ENVIRONMENTAL BLOCKADES

Since the 1970s, environmental blockades disrupting the exploitation and destruction of forests, rivers, and other biodiverse places have been one of the most attention-grabbing and contentious forms of political action. This book explores when, where, and why environmental blockading and its associated tactics first arose.

The author explores a broad range of questions, including how did tactics and practices first developed and popularised during environmental blockades come to feature regularly in animal rights, peace, refugee, and other campaigns? What are blockaders hoping to achieve? How have such blockades and tactics shaped government policy, the culture of modern politics, and popular understandings of ecology, colonialism, and activism? This book offers the first comprehensive history and analysis of environmental blockading in three key countries: Australia, the United States, and Canada. As the first places to experience sustained protest cycles which fully established, promoted, and developed the environmental blockading repertoire as an ongoing strategic option for movements nationally and internationally, these campaigns were central in creating a new approach to conservation issues. They also played a leading role in making obstructive direct action a regular part of political campaigning, as seen in the form of the Extinction Rebellion (XR), alter-globalisation, climate justice, and other movements.

This book draws on rigorous archival research including sources ranging from personal diaries, campaign minutes, and video footage through to police reports and newspaper articles, as well as interviews with more than 30 protest leaders and campaigners. It will be of great interest to students and scholars in the fields of sociology, political science, history, green criminology, and interdisciplinary environmental studies.

Iain McIntyre is a researcher, writer, lecturer, and community radio broadcaster based in Melbourne, Australia. He is the author of a number of books on history, social movements, and popular culture. He is a regular contributor to commonslibrary.org, a web repository which collects and distributes the key lessons of progressive movements across Australia and around the globe.

TRANSFORMING ENVIRONMENTAL POLITICS AND POLICY

Series Editors Timothy Doyle, Keele University, UK and University of Adelaide, Australia and Philip Catney, Keele University, UK

Series Blurb

The theory and practice of environmental politics and policy are rapidly emerging as key areas of intense concern in the first, third and industrializing worlds. People of diverse nationalities, religions and cultures wrestle daily with environment and development issues central to human and non-human survival on the planet Earth. Air, Water, Earth, Fire. These central elements mix together in so many ways, spinning off new constellations of issues, ideas and actions, gathering under a multitude of banners: energy security, food sovereignty, climate change, genetic modification, environmental justice and sustainability, population growth, water quality and access, air pollution, mal-distribution and over-consumption of scarce resources, the rights of the non-human, the welfare of future citizens – the list goes on.

What is much needed in green debates is for theoretical discussions to be rooted in policy outcomes and service delivery. So, while still engaging in the theoretical realm, this series also seeks to provide a "real world" policy-making dimension. Politics and policy-making is interpreted widely here to include the territories, discourses, instruments and domains of political parties, non-governmental organizations, protest movements, corporations, international regimes, and transnational networks.

From the local to the global – and back again – this series explores environmental politics and policy within countries and cultures, researching the ways in which green issues cross North–South and East–West divides. The 'Transforming Environmental Politics and Policy' series exposes the exciting ways in which environmental politics and policy can transform political relationships, in all their forms.

Environmental Citizenship in the Indian Ocean Region
Edited by Benito Cao

For more information about this series, please visit: www.routledge.com/ Transforming-Environmental-Politics-and-Policy/book-series/TEPP

ENVIRONMENTAL BLOCKADES

Obstructive Direct Action and the History of the Environmental Movement

Iain McIntyre

Routledge
Taylor & Francis Group

LONDON AND NEW YORK

First published 2021
by Routledge
2 Park Square, Milton Park, Abingdon, Oxon OX14 4RN

and by Routledge
605 Third Avenue, New York, NY 10158

Routledge is an imprint of the Taylor & Francis Group, an informa business

© 2021 Iain McIntyre

British Library Cataloguing-in-Publication Data
A catalogue record for this book is available from the British Library

Library of Congress Cataloging-in-Publication Data
A catalog record has been requested for this book

ISBN: 978-0-367-48054-7 (hbk)
ISBN: 978-0-367-48055-4 (pbk)
ISBN: 978-1-003-03778-1 (ebk)

Typeset in Bembo
by Newgen Publishing UK

I would like to thank Professor Sean Scalmer and Associate Professor Sara Wills for their support, encouragement and guidance. Thanks are also due to all those who gave time to be interviewed and made available primary sources and other materials. I acknowledge the support of the Australian Government Research Training Program Scholarship, the Dennis-Wettenhall prize, the Gilbert Postdoctoral Career Development Fellowship, and the University of Melbourne in producing this work. Thanks, as always, to my family.

CONTENTS

ACKNOWLEDGEMENT

Cover image: The Biscuit (1990). This photograph was taken by Orin Langelle during a blockade in Illinois's Shawnee Forest and is used with his permission. The car barricade was a replica of one used in the then-ongoing "Oka Crisis" in Canada to show solidarity with the Kanesatake Mohawk community. © Orin Langelle.

ABBREVIATIONS

ABC	Australian Broadcasting Commission
ACF	Australian Conservation Foundation
ALP	Australian Labor Party
AWA	Alberta Wilderness Association
BC	British Columbia
BLF	Builders Labourers Federation
BLM	US Bureau of Land Management
CAFNEC	Cairns and Far North Environment Centre
CAMP	Campaign Against Marijuana Planting
CFAG	Cathedral Forest Action Group
CRAG	Cedar River Action Group
EB	Environmental Blockading
EIS	Environmental Impact Study
EF!	Earth First!
EFer	Earth Firster
EFDAF	Earth First Direct Action Fund
EFJ	*Earth First! Journal*
EFNAG	Earth First! Nomadic Action Group
EPIC	Environmental Protection and Information Centre
FC	NSW Forestry Commission
FOCS	Friends of Clayoquot Sound
FOSP	Friends of Strathcona Park
FS	US Forest Service
GWTC	Gitxsan-Wet'suwten Tribal Council
HEC	Tasmanian Hydro-Electric Commission
IPS	Island Protection Society
IWW	Industrial Workers of the World

MHSMAG	Middle Head Sand Mining Action Group
MNS	Movement for a New Society
NAG	Nightcap/Nomadic Action Group
NDP	New Democratic Party
NSW	New South Wales
NVDA	Non-Violent Direct Action
NVT	Non-violence trainers
NZ	New Zealand
ODA	Obstructive Direct Action
OMNR	Ontario Ministry of Natural Resources
ONV	Orthodox Non-violence
PNW	Pacific Northwest
RACE	Regional Association of Concerned Environmentalists
RCMP	Royal Canadian Mounted Police
SLAPP	Strategic Lawsuit Against Public Participation
SWAAT	Save Wilderness At Any Time team
TAA	Teme-Augama Anishnabai Association
TNFAG	Terania Native Forest Action Group
TRO	Temporary Restraining Order
TWS	Tasmanian Wilderness Society/The Wilderness Society
TWS-C	Temagami Wilderness Society (Canada)
UA	Upriver activists
VOT	Voices of Temagami
WAG	Wilderness Action Group

INTRODUCTION

On 17 October 2018 staff at Greenpeace's London headquarters were surprised to find themselves the target of a sit-in by members of the hitherto little known Extinction Rebellion (XR). Explicitly aiming to differentiate the new from the old, the occupiers lauded Greenpeace's pioneering work in decades past while appealing for it to undertake a major shift in moving away from reformism and professionalised activism to join a "new journey" of "mass civil disobedience". In the face of "ecological and societal collapse" XR called for a "mobilisation of war time like proportion" to create a rapid transition to a grassroots democracy via a "Citizen's Assembly" tasked with addressing catastrophic climate change.[1] A fortnight later XR unveiled its "Declaration of Emergency" to a crowd of over 1,000 people. The movement's strategy of disrupting everyday life in order to highlight the climate crisis was immediately enacted as the crowd moved to blockade the road outside the UK Houses of Parliament.[2] Over the next two weeks, further protests including dropping banners from Westminster Bridge and closing down access points to a government ministry, the Prime Minister's residence, and the Brazilian embassy via means such as activists chaining themselves together and gluing their hands to doors and gates.[3]

These actions were calculated to generate momentum and publicity ahead of "Rebellion Day", 17 November 2020. Having declared "We are bold. We will not hide. We are all in open rebellion", XR notified authorities and the public that it would close down five major bridges. Come the day, approximately 6,000 people did so. As they listened to speeches and danced to music pumped from sound systems, the mass occupation steadily caused traffic gridlock across London.[4]

This proved to be the prelude to further protest across London as well as other British cities, much of which involved "swarming roadblock" actions during which small groups occupied multiple intersections for short periods. Dispersing ahead of the arrival of police they then moved on to disrupt other sites. Despite making arrests, police found themselves unable to effectively counter the blockaders' tactics and were reduced to advising motorists to avoid London.[5]

The impact of these actions, boosted through associated media attention, rapidly propelled XR into a global movement with over 1,000 chapters arising globally to carry out actions involving hundreds of thousands of people. The means of disrupting traffic, entry to offices, and other activities have ranged from roadblocks involving crowds sitting and standing in the road way to individuals locking their arms into modular boxes and sitting precariously atop trees and wooden tripod structures.[6] *Commenting on the effectiveness of mass mobilisations, early XR strategist Roger Hallam argued, "They can't do anything about it unless they start shooting people, and presumably they won't do that."*[7]

The dangers of industrially induced climate change first came to international prominence during the 1990s. Since that time, activists had employed a variety of means to raise awareness and pressure governments, companies, and civil society to remedy the issue. XR successfully emphasised the urgency of the matter and focused widespread public attention upon it, through the use of disruptive activity, which physically impeded movement and work by others, a practice that can be termed Obstructive Direct Action (ODA).

Many ODA tactics involve the creation of what theorist Brian Doherty describes as "manufactured vulnerability".[8] By placing their bodies in the way of their opponents, tactics such as sit-ins[9] have long been used by activists to place "responsibility for the blockaders' safety in the hands of the authorities."[10] Techniques such as being chained together or attaching body parts to entrances extend activists' ability to minimise arrests while creating "a sustained performance of their own moral commitment for a media and public captured by the epic quality of the confrontation".[11]

Although largely reported as novel and surprising phenomena, and in some cases treated by the activists involved as such, the forms of action on display during XR and other contemporary ODA-based campaigns have deep roots in earlier political action. Knowledge, forms, and approaches drawn from peace, civil rights, environmental, labour, and other movements all exert influences on contemporary activism.

This book concerns itself with a series of events which played a critical role in establishing and shaping ODA as a key response to environmental dangers from 1979 onwards. The campaigns it covers involved environmental blockading (EB). This practice uses ODA to disrupt logging, clearing, and other activities that threaten to transform biodiverse areas such as forests, beaches, and rivers.[12] Such ODA-based campaigns allow activists to physically "defend environmental sites threatened by economic 'development' and to confront and expose those economic and political groups whom they [see] as perpetuating environmental degradation."[13]

Although its contentious nature has ensured that it has not achieved parity with conventional forms of lobbying in developed countries, the threat and use of environmental blockading has nevertheless become a regular form of action to which environmental, First Nations, and other activists turn. The protest waves covered in this book occurred in Australia, the United States, and Canada between 1979 and 1997. As discussed below, these countries and this period played a particularly critical role in establishing environmental blockading.

The influence of environmental blockading has extended beyond the protection of biodiverse places. Contemporary campaigns addressing other issues, ranging from climate change to student rights, regularly draw on ODA tactics and approaches that were innovated, modified, developed, and/or popularised during environmental blockades. Sustained campaigning in Australia, Canada and the United States from 1979 to 1990 generated and crystallised practices and approaches, on the side of activists, police, and others, that have since been applied to a range of issues, nationally and globally.

Much writing concerning ODA-based movements has tended to be ahistorical, failing to address the question of how tactical innovations emerge and come to be nationally and internationally established and diffused. This study addresses gaps in the historical record regarding specific events and protest waves to provide a fuller understanding of the nature and development of environmental movements as well as their transnational dimensions. It not only explores the roots of contemporary ODA, but also furthers understanding of it by identifying and analysing approaches, tactics and dynamics that persist through to the present.

In documenting and analysing the period, this book addresses a number of under-researched areas within environmental and social movement scholarship and furthers understandings concerning "repertoires of contention". This term, originally coined by Charles Tilly, denotes "the distinctive constellations of tactics and strategies developed over time and used by protest groups to act collectively in order to make claims on individuals and groups".[14] Based on observations regarding French and British popular contention from the seventeenth to the twentieth centuries, Tilly developed the concept to describe how, like "troupes of street musicians … general participants in uprisings and local struggles followed available scripts, adapted those scripts, but only changed them bit by bit".[15] The concept, and its focus on how activists build their means of contention on existing bodies of knowledge, is intrinsic to many studies of social movements. Many areas, including environmental blockading, remain under-researched, including aspects regarding how innovation takes place, why it emerges during certain protest events, and why certain tactics and approaches endure and are expanded upon.[16]

Scholars have noted that innovation within social movements takes a variety of forms, including the creation of novel forms of contention and the application of existing forms to new settings and issues. Tactics may also be innovatively combined or reintroduced.[17] The importance of certain periods to the evolution of repertoires, as well as the rate at which change occurs, has emerged as a key point of debate and analysis.[18] In relation to such periods Tarrow defines "cycles of contention", also referred to as "protest waves", as phases "of heightened conflict across the social system with a rapid diffusion of collective action frames; a combination of organised and unorganised participation; and sequences of intensified information flow and interaction between challengers and authorities." Within these are what Sewell has termed "transformative events", critical turning points that dramatically effect mobilization and movement effectiveness.[19]

In providing a history of such transformative events and waves, this book builds on and contributes to understandings regarding the nature and development of tactics and strategies as well as the internal dynamics of the networks in which they are inculcated. It combines a focus on key points of rapid innovation with an analysis of longer term processes of change and embedding.

The book also addresses gaps in the historical record by describing the key Australian, Canadian, and American events that generated and employed EB tactics. In doing so, it contributes to a fuller understanding of the nature and development of environmental movements in these nations as well as their transnational dimensions. Previous studies have tended to focus on formal groups operating within institutional frameworks. Where research has detailed EB campaigns, it has rarely addressed the question of how tactical innovations emerged and came to be nationally and internationally established and diffused.

Australia, Canada, and the United States as prime early innovators

Australia, the United States, and Canada have been chosen as the focus of this study because they were the first three countries to nationally establish EB as a repertoire of contention. In each the repertoire was applied in various regions and eventually came to form a distinct part of their national political culture.

Initial research involved scouring a wide range of primary and secondary sources, including searches of mainstream and activist media archives and indexes as well as academic studies, to yield a timeline of 392 events and campaigns between 1973 and 1997 that come under the definition of EB. The year 1997 was chosen as the end date for the timeline on the basis that this was the year by which detailed activist guides to environmental blockading tactics had first been produced in the United Kingdom, the United States, and Australia, thereby codifying and documenting much of the innovation that had occurred up until that point.[20]

The timeline showed that of the campaigns and protest events that occurred during this time, 55 took place in Australia, 111 in the United States, 45 in the UK, 46 in Canada, and 62 in Malaysia.[21] Smaller numbers also took place in India, New Zealand, Finland, Norway, Germany, the Philippines, Poland, Russia, Belgium, the Netherlands, France, and Papua New Guinea. Because of a lack of reporting, both in English and in general due to the remote location of the protests, numbers are harder to quantify for Brazil, but at least 45 blockades had taken place by 1988, with more following until 1994.[22]

Among a variety of patterns, the timeline demonstrates that Australia, the United States, and Canada were the first countries in which the EB repertoire came to be used on a regular basis across a variety of regions. Large numbers of blockades took place in Malaysia and Brazil, but in both cases, they were primarily located in one or two regions and employed a limited body of tactics. Although they have received greater academic attention, activists in the UK did not consistently employ the repertoire until 1992.

A transformative event, both for this period and today, occurred during August 1979 in a remote and previously largely unknown corner of northern New South Wales (NSW), Australia. Having spent five years lobbying politicians and lodging public submissions, activists there turned to direct action to save rainforest from being logged at Terania Creek. For over four weeks, they delayed work by placing bodies, objects, and vehicles upon roads, by digging ditches and flooding areas, and by occupying and connecting trees with steel cables. As pressure mounted, the NSW government called a halt to logging and within three years the area was protected as part of a new series of national parks.[23]

Although Australian activists were largely unaware of them, environmental blockades had been employed in other places prior to this.[24] As part of what became known as the Chipko movement, women and children in Uttarakhand, India, initially protected trees in 1974 by hugging them before being reinforced by wider village communities, who occupied work sites for four days.[25] Activists in New Zealand saved old growth forest from logging at Pureora in 1978 by scaling and occupying trees.[26] In the summer of 1979, Finnish environmentalists occupied work sites and chained themselves to machinery in an attempt to prevent the drainage of Lake Koijarvi.[27] Following these protests the activists involved returned to the use of other protest means and did not employ EB consistently until later decades.

In contrast, Brazilian rubber tappers based in Xapuri and Brasilia repeatedly sought to protect Amazonian rainforests and their livelihoods through the use of *Empates* or "stand-offs" from 1973 until 1994. Defying heavy repression, including the murder of activists, community members occupied worksites, laid in the path of bulldozers, and stood in front of trees and chainsaws to prevent clearing for ranches. These actions constituted the first sustained use of environmental blockading. They generally disrupted operations and in some cases saved areas altogether. Although relatively unknown internationally in the 1970s, by end of the following decade, the actions of the rubber tappers had generated enough regional and national support, and global attention, to lead to the creation of "extractive reserves", which combined environmental protection with small scale, sustainable resource use.[28]

Tactical innovation associated with the *empates* was relatively limited and the use of the strategy largely confined to one area.[29] In contrast, within Australia a series of blockades followed Terania Creek at NSW's Nightcap rainforest and Middle Head beach, Tasmania's Franklin river, Queensland's Daintree rainforest and Victoria's Errinundra Plateau. Over a five-year period, these formed part of a protest wave, which successively generated and shifted tactics across the nation, creating cohorts of protesters experienced in using them. Subsequent waves of blockading from the late 1980s onwards further enlarged tactical repertoires and spread the use of the strategy to other parts of Australia.[30]

The template and tactics developed at EB events in Australia from 1979 to 1983 also emerged in the United States and Canada in a process that combined domestic innovation and dissemination with transnational diffusion. Citing Australian precedents US activists also drew on the experiences of local anti-nuclear, civil rights and other social movements to undertake their first forest blockade at Bald Mountain in Oregon during 1983. As part of the rise of the Earth First! network, further actions followed throughout the country, building movement capacity to a point where sporadic, short-lived ODA was joined by sustained occupations and campaigns.[31]

In 1984 an alliance between the Nuu-chah-nulth community and environmental activists successfully turned loggers away from Meares Island in Canada's Clayoquot Sound by blocking boats and occupying a landing bay in order to protect old-growth forest and assert Aboriginal sovereignty. At times during the 1980s, conservationism in Australia and the United States also involved Indigenous communities and became interconnected with concerns regarding land claims and the protection of culturally significant sites. It never did so to the degree that Canadian campaigns, particularly in British Columbia, would between 1984 and 1989. There a series of campaigns against forestry, mining and other activities were typically led by First Nations communities or involved alliances with non-Indigenous environmentalists.[32]

By the end of the 1980s environmental blockading had become embedded as a strategic option in all three countries. The term "embedding" refers to the process by which a political strategy becomes established to the point where it is a regularly considered and recurring response to grievances. As will be discussed in Chapter 7, the repertoire would become progressively more sophisticated and transnationally entwined in the years that followed. For now, it is important to underscore that these initial periods of sustained action were crucial for each country, and over time globally, as for the first time they firmly established a model of environmental blockading and forms of ODA that others would use and build upon.

Commonalities between Australia, the United States, and Canada

While each country features distinctive societies and political and economic structures, it is important to acknowledge that Canada, Australia, and the United States have much in common. The degree to which these led to shared patterns of activism will be explored in further chapters, but some broad characteristics are noted here.

All three are settler nation states created by colonists who primarily framed the places they invaded as "human free" and devoid of culture or development. Although a variety of relationships were formed between settler and Indigenous populations, colonisation generally involved the displacement and dispossession of the latter through the instalment of new economies, customs, and geographies. Following what Wolfe terms the "logic of elimination" this in some cases included the "summary liquidation of Indigenous people" alongside widespread "child abduction, religious conversion, resocialization in total institutions such as missions or boarding schools, and a whole range of cognate biocultural assimilations", all of which were resisted in varying ways.[33]

Each country has generally been considered as belonging to the "British World" or "Anglosphere", an imagined community of English-speaking states in international society which have historically identified with each other and drawn on traditions and values associated with Britain and its former empire.[34] All featured capitalist economies in which, as part of an international trend towards neo-liberalism, state intervention was beginning to be scaled back during the 1980s. Although some sections of each country are covered by agreements with Indigenous peoples regarding land rights and treaties, the nature of these, and sovereignty concerning other regions, remain contested. All three countries are representative democracies based on a federal model in which a central government and those in provinces, states, and other localities hold differing, but often overlapping and conflicting, powers regarding business, policing, and land use. Each has a level of tolerance for, and political inclusion of, protest. This, with some notable exceptions, has allowed social movements to operate openly and employ ODA without the threat of major physical harm to participants. To varying degrees policies and practices regarding land use, policing and other factors relevant to blockading were subject to international and cross-national treaties, law and trade.[35]

While these countries have industrialised economies, they also include large areas of undeveloped or lightly developed land. They all include regions largely dependent on extractive industries and markets for profit making. Although some degree of logging, mining, and other activities take place on privately owned property, each has significant amounts of land which remain under at least nominal government control and are leased out to private interests for exploitation.[36]

Despite significant national and regional differences in structures and processes all three countries had political and administrative policy regimes that prioritised profit making based on resource extraction and development over Indigenous, heritage, ecological, recreational, and other values during the 1970s and 1980s. Each allowed for public input regarding land use, but in terms of effecting policy change and challenging decisions, government bodies and regulators were largely closed to dissenting opinions. At the same time, each nation recognised that certain places deserved protection and had established a network of national parks and other areas that were zoned in such a way as to limit or preclude residency and development.[37]

As part of trends occurring transnationally, new environmental movements emerged in each country during the 1960s and 1970s to challenge existing policies and concepts regarding ecology and land use. This "second wave"[38] of environmentalism, which primarily focused on issues related to pollution, energy, and toxic materials, had a political and cultural impact to the degree that public attitudes concerning the environment shifted and new legislation was enacted concerning public and environmental health, and in some cases wildlife and biodiverse places. Although overall activism subsided, each nation experienced an increase in interest concerning the protection of biodiverse places during the 1970s and 1980s. This was in part due to the international rise of new cultural and scientific understandings regarding ecosystems. Grievances also stemmed from marked increases in exploitation and the movement of extractive activities into areas that had been previously considered safe or too remote to be utilised. While these campaigns often had some base in the areas immediately affected, they also drew support from urban communities whose knowledge and attraction to nature were largely drawn from images, aesthetics, and recreation rather than direct experience and work.[39]

At the same time, many of those working in resource industries, particularly logging, were enduring periods of global and local market instability, technological change, and restructuring. The disquiet related to resulting job losses and precarity often came to be focused upon environmentalists and urban populations rather than employers.[40]

Conservation movements in each country were made up of loose and heterogeneous alliances of local residents, recreational users, and environmental advocacy groups. The latter included internationally, nationally, and regionally based organisations concerned with multiple issues as well as locally based and focused groups and informal networks. These engaged in campaigns, educational activities, and interaction with the mainstream media, but none had held environmental blockades until the end of the 1970s.[41] As will be demonstrated in later chapters the use of blockades in each country would play a key role in introducing issues, topics, and scientific knowledge involving concepts such as "biodiversity", "rainforest", "ecological systems", and "old growth forest" into public discourse. In doing so they helped to delegitimise existing policy regimes and facilitated later waves of environmentalism.

Tactical innovation, choices, and diffusion

In providing an account of how the EB repertoire came to be formed, used, and embedded in each country, as well as transnationally, this book addresses a range of areas regarding social movement tactics and strategy. Mayer, et al. define strategy as involving "defining, interpreting, communicating and implementing a plan of collective action that is believed to be a promising way to achieve a desired alternative future in light of circumstances."[42] One key element, alongside choosing targets, organisational forms, etc., within this process is selecting, inventing, and implementing the specific tactics to be used in particular situations. Social movement scholars have long debated the degree to which culture, identity, and emotion, as well as calculations regarding efficiency and efficacy, affect strategic and tactical choices as well as what is understood to be effective and efficient.[43] This book interrogates and adds to understandings of the factors that influence choices and the conditions which influence the rate and direction of innovation, both during the period under study and more generally. It also pays close attention to the interaction of differing groups of activists with opponents, authorities, and other stakeholders, thereby avoiding what McAdam and Tarrow have criticised as an "excessively 'movement-centric'" approach.[44]

The study also contributes to understandings of how national and cross-national tactical diffusion take place. Michaelson defines diffusion as the "process by which any innovation (new idea, activity or technology) spreads through a population."[45] According to Soule, most social movement studies models of diffusion include four elements: "a transmitter, an adopter, an innovation that is being diffused, and a channel along which the item may be transmitted".[46] In terms of types of innovation, Kriesi, et al. distinguish between two major areas: the use of collective action and protest in general, and particular features of protest (goals, issues, organisational structures, action forms).[47] For such innovations to be shared they must be, in Tilly's words, "modular", that is "transferred easily from place to place, issue to issue, group to group".[48]

Theorists generally suggest that diffusion occurs along two sets of channels: those that are direct/relational, such as face-to-face meetings, correspondence, and workshops, and those that are indirect/non-relational, such as media depictions. In cases where "scale shift" occurs, that is when movements increasingly coordinate and bridge their activities, Tarrow, Tilly and McAdam, argue that a third process comes in to play, that of "brokerage", in which transfers depend on the deliberate and conscious linking of previously unconnected social actors. The nature of these channels, and the point at which movements pick them up, in turn affects the speed, influence, and adaptation of the ideas being diffused.[49]

Activism and protest in different locales can be triggered by similar causes, but all require diffusion and brokerage to take place if they are to expand beyond localised incidents to become part of larger protest waves.[50] A criticism levelled at many social movement scholars' treatment of diffusion is that they incorrectly suggest "objects of diffusion are easily transferable and translatable" and assume "receivers will simply adopt an idea or practice when it is seen as appropriate or useful."[51] According to Scalmer, a more complex and nuanced approach involves attention to on how local movements "translate", rather than transmit, passively receive, and automatically apply knowledge. He argues that this process, in which actors "learn about, connect with, and incorporate new forms of collective claim making... from one context to another" is a "historically variable, enculturated process, that rests upon sustained intellectual labor".[52]

By focusing upon those who created innovations and those who first adopted and adapted them, this book demonstrates that diffusion practices are active, contextual, and rarely automatic or immediate. It also explores how under similar conditions tactics can emerge separately and in parallel with only weak, or no, levels of diffusion involved.

Obstructive Direct Action

Environmental blockades are primarily predicated upon using what participants and others have described as "direct action". Direct action is a practice that is extensively covered in social movement literature, yet the exact meaning and parameters of the term are rarely discussed. Where definitions are provided, they are often vague and/or allow for such a wide range of activities as to make the term indistinguishable from "protest" in general.

A cross-section of the relevant literature indicates two main uses of "direct action". The traditional sense of the term stems from the nineteenth-century syndicalist movement, which aimed to abolish capitalism via industrial action and other activities under workers' control such as boycotts and cooperatives. These means were posited as an alternative to parliamentary and electoral strategies.[53] The term had a clearly prefigurative sense in that present-day tactics and strategies were to reflect the society that the movement wished to achieve in the future. This sense has since been widened by anarchist and autonomist movements who use "direct action" to denote activities through which "individuals assert their ability to control their own lives and participate in social life without the need for mediation or control by bureaucrats or professional politicians."[54]

The second broad use of "direct action" is to describe the means through which social movements create "disruption", a term which itself indicates the interruption of "routine activities of opponents, bystanders or authorities".[55] The forms of contention indicated by this use of "direct action" can include but are not limited to prefigurative ones. Indeed, their primary aim is often to influence institutions, corporations, and/or government bodies to change their policies rather than to remove them from the equation.[56] Such activities can be as varied as rioting, the use of provocative symbols, and the refusal to play customary social roles. Actions may or may not be construed as "violent", but stop short of armed insurrection and guerrilla warfare, strategies which influential Italian sociologist Melucci defines as the "deliberate and continuous use of violence and a military-type organisation".[57]

Movements have long engaged in disruptive direct action involving the use of barricades and other physical installations, as well as their bodies, to interfere with activities they are opposed to and obstruct the movement of people, resources, and vehicles connected to them. The types of action and dynamics of organisation involved in such physical obstruction differ from non-obstructive kinds of disruptive activity such as the singing of banned songs and the wearing of nonconformist clothing. To distinguish physically obstructive forms of disruptive "direct action" from others this study employs an original term, Obstructive Direct Action (ODA).[58]

In addition to "direct action", activists and others have also regularly described environmental blockading activities as "civil disobedience" and "nonviolent direct action" (NVDA). "Civil disobedience" is defined by political philosopher John Rawls as the "public, nonviolent, conscientious, yet political act contrary to law usually done with the aim of bringing about a change in the law or policies of the government".[59] As will be discussed below and in subsequent chapters, "nonviolence" and NVDA are terms that cover a gamut of means and ethics. These extend from minimal definitions eschewing the instigation of physical violence through to those that reject any form of sabotage, secrecy, and coercion. Where definitions refer to obstructive activities both concepts can be seen as crossing over with ODA. However, as ODA also encompasses legal forms of physical block-ading and those which may be construed as "violent", it is the primary term used throughout this book.

Environmental blockading and its key tactical forms

As a sub-set of the broader ODA repertoire, environmental blockading, as defined in this study, takes place at "the point of destruction"'. This term denotes locations where logging, clearing, or other development activities are threatening to directly destroy or significantly and irrevocably alter an area that has not previously been subject to such activity. Traditional ODA tactics used by environmental blockaders, such as occupations and barricades, as well as means innovated by them, are also utilised at other locations. These include the "point of production" (workplaces engaged in environmentally destructive activities), "point of transportation" (places from and through which environmentally harmful products are moved), "point of decision" (summits, parliaments, corporate offices, etc.), and "point of consump-tion" (retailers and other places from which environmentally harmful products are sold).[60] As my research indicates that campaigns focused on the "point of destruc-tion" were a key locus of tactical innovation and diffusion during the 1979–1990 period, I primarily focus upon them. Their influence upon campaigns elsewhere will be canvased in Chapter 7.

Drawing on interviews, activist manuals, and other primary sources, the tactics employed during environmental blockading can be placed into four categories. Depending on the conditions of protest and beliefs of those involved, these may be combined. The first, "soft blockades", involves people using only their bodies, rather than devices or barricades, to occupy a site and/or block work and the passage of goods and equipment. Mobile forms of action, such as hiding in or running through an area to be cleared, also fall within this category.[61]

A second group of tactics, "barricades", involves the obstruction of roads and other key points through the placement of debris, boulders, cars, and other objects. A third category, "enhanced vulnerability",[62] sees activists place their bodies in devices to obstruct their opponents by both blocking and occupying a space while amplifying threats to their well-being. Combining elements of the first two cat-egories, this can involve static techniques of manufactured vulnerability, such as locking a body part to opponent's equipment to prevent its use or sitting in a tree to prevent it from being cut.

A final category includes tactics based upon "sabotage". The evolution of tactics involving the major destruction of opponents' equipment and resources, often referred to in activist circles as "monkeywrenching" or "pixieing", largely falls out-side the purview of this book. As they attract heavy penalties, and are of a conten-tious nature, such activities are generally carried out by individuals and small groups on a clandestine and fleeting basis rather than as a regular and core part of block-ading. More typically the tactics of sabotage involved in EB have focused upon cutting through fences, eroding and removing sections of a road, or slowing work by removing survey stakes. Despite major sabotage being largely divorced from EB during this period, questions regarding what constituted acceptable levels of sabotage and how activists defined it became key issues within some campaigns – especially in the United States – and this book canvases such debates.

Outcomes and goals of ODA

By employing ODA, activists – deliberately or otherwise – create a range of outcomes beyond simply physically impeding activities to which they are opposed. As will be seen in the chapters that follow, activist goals, often simultaneously pursued, form an important consideration in shaping which tactics are chosen and how they are modified. A variety of theorists, including those adhering to Resource Mobilisation and Political Process Theory approaches, have long focused on the ways in which activists attempt to mobilise resources efficiently in order to achieve desired ends at lowest cost.[63] Such a focus has been criticised for reducing choices to rational calculations, thereby ignoring the role of emotions, culture and ethics, as well as how these influence the ways in which goals are defined and understood.[64] While acknowledging that activists do act instrumentally, and that the ends they explicitly seek are of great importance, this study concerns itself with a wide range of factors influencing tactical choice.

First and foremost, blockades allow campaigners to publicly and clearly demonstrate that organised opposition exists and that those involved are committed to their cause. Such action inevitably includes personal, expressive, and symbolic dimensions. Blockades have been used by some First Nations communities to express connection to land and sovereignty over it. For other activists, such as those from religious backgrounds, "bearing witness", that is "recording ethical or moral opposition or disapproval by calling attention to an event through a person's presence", will be the primary reason for undertaking ODA.[65] For Gandhian-influenced activists "psychological intervention" to convert opponents and show commitment via deliberate suffering is a further expressive and strategic dimension.[66]

ODA shares benefits with other forms of disruptive activity by allowing protesters to draw attention to their concerns. The primary vehicle for doing so is generally via the mainstream media. Maddison and Scalmer identify three main activist positions regarding mass media: "separate camp", in which activists reject involvement with journalists and focus on their own media networks; "modernising accommodation", in which they tailor, and often moderate, their activities to gain coverage which portrays them in a favourable light; and "expressive militancy", in which they employ confrontational tactics in the knowledge they are likely to generate coverage, albeit of an often negative nature.[67]

ODA can also give activists the power to coerce their opponents and supporters via the imposition of financial and political costs. Turner and Killian define coercion as "the credible threat of punishment in case the target group does not support the movement goals", and distinguish it from other means, with which it can overlap, such as persuasion, facilitation, and bargaining.[68] In terms of protest activity, Michael

Albert argues: "At a given moment, elite policy-makers have a whole array of priorities. Change occurs… when movements raise the social costs that policy-makers are no longer willing to endure and which they can only escape by relenting to movement demands."[69]

Further to this, disruptive activities can broaden the field of conflict, critique capitalism and other social and economic relations, and expose the state's repressive nature and complicity with destructive or anti-social activities. They can also temporarily delay an activity while lobbying, litigation, and other strategies progress. Moreover, disruption can allow confrontational activists to act as a "radical flank", thereby bestowing legitimacy upon moderate activists and allowing them to present their demands and behaviour as a serviceable alternative.[70] As previously mentioned ODA may also have a pre-figurative element in demonstrating that people can directly deal with an issue without recourse to mediating bodies.

Although disruptive activities can give "weak actors leverage against powerful opponents", Tarrow argues that they are not commonly employed as they rely upon "a high level of commitment, on keeping authorities off balance and on resisting the attractions of both violence and conventionalisation".[71] Despite its costs, activists have nevertheless consistently prioritised the use of disruption. As will be demonstrated in the chapters to come, ODA has generally run parallel to and complemented other strategies, particularly when activists have used it to buy time and garner media attention. It has less often been viewed as the primary vehicle for permanently ending an activity or as the sole means of exerting coercion.

"Ecologies of protest action", "normative protester behaviour", "protest policing", and "Green Criminology"

This book's approach blends a range of social movement theories to combine an understanding of specific external factors affecting mobilisation while also emphasising their changing nature and the way in which activists, consciously or otherwise, interpret and respond to them. It also builds on insights developed by Feigenbaum, Frenzel, and McCurdy in their 2013 study *Protest Camps*. The authors define the use of "protest camps" as "a place-based social movement strategy that involves both acts of ongoing protest and acts of social reproduction needed to sustain daily life".[72] They argue that such camps often serve as critical nodes of tactical, ideological, and infrastructural development due to the lengthy periods of sustained protest and close interaction between protesters involved as well as that with opponents, authorities, and their activities.[73]

As this study demonstrates, these observations can be extended to environmental blockades. Some of the events analysed involved protest camps, while in others a sizeable number of protesters travelled to and from local residences and other locations rather than living on site. In all cases, the experience of blockading – the sustained, close, and intense level of protest and associated work and living and the engagement with the biodiverse environments involved – made these events key points of tactical innovation. The periods of waiting for actions to begin or opponents to react also allowed for extended debate and discussion. Combined with episodes of intense action such factors further enabled political and protest tendencies to form as individuals and groups bonded as well as bickered. Similar observations can be made for other forms of action when they are intense and sustained, such as picket lines, occupations, and insurrections.

The degree to which sustained protest allows for discussion and innovation is subject to cultural and contextual factors as well as incidents, dynamics, and geographies particular to events. Feigenbaum et al. use the term "ecology of protest action" to encapsulate the "entangled ways in which objects, people and environments come together in protest action".[74] Elements making up such "ecologies" include the broader political framework and situation, individual and collective emotional responses to events, previous experiences of protest and policing, and the physical abilities of those involved. Further to this are factors such as terrain and geography, the strategies and emotions of opponents, the objects and resources available and used to both generate and prevent obstruction, and the gender, cultural, and ethnic differences between and within parties.[75]

Feigenbaum et al.'s emphasis on the complexity of factors driving protest forms and tactical choice is intended to redress what they see as a propensity among activists and theorists to reduce tactical choices to binaries concerning "violence" and "non-violence" as well as to ignore or downplay the malleable nature of such concepts and the range of non-ideological influences upon them.[76] Some previous studies, as well as much activist discourse, have tended to condense the discussion of tactics in this way, elevating specific definitions of "non-violence", and lending them an enduring importance beyond that they may have had in particular instances.[77]

Despite being one among a range of tactical influences, differing interpretations of "non-violence" have nevertheless at times constituted one of the key disputes in environmental blockading. Almost all of the organisers of Australian, Canadian, and American environmental campaigns covered in this study, and the majority since, have defined their events as "peaceful" or "non-violent", in the minimal sense of what Amory Starr describes as "eschew[ing] offensive violence against persons".[78]

Despite acknowledging that they are far from the only determinant of tactics and strategy, this study employs the original term "normative protest behaviour" to describe and emphasise the range of attitudes and actions that activists have deemed ethically and strategically acceptable at different times. As they have usually done so under the banner of "non-violence", it also provides an analysis of the evolving and varying meanings associated with that concept within and across protests.

Among the range of factors already canvassed, the book also pays close attention to the question of protest policing. Running the gamut of brutal repression through to non-interference, the dynamic interplay of tactics between authorities and protesters clearly shapes and limits the choices available to both. Work by theorists such as della Porta, Reiter, and others postulate that responses to protest from authorities are shaped by a range of factors including the organisational structures and political character of the state, police, and relevant bureaucracies.

Opinions among state actors as well as broader publics regarding policing and protest in general, as well as the specific campaign issues, also play a role. Further to this the occupational culture of the police and the nature of specific and historical interactions with protesters combine with perceptions of external realities, such as those involving the behaviour and nature of crowds, to shape and determine officers' individual and collective responses.[79]

The history of protest policing and EB, outside of specific events and periods, remains largely under-theorised. As will be demonstrated, authorities experimented with a variety of responses that included enough tolerance to allow activists to, at the very least, initiate blockades and employ tactics, which involved manufactured vulnerability.

During this period, the responses of police, legislatures, and bureaucracies were largely reactive. This study provides a history and analysis of how many now typical counter-strategies, including changes to trespass laws, bail conditions, injunctions and fines, and approaches to negotiation and arrests, were first developed and in turn affected the tactics of those they targeted. The evolution and effects of responses by workers and communities at the point of protest are similarly canvassed, as is the development of broader media and campaigning counter-strategies by affected industries.

In addressing such questions, this book also contributes to and draws on the sub-discipline of Green Criminology. This involves "the study of environmental harms and crimes, and their effect upon the planet, including both humans and non-humans".[80] Much of the work within this field addresses how criminal justice systems and differing communities and groups define and deal with activities that damage the environment and, by association, human health and well-being. The competing discourses and reactions to such "crimes" are analysed in relation to political, economic, cultural, legal, and other factors. Alongside criminology, more broadly, the field is also concerned with the policing and regulation of protest.[81]

This study contributes to green criminological scholarship by analysing blockader motivations for undertaking civil disobedience and other illegal acts. These are discussed in light of varying conservationist and Indigenous communities' ethical and ideological understandings of crime and injustice, and the ways in which they dispute, ignore, obey, and make use of existing law and legal practices. Further contributions concern the analysis of the entwinement of patterns of law and policing with those of ODA, as well as campaigners' role in "policing" and countering environmental harms and forcing courts and authorities to enforce or change existing laws and regulations relating to land use.

Methodological framework

Environmental Blockades combines the use of social movement theory regarding the context and emergence of ODA with an approach that Bevington and Dixon describe as "direct engagement". This involves "putting the thoughts and concerns of movement participants at the centre of the research agenda and showing a commitment to producing accurate and potentially useful information about the issues that are important to activists".[82] In line with this, I primarily reflect on and analyse the issues and opinions raised by activists rather than those of politicians, media commentators, and police. In part this is because many of the questions the book deals with – the forms, dynamics, and efficacy of ODA – are primarily of interest to social movement actors and have mainly been debated and written about by them.

I also concentrate on activist accounts in an attempt to bridge a gap between the concerns of academic and activist intellectuals that has been identified in a series of debates, books, and articles. Academic work on social movements has been characterised as being of a generally abstract, rarefied, and complex nature, seldom engaging with the practical, immediate concerns of the activists being studied.[83] These concerns have been described as in short, "what works and how".[84] Activists in turn have been criticised for being reactive and indifferent "to thinking through the strategic implications and broader meanings of their efforts".[85] Such deficiencies are said to weaken movement efficacy, with Meyer arguing that in lieu of detailed analysis activists often resort to "habits and belief, familiar routines and well established scripts for action that may not have ever been particularly effective or – even if optimal at one point – are less adapted to current circumstances".[86] By combining what McAdam describes as an analysis of "the role of system level factors in either facilitating or constraining movement activists" with accounts of "the lived experience of activism and everyday strategic concerns of movement groups", this book aims to inform the theoretical and practical concerns of both audiences.[87]

Chapter outline

Each of the chapters in the book deals with a key phase in the development of EB tactics and focuses on those campaigns that invented and promoted new methods or introduced them to a new region or country. The patterns, dynamics, and outcomes of organisation and innovation within each country are examined along with levels and methods of diffusion, and the relationships to workers, media, politicians, and other actors that developed.

The first chapter provides a history of the seminal 1979 Terania Creek campaign. It discusses the grievances involved and its tactical unfolding in the context of forestry practices and environmental activism of the time as well as protest policing and the culture of Northern NSW's "New Settler" movement. The impact of the campaign in terms of pioneering the EB template and introducing practices within it concerning organisation, infrastructure, tactics and interpretations of appropriate protester behaviour is emphasised.

The second chapter examines six Australian campaigns that followed between 1980 and 1984. It canvasses the introduction and expansion of new tactics as well as the way in which activists debated and applied guidelines regarding normative protester behaviour. Particular attention is paid to evolving patterns of protest policing as well as how the introduction of actors influenced by overseas models of action, and involved in formal environmental organisations, intensified and solidified processes of tactical, strategic and ideological differentiation.

The third chapter explores the emergence of environmental blockading in the United States from 1983 to 1986. This is linked to the foundation of the US Earth First! (EF!) network and the evolution of its experimental culture as a response to conventional conservation strategies. It analyses the role of direct connections and international publicity surrounding Australian campaigns in inspiring American activists to undertake their own blockades as well as the domestic roots of tactics and organisational forms.

The fourth chapter considers how environmental blockading became embedded as an action template in the United States and helped make the protection of old-growth forests a major national issue from 1987 to 1990. It demonstrates that 1990 was the point at which the results of the previous six year's capacity building enabled EF! to launch its first long term protest camps and mass forest defense campaign. The chapter also examines how shifts in the goals of activists and the role they assigned to ODA led to a major split in the network regarding strategy and approaches to normative protester behaviour emerged to exert an enduring impact upon contemporary activism.

Notes

1 Extinction Rebellion Occupy Greenpeace Offices, Available [Online]: https://risingup. org.uk/extinction-rebellion-occupy-greenpeace-offices (Accessed 1 September 2020).
2 Over 1,000 people block Parliament Square, Available [Online]: https:// extinctionrebellion.uk/2018/10/31/over-1000-people-block-parliament-sq-to-launch-mass-civil-disobedience-campaign-demanding-action-on-climate-emergency/ (Accessed 1 September 2020).
3 Extinction Rebellion says, 'We're Fucked," block access to Downing Street, Available [Online]: https://extinctionrebellion.uk/2018/11/14/breaking-london-extinction-rebellion-says-were-fucked-block-access-to-downing-street/ (Accessed 1 September 2020); Extinction Rebellion protests block London bridges, Available [Online]: www. bbc.com/news/uk-england-london-46292819 (Accessed 1 September 2020).
4 Extinction Rebellion – First time in living memory central London's bridges blocked by protest group, Available [Online]: https://extinctionrebellion.uk/2018/11/17/update-extinction-rebellion-first-time-in-living-memory-central-londons-bridges-blocked-by-protest-group/ (Accessed 1 September 2020).
5 Simon Murphy, "Environmental Protesters Block Access to Parliament Square," *Guardian*, 24 November 2018, Available [Online]: www.theguardian.com/environment/ 2018/nov/24/environmental-protesters-block-access-to-parliament-square-extinction-rebellion (Accessed 1 September 2020).
6 Sam Knights, "The Story So Far" in *This Is Not A Drill: An Extinction Rebellion Handbook*, ed. Extinction Rebellion (London: Penguin, 2019), 16–21; Diane Taylor, "Extinction Rebellion Activists Claim Victory in HS2 Tree Protest," *Guardian*, 27 April 2019, Available [Online]: www.theguardian.com/environment/2019/apr/27/extinction-rebellion-activists-scale-trees-in-anti-hs2-protest (Accessed 1 September 2020); India Block, Modular Boxes used by Extinction Rebellion are Protest Architecture, Available

[Online]: www.dezeen.com/2019/10/17/extinction-rebellion-protest-architecture/ (Accessed 1 September 2020).

7 Matthew Taylor and Damien Gayle, "Dozens Arrested After Climate Protest Blocks Five London Bridges", *Guardian*, 17 November 2018, Available [Online]: www.theguardian. com/environment/2018/nov/17/thousands-gather-to-block-london-bridges-in-climate-rebellion." (Accessed 1 September 2020).

8 Brian Doherty, "Manufactured Vulnerability: Protest Camp Tactics," in *Direct Action in British Environmentalism*, ed. Benjamin Seel, Matthew Paterson, and Brian Doherty (London, New York: Routledge, 2000), 70.

9 These were limited to protesters statically occupying a space until being removed by their opponents and did not involve people sitting in trees, attaching their bodies to objects, etc. Gene Sharp, *The Politics of Nonviolent Action* (Boston, MA: Extending Horizons Books, 1973), 371–74.

10 Doherty, "Manufactured Vulnerability," 70.

11 Doherty, "Manufactured Vulnerability", 70.

12 In this study, biodiverse areas are defined as those that have not been subject to resource extraction or forms of development that have otherwise majorly transformed sections of land since colonisation. Despite it being in common usage the thesis will not employ the term 'wilderness' to describe such areas, because the use of the term typically suggests that they are closed ecological networks devoid of culture and human interference. The concept of wilderness has been used by some Indigenous activists, but more typically has been criticised by them for feeding into processes that deny and erase First Nations ownership and understandings. Although the study will not use the term to denote these places, it will discuss how understandings of "wilderness" informed activist practice and debates. William Cronon, "The Trouble with Wilderness: Or, Getting Back to the Wrong Nature," *Environmental History* 1, no. 1 (1996): 7–25.

13 Benjamin Seel and Alexandra Plows, "Coming Live and Direct: Strategies of Earth First!," *Direct Action in British Environmentalism*, 115.

14 Verta Taylor and Nella Van Dyke, "'Get up, Stand Up': Tactical Repertoires of Social Movements," in *The Blackwell Companion to Social Movements*, ed. David Snow, Sarah Soule, and Hanspeter Kriesi (Malden, Oxford, Carlton: Blackwell Publishing, 2004): 265.

15 Charles Tilly, *Contentious Performances* (Cambridge: Cambridge University Press, 2008), xiii.

16 Wang and Soule, "Tactical Innovation in Social Movements: The Effects of Peripheral and Multi-Issue Protest," 518–20; Frances Polletta, *It Was Like a Fever: Storytelling in Protest and Politics* (Chicago: University of Chicago Press, 2006), 55; Jeff Larson, "Social Movements and Tactical Choice," *Sociology Compass* 7 (2013): 874.

17 Dan Wang and Sarah Soule, "Tactical Innovation in Social Movements: The Effects of Peripheral and Multi-Issue Protest," *American Sociological Review* 81, no. 3 (2016): 520.

18 Tilly, *Contentious Performances*, xiii–xv, 4–29; Donatella Della Porta, "Eventful Protest, Global Conflicts," *Distinktion* 17(2008): 29–32, 48–50; Sidney Tarrow, *Power in Movement: Social Movements and Contentious Politics*, Third ed. (Cambridge: Cambridge University Press 2011), 140–45; William H. Sewell Jr, *Logics of History: Social Theory and Social Transformation* (Chicago, IL: University of Chicago Press, 2009), 6–14.

19 Ibid; 225–227.

20 Although certain patterns are clear by 1990, the timeline includes another seven years of information to establish the influence of earlier periods of embedding. The timeline and further information regarding sources and parameters can be found at Iain

McIntyre, Environmental Blockading Timeline, 1974–1997 Available [Online]: https://commonslibrary.org/environmental-blockading-in-australia-and-around-the-world-timeline-1974-1997/ (Accessed 8 November 2020).

21 Ibid. These figures are based on counting blockades which took place in distinctly separate locations. If a single campaign involved a number of actions that all targeted the same mining, clearing, or logging operation, or neighbouring ones, then they have not been counted separately.

22 Although the timeline is primarily based on materials produced in industrially developed and English-speaking nations, over-representation of these has been mitigated as groups based in Australia, Canada, New Zealand, Europe, and the United States provided support to activists in non-English speaking and developing countries and activist sources regularly carried and highlighted news from them. This was particularly the case with magazines such as the World Rainforest Report, Survival International and New Internationalist.

23 Vanessa Bible, "Aquarius Rising: Terania Creek and the Australian Forest Protest Movement" (Honors thesis, University of New England, 2010), 36–50.

24 John Seed recalled in an interview with Christopher Rootes that some involved in the Terania Creek campaign had travelled to India and were familiar with the Chipko campaign. However, other blockaders involved appear not to have known of this precedent and describe the blockade as "spontaneous". For more discussion of this refer to Chapter One. Christopher Rootes, "Exemplars and Influences: Transnational Flows in the Environmental Movement," *Australian Journal of Politics and History* 61, no. 3 (2015), 429; Dudley Leggett, Interviewed 5 February 2015; Nan Nicholson, Interviewed 3 August 2015; Lisa Yeates, Interviewed 29 January 2015.

25 Rik Scarce, *Eco-Warriors: Understanding the Radical Environmental Movement* (Chicago, IL: Noble Press, 1990), 157–58.

26 Roger Wilson, *From Manapouri to Aramoana* (Auckland: Earthworks Press, 1982), 120–24.

27 Tino Järvikoski, "Alternative Movements in Finland: The Case of Koijarvi," *Acta Sociologica*, January (1981), 313–15.

28 Andrew Revkin, *The Burning Season: The Murder of Chico Mendes and the Fight for the Amazon Rainforest* (Washington, DC: Island Press, 2004), 123–45; Luis Barbosa, *The Brazilian Amazon Rainforest: Global Ecopolitics, Development and Democracy* (Lanham, MD: University Press of America, 2000), 116–31; Gomercindo Rodrigues, *Walking The Forest with Chico Mendes: Struggle for Justice in the Amazon* (Austin: University of Texas, 2007), 100–22.

29 Ibid.

30 Ian Cohen, *Green Fire* (Sydney: Angus & Robertson, 1997), 31–209.

31 Anthony Silvaggio, "The Forest Defense Movement 1980–2005: Resistance at the Point of Extraction, Consumption, and Production" (PhD thesis, University of Oregon, 2005), 122–24, 75.

32 Nicholas Blomley, "'Shut the Province Down': First Nations Blockades in British Columbia, 1984–1995," *BC Studies: The British Columbian Quarterly*, no. 111 (1996).

33 Patrick Wolfe "Settler Colonialism and the Elimination of the Native," *Journal of Genocide Research* 8 no.4 (2006): 387–409.

34 The core members are said to be Australia, Canada, New Zealand, the United Kingdom and the United States. Srdjan Vucetic, *The Anglosphere: A Genealogy of an Identity in International Relations* (The Ohio State University, 2008): 1–6; Carl Bridge and Kent Fedorowich, "Mapping the British World," *The Journal of Imperial and Commonwealth History*, 31 no. 2 (2003): 1–15.

35 Chrissa Scholtz, *Negotiating Claims: The Emergence of Indigenous Land Claim Negotiation in Australia, Canada, New Zealand and the United States* (New York: Routledge, 2006), 15–16.

36 Thomas Dunlap, *Nature and the English Diaspora* (Cambridge: Cambridge University Press, 1999), 13–15, 165–217; Michael Howlett and Jeremy Rayner, "The Business and Government Nexus: Principal Elements and Dynamics of the Canadian Forest Policy Regime," in *Canadian Forest Policy: Adapting to Change*, ed. Michael Howlett (Toronto: University of Toronto Press, 2001), 29–43.

37 Dunlap, *Nature and the English Diaspora*, 118–36; Jeremy Wilson, *Talk and Log: Wilderness Politics in BC, 1965–96* (Vancouver: University of British Columbia Press, 1998), 48–71, 208–15.

38 The "first wave" of environmentalism is generally dated from the late nineteenth century. It focused on environmental degradation and alternatives to it and led to the creation of national parks in a number of countries. Andrea Olive, *The Canadian Environment in Political Context* (Toronto: University of Toronto Press, 2016), 83–97.

39 Dunlap, *Nature and the English Diaspora*, 275–304; Philip Van Huizen, "'Panic Park': Environmental Protest and the Politics of Parks in British Columbia's Skagit Valley," *BC Studies*, no. 170 (2011): 72; Libby Connors and Drew Hutton, *A History of the Australian Environmental Movement* (Cambridge: Cambridge University Press, 1999), 117–27.

40 Roger Hayter and John Holmes, "The Canadian Forest Industry: The Impacts of Globalization and Technological Change," in *Canadian Forest Policy: Adapting to Change*, 127–45; Sierra Braggs, "Earth First!: The Rise of Eco-Action" (Master's thesis, Humboldt State University, 2012): 22–25.

41 Connors and Hutton, *A History of the Australian Environmental Movement*, 117–27, 56–68, 223–39; Anthony Silvaggio, "The Forest Defense Movement," 57–83, 113–40.

42 Gregory M. Maney et al., "Introduction," in *Strategies for Social Change*, ed. Gregory M. Maney et al. (Minneapolis: University of Minnesota Press, 2012), xviii.

43 James Jasper, *The Art of Moral Protest: Culture, Biography, and Creativity in Social Movements* (Chicago: The University of Chicago Press, 1997), 5–43, 229–50; John D. Mccarthy and Mayer N. Zald, "Resource Mobilization and Social Movements: A Partial Theory," *American Journal of Sociology* 82, no. 6 (1977): 1212–38; Larson, "Social Movements and Tactical Choice", 866–74.; David Meyer, "Protest and Political Opportunities," *Annual Review of Sociology* 30 (2004): 125–41.

44 Doug McAdam and Sidney Tarrow, "Ballots and Barricades: On the Reciprocal Relationship between Elections and Social Movements," *Perspectives on Politics* 8, no. 2 (2010): 529.

45 Alaina Michaelson, "The Development of a Scientific Speciality of Diffusion through Social Relations: The Case of Role Analysis," *Social Networks* 15 (1993): 217.

46 Sarah Soule, "Diffusion Processes within and across Movements," in *Blackwell Companion to Social Movements*, ed. David Snow, Sarah Soule, and Hanspeter Kriesi (Oxford: Blackwell, 2004), 295.

47 Hanspeter Kriesi et al., *New Social Movements in Western Europe: A Comparative Analysis* (Minneapolis: University of Minnesota, 1995).

48 Charles Tilly, *Regimes and Repertoires* (Chicago, IL: University of Chicago, 2006). 42.

49 Doug MacAdam, Sidney Tarrow, and Charles Tilly, *Dynamics of Contention* (Cambridge: Cambridge University Press, 2001). 142, 57.

50 Tarrow, *Power in Movement*, 205–6.

51 Conny Roggeband, "Translators and Transformers: International Inspiration and Exchange in Social Movements," *Social Movement Studies* 6, no. 3 (2007): 246.

52 Sean Scalmer, "Translating Contention: Culture, History, and the Circulation of Collective Action," *Alternatives* 25, no.4 (2000): 492–493; Sean Scalmer, *Gandhi in the West: The Mahatma and the Rise of Radical Protest* (Cambridge: Cambridge University Press, 2011).

53 Voltairine De Cleyre, *Direct Action* (New York: Mother Earth Publishing Association, 1912). 1–19.

54 Reclaim The Streets, Propaganda, Available [Online]: http://rts.gn.apc.org/prop15.htm (Accessed 10 October 2010).

55 Tarrow, *Power in Movement*, 97.

56 Damian Grenfell, "The State and Protest in Contemporary Australia: From Vietnam to S11" (PhD thesis, Monash University, 2001), 1–2.

57 Alberto Melucci, *Challenging Codes: Direct Action in the Information Age* (Cambridge: Cambridge University Press, 1996). 378–79.

58 Sharp characterises such obstructive actions as "physical intervention", but his term only allows for the use of human bodies and does not include barricades and other physical installations. Gene Sharp, *The Politics of Nonviolent Action* (Boston: Extending Horizons Books, 1973), 371–89.

59 John Rawls, *A Theory of Justice* (Cambridge, MA: Belknap Press, Revised edition, 1999), 320.

60 This is a partially original expansion of Reinsborough's categories. Patrick Reinsborough, "Post Issue Activism", in *Globalize Liberation: How to Uproot the System and Build a Better World*, ed. David Solnit (San Francisco: City Lights Books, 2004), 183–85.

61 "Soft blockades" is a term that does not appear to have been used prior to 1990, but has become part of activist parlance since. Lesley Wood, *Direct Action, Deliberation and Diffusion: Collective Action After the WTO Protests in Seattle* (Cambridge: Cambridge University Press, 2012), 36.

62 "Enhanced vulnerability" is an original derivation of Doherty's term "manufactured vulnerability" with the word enhanced used to distinguish these tactics from soft blockading. Doherty, "Manufactured Vulnerability", 70.

63 William Gamson, "Reflections on the Strategy of Social Protest," *Sociological Forum* 4, no. 3 (1989): 455–66; Mccarthy and Zald, "Resource Mobilization and Social Movements," 1212–38; Charles Tilly, *From Mobilization to Revolution* (Reading: Addison-Wesley, 1978), 12–50.

64 James Jasper and Francesca Polletta, "Collective Identity and Social Movements," *Annual Review of Sociology* 27 (2001): 292–300.

65 Ana Acosta, "Quakers, the Origins of the Peace Testimony and Resistance to War Taxes," in *War and Peace: Essays on Religion and Violence*, ed. Bryan Turner (London: Anthem Press, 2013), 113.

66 Sharp, *The Politics of Nonviolent Action*, 359–67; Anthony Silvaggio, "The Forest Defense Movement," 91–94, 106–10.

67 Sarah Maddison and Sean Scalmer, *Activist Wisdom: Practical Knowledge and Creative Tension in Social Movements* (Sydney: UNSW Press, 2005), 215–21.

68 Ralph Turner and Turner Killian, *Collective Behaviour* (Sydney: Prentice-Hall, 1987). 219.

69 Quoted in Cynthia Kaufman, *Ideas for Action: Relevant Theory for Change* (Cambridge: South End Press, 2003), 292.

70 Jo Freeman, *The Politics of Women's Liberation a Case Study of an Emerging Social Movement and Its Relation to the Policy Process* (New York: D. McKay Publications, 1975), 236; Verity Burgmann, "The Importance of Being Extreme," *Social Alternatives* 37 no.2 (2011), 10–12.

71 Tarrow, *Power in Movement*, 98.

72 Anna Feigenbaum, Fabian Frenzel, and Patrick McCurdy, *Protest Camps* (London: Zed Books, 2013), 70.

73 Ibid., 116–18.

74 Feigenbaum, Frenzel, and McCurdy, *Protest Camps*, 127.

75 Ibid., 125–34.

76 Ibid., 125.

77 Thomas Weber and Robert Burrowes, "Nonviolence: An Introduction," *Peace Dossier* no. 27 (1991): 1–10.

78 Amory Starr, *Global Revolt: A Guide to the Movements against Globalization* (London, New York: Zed Books, 2005), 134.

79 Donatella della Porta and Herbert Reiter, *Policing Protest: The Control of Mass Demonstrations in Western Democracies* (Minneapolis: University of Minnesota Press, 1998), 9–25; David Schweingruber, "Mob Sociology and Escalated Force: Sociology's Contribution to Repressive Police Tactics," *The Sociological Quarterly* 41, no. 3 (2000): 383–86; Jan Terpstra, "Policing Protest and the Avoidance of Violence: Dilemmas and Problems of Legitimacy," *Journal of Criminal Justice and Security* 8, no. 3–4 (2006): 207–11.

80 John Cianchi, *Radical Environmentalism: Nature, Identity and More-Than-Human Agency* (Basingstoke: Palgrave-Macmillan, 2015), 3.

81 Rob White and Diane Heckenberg, *Green Criminology: An Introduction to the Study of Environmental Harm* (London: Routledge, 2014), 22–36; Ragnhild Aslaug Sollund, "Introduction: Critical Green Criminology – An Agenda for Change," in *Green Harms and Crimes: Critical Criminology in a Changing World*, ed. Ragnhild Aslaug Sollund (Basingstoke: Palgrave Macmillan, 2015): 3–13; Michael Lynch et al., *Green Criminology* (Oakland: University of California Press, 2017), 1–45; Avi Brisman and Nigel South, "A Green-Cultural Criminology: An Exploratory Outline," *Crime, Media, Culture: An International Journal* 9, no. 2 (2013): 115–25.

82 Douglas Bevington and Chris Dixon, "Movement-Relevant Theory: Rethinking Social Movement Scholarship and Activism," *Social Movement Studies* 4, no. 3 (2005): 200.

83 Maddison and Scalmer, *Activist Wisdom*, 37–39.

84 David Meyer, "Scholarship That Might Matter," in *Rhyming Hope with History: Activists, Academics and Social Movement Scholarship*, ed. David Croteau and William Hoynes (Minneapolis: University of Minneapolis Press, 2005), 201.

85 Marcy Darnovsky, Barbara Epstein, and Richard Flacks, "Introduction," in *Cultural Politics and Social Movements*, ed. Marcy Darnovsky, Barbara Epstein, and Richard Flacks (Philadelphia: Temple Press, 1995), xvi.

86 Meyer, "Scholarship That Might Matter," 196.

87 Doug McAdam, "The Framing Function of Movement Tactics: Strategic Dramaturgy in the American Civil Rights Movement," in *Comparative Perspectives on Social Movements: Political Opportunities, Mobilizing Structures and Cultural Meanings*, ed. Doug McAdam, John McCarthy, and Mayer Zald (Cambridge: Cambridge University Press, 1996), 339.

1

"THE FOREST IN QUESTION WAS REGARDED AS TERRITORY"

The Terania Creek campaign and the creation of the environmental blockading template

As discussed in the Introduction, Australia's first forest blockade took place at Terania Creek in 1979. This campaign was a catalytic event in an Australian cycle of protest that extended between 1979 and 1984, popularising environmental blockading in the process. It pioneered a new combination of consensus decision-making, protest camps, and normative protester behaviour. And it had a global influence on environmental struggle.

This chapter will explore and analyse the unfolding of this pioneering campaign as well as the conditions under which it developed. It will first consider the grievances that gave rise to the protest and the political context it emerged from before providing a history of the blockade itself. As a case study presenting many dynamics which remain current in blockading it introduces a number of processes and factors shaping tactical choice and innovation.

The founding of the Terania Native Forest Action Group

The Terania Creek campaign and the forms of protest it developed grew out of a specific community's concern with the natural environment. It was shaped by members' pre-existing attitudes towards politics, protest, and the law, and dissatisfaction with contemporary life. From the early 1970s onwards up to one thousand people moved to rural northern New South Wales (NSW) in response to disenchantment "with the technocratic, economic, and political realities of the city, and unease with the social divisiveness and environmental impact of modernity".[1] A distinctive set of lifestyles, labelled the "New Settler" movement, emerged, combining economic and cultural alternatives involving communes, spirituality, self-sufficient agriculture, and environmentally sustainable technology. This reflected a global phenomenon as tens of thousands of members of radical and countercultural movements in Western countries left cities in attempts to explore alternative lifestyles and more immediately implement cultural and economic change.[2]

The campaign to prevent logging at Terania Creek, a 700-hectare remnant of rainforest in rural northern NSW, began in 1974 when two NSW Forestry Commission (FC) workers were discovered working on roads in the state government–owned forest, neighbouring land owned by Nan and Hugh Nicholson. When they learnt that the FC intended to clear fell and burn the forest before replanting it with Flooded Gum, they and others formed the Channon Resident's Group, later renaming it the Terania Native Forest Action Group (TNFAG).[3]

The form of organisation that these activists drew upon, that of a "resident action group", had been popularised in Australia during the late 1960s and early 1970s. Such semi-formal political bodies were based in a specific neighbourhood or area, organised on a "grassroots" rather than party political basis, and campaigned against activities by municipal and state governments and private companies that were deemed to be a threat to the social and environmental health of local communities. Their form of activity and organisation drew on examples from the United States and the United Kingdom, as well as Australian traditions of community organising during the 1930s and 1950s, and extended the existing protest repertoire through connections to the "green bans" movement.[4] 'As a modular form of organisation, it was easily extended from its urban origins to rural campaigns such as that at Terania Creek.

While the countercultural nature of the New Settler community meant that it largely eschewed the accumulation of income and capital, significant sections of its membership were highly resourced in other ways. A number of those involved in Terania Creek and the campaigns that followed had developed skills in public relations, journalism, research, and other areas through the careers they had pursued before "dropping out" and new ones were learnt as required. Many of these individual's pre-existing political and media contacts were reactivated and exploited as the group engaged in scientific research, lobbying, and a media campaign. Although involvement fluctuated, by 1979 TNFAG numbered around 30 members with a core of ten who had developed substantial campaigning skills.[5]

Although concerned about road safety due to the presence of logging trucks, TNFAG was primarily troubled with what they saw as the unnecessary destruction of a beautiful area. As with many involved in future blockades those living close to Terania Creek had formed an intimate connection with the ecosystem and perceived that it had an intrinsic worth enhanced by the degree to which it had remained undeveloped. Over the decades such values and emotional attachments would drive many activists to engage in long and exhausting campaigns as well as put their bodies at risk. As Nan Nicholson recalled in an unpublished manuscript:

> The forest in question was regarded as territory, producing very strong protective instincts in those who lived next door or very close at hand… No political reasons or ideology could have given such impetus to opposition as the sight of such a valley night and morning.[6]

The political and bureaucratic context of the campaign

TNFAG's opposition to logging formed part of growing scientific criticism regarding bureaucratic administration and government policies. The body responsible for overseeing the exploitation of NSW's publicly owned forests, the Forestry Commission, was typical of many around Australia and the world at the time. Its critics considered it to favour commercial value, employment, and sawmiller' profits over biodiversity. Policies giving precedence to logging over conservation in turn led to forms of regeneration that prioritised individual species over the conservation and maintenance of entire ecosystems.[7]

Activism concerning forest issues was led by a new wave of conservation organisations which had emerged in the 1970s, such as the Colong Committee and Total Environment Centre (TEC), or developed out of older ones such as the Australian Conservation Foundation (ACF). Dominated by a core of fulltime activists, these bodies, unlike their polite forebears, were willing to engage in open, albeit civil, conflict with their opponents. Engaging in public protest they focused on winning campaigns by lobbying business and government, building political alliances, and mobilising public support via the media.[8]

Sydney-based organisations had been campaigning to preserve the remaining 253,000 ha of rainforest in the state, 12.5 per cent of which had not been logged or developed since colonisation.[9] Their aim was to protect large swathes of forest in national parks, in which logging and extractive activities would be banned. Although the FC remained unswayed, political opportunities had begun to open up with the election of the Australian Labor Party (ALP) to the State Government of NSW on a strong environmental platform in 1976. Although the party was divided over its policies, activists enjoyed considerable support in cabinet and the Premier's Office.[10]

As the ALP had begun establishing inquiries, rescinding FC decisions, and negotiating industry compensation in order to create new national parks, the major organisations considered Terania Creek, sections of which had been previously logged, as too insignificant to prioritise. Privately these increasingly professionalised organisations doubted the ability of what they viewed as the "hippies" associated with TNFAG to run an effective campaign and advised their ALP allies that the issue was unlikely to develop into one of concern to them or the public.[11]

Despite this, and the fact that the NSW Forestry Act lacked formal public input processes, TNFAG pressed their case by writing over 150 letters to interested parties. They also presented submissions to politicians, held public meetings and protests, and produced Australia's first environmental TV advertisement.[12]

Faced with opposition, the FC implemented compromises regarding transportation and the species to be cut. For a bureaucracy little used to negotiation, these may have seemed major concessions, but they were rejected by TNFAG. In line with ALP policy and draft legislation, the group demanded an Environmental Impact Statement (EIS) be issued before logging could commence. The request was denied and by February 1977 key conservation allies Premier Wran and Environment

Minister Paul Landa had made it clear that they would not risk a confrontation with pro-logging forces within their party. Despite the minimal profit of $9,000–15,000 expected on logging, the FC provided Lismore company Standard Sawmills with a licence to begin operations.[13]

From TNFAG's perspective the primary issues for the FC concerned the wider consequences for the timber industry and a desire to maintain its decision-making dominance. Added to this were gender dynamics. Although women, mainly the wives of loggers and contractors, would later take a leading role in pro-logging groups such as Ladies for Environmental Awareness in Forests (LEAF), none held official roles in the industry. As Nan Nicholson affirms, "It was clear it was about emotions, it was about them being challenged. The other point was that our group was half women and they'd never had women taking the initiative."[14]

The creation of a protest camp and other improvisations

Logging would not begin until two years later. By May 1979 TNFAG was concerned enough to raise the possibility of Obstructive Direct Action (ODA), with spokesperson Keith Bashford telling the *Australian*, "We hoped it would not get this far, but if it comes to that, we will lie down in front of the trucks and bulldozers."[15]

With TNFAG continuing to hope that the government would shift its position, no specific response had been planned by the time logging became imminent. Nevertheless, the combination of a strong grievance, isolation from more moderate conservationists, and the basis of TNFAG in a community formed around alternative values soon saw a new strategic response emerge.[16]

On Sunday, 12 August 1979, TNFAG called a protest meeting at Terania Creek. This came at the end of the nearby Channon market day, a popular event among the region's New Settlers, and drew around 100 participants. When the initial mass meeting, which had been directed from a stage, began to lose focus a circle was formed and a form of consensus decision making introduced.[17] As Jasper has observed, the organisational forms groups innovate and adopt are "never simply the most efficient way to solve some problem", but are also rooted in the culture and values of those involved, forming "symbols and myths designed to send messages to various audiences".[18] Reflecting critiques of wider societal power relations and an interest in direct democracy that already existed amongst the New Settler movement, consensus decision making also had the practical aim of fostering inclusivity.

The processes used at Terania Creek were not as highly formalised as those, largely codified by the US organisation Movement for a New Society (MNS), widely adopted among Australian peace and environmental activists during the 1980s. Instead, as with much during the protest, they evolved as events unfolded. Meetings involved people passing a conch shell to indicate whose turn it was to speak, with unofficial and rotating facilitators managing the process and helping bring the group to an agreement. Singing, the chanting of "Om" and other spiritual practices were closely entwined with the process.[19]

Forms of consensus decision making remain commonly used in many blockades. Critics argue that they only work in small, homogenous groups, are time-consuming, allow minorities to block decisions, and are open to manipulation by popular or unscrupulous participants.[20] Such issues were occasionally apparent at Terania Creek, although they would not be widely aired within Australia until later campaigns.

A greater issue during Terania Creek, and one that has recurred nationally and internationally, arose around the enforcement of agreements when individuals or groups failed to recognise or abide by them or participate in meetings. The lack of precedents, the fact that involvement in the protest was open to anyone who turned up, and that there had been little time available to establish binding norms, intensified the heterogeneity of opinions and positions concerning appropriate tactics and behaviour.[21]

In addition to these pressures was the intense emotional identification with the forest that many already possessed or developed through living in it and directly witnessing its destruction. In his study of Tasmanian environmentalists during the 2010s, Cianchi observes "nature is more than a rationale for action", but acts as "an active contributor to activist self-identity and culture".[22] For many activists he found that a "pivotal or epiphanic experience" regarding "the transcendent power of nature" as well as "connection and communication" with it became "fundamental to their self-identity, their motivation to defend nature and their willingness to engage in direct action".[23] As will be seen, this has been a key factor throughout the history of environmental blockading.

Cianci observes that while his interviewees touched on spiritual dimensions they were reluctant to discuss them in depth for fear of appearing "hippy" or "foolish or unprofessional, and therefore someone who is not taken seriously by other activists, foresters, police or media".[24] Some environmentalists, particularly in professionalised organisations, were similarly concerned with being associated with alternative subcultures and spirituality during earlier periods. Others have proudly expressed their connection to subcultures and nature through music, art, clothing, writing and ritual. For those Indigenous communities who sought to protect biodiverse places, particularly in Canada, cultural expressions of connection to land, and sovereignty regarding it, were and remain central to campaigning.

Theorists such as Jasper have emphasized the importance of "moral shocks", images or events that create outrage thereby mobilising new participants and radicalising and reinforcing the commitment of existing ones.[25] Indirect exposure to the damaging of biodiverse places, often due to media coverage of protest within them, typically acts as a moral shock for audiences during environmental campaigns. The combination of directly bearing witness to destruction and the state's complicity in it has also had a consistently powerful effect on blockaders. At Terania Creek, as elsewhere, a mixture of emotions and experiences deepened commitment to the campaign and marked for many the beginning of lifelong involvement in environmental activism. The urgency and militancy that such feelings engendered would drive the protest to use ODA, and some among it to break consensus regarding acceptable tactics.[26]

During Terania Creek immediate interventions sometimes served to convince breakaway factions to adhere to group decisions. Where this was not possible, actions that were contrary to the stated aims of the majority occasionally occurred. Some future campaigns, beginning with that against the Franklin Dam, have endeavoured to use binding rules and other means to prevent such transgressions. More commonly, as at Terania Creek, key activists seek to maintain consensus regarding acceptable behaviour by reiterating agreements and discussing them with individuals believed likely to break them. Where this fails and actions are undertaken that are viewed as harmful and divisive, public apologies and denunciations are often issued.[27]

Out of the initial rally and meeting came further improvisations. It was agreed that people would remain on site to monitor and protest logging and that a camp would be set up on the Nicholson's land. People divided themselves into teams responsible for various tasks and within days basic requirements concerning food, sanitation, accommodation, legal support, childcare, and entertainment had been met for hundreds of people. The creation of the camp would later prove decisive as its location next to the logging site allowed blockaders immediate access to the point of protest.[28]

At this time protesters were unaware that the camp would remain in place for almost a month. Protest camps had been employed at sustained anti-nuclear and environmental protests in the United States and Germany and as part of two three-day occupations at a proposed aluminium smelter site earlier in 1979 at Wagerup, Western Australia.[29] Those present at Terania Creek were largely or wholly unaware of such precedents. Instead, the experimental, frugal, and communal nature of the New Settler movement was applied to planning and carrying out environmental campaigning.[30]

While day-to-day decisions regarding protest tactics and camp organisation would flow from the circle meetings and teams, a division of labour soon emerged regarding the direction and implementation of the broader political and media aspects of the protest. These duties were largely performed by a core of 10–15 people based inside the Nicholson's single-room home. Membership of this group appears to have mainly been self-selected and involved those who had been carrying out these roles in the campaign for years and already developed relevant contacts and skills.[31]

Many, although not all, later blockades would echo this allocation of tasks. Alongside other elements of organisation the way in which blockaders and political campaigners relate, and the degree to which they play separate roles, forms an important influence upon tactical and strategic dynamics. Although later campaigns would see tensions, during the Terania Creek campaign little dissent concerning the actions of the informal leadership appear to have arisen. This can be attributed to the fact that major decisions were ratified at circle meetings and the office made easily accessible. The localised nature of the campaign, which meant many involved knew each other personally or by reputation, and the respect that TNFAG members had built up over time also contributed to the level of trust.[32]

Although logging was yet to commence, and the tactics against it yet to emerge, the camp resolved it would be "non-violent" from its inception.[33] This adoption of "non-violence" as a key determinant regarding appropriate tactics and normative protester behaviour was another campaign theme that would be commonly followed in the future. The meaning of "non-violence", however, has been highly variable.

Some accounts of Australian blockading have tended to stress the primacy of "Orthodox Non-violence" (ONV) as an approach. ONV is a term associated with a set of practices that were first advocated by Gandhi and further refined through pacifist, anti-nuclear and civil rights movements from the 1950s onwards. Processes, protest forms, and ethics developed by MNS, and codified in a manual it produced in 1977, formed the basis of many ONV activists' practice in Australia and elsewhere. Adherents believe that any form of physical and verbal aggression, property damage, or secrecy dehumanises all involved and creates an inconsistency between means and ends. Individuals should not resist or avoid arrest and groups should adopt a strategy of fully briefing and engaging in dialogue with the police, media and other parties before engaging in protest or civil disobedience. ONV proponents use protest as a radical form of "bearing witness" and aim to convert their opponents, state actors, and the public at large by reaching out to their humanity via dialogue and a willingness to suffer for their beliefs. As part of their actions, they emphasise the importance of workshops, role plays, and training aimed at preparing activists for actions and minimising unpredictable situations, which to their thinking often result in violence.[34]

Although it would become popular among certain organisations and networks, ONV, as this book demonstrates, has never achieved a wholly dominant position within Australian and other countries' environmental movements, except for short periods and specific campaigns. Instead, fluid and less defined approaches have usually been employed, even when organisers and official campaign material have stated otherwise. At Terania Creek ONV was an influence on some activists but did not constrain the campaign's approach and methods to the degree it would to at the Franklin River blockade three years later.[35] Instead of taking its cues from overseas models, as that protest largely did, Terania Creek was an example of a domestically created approach to normative protest behaviour that was far less prescriptive.

At Terania Creek, the initial consensus concerning protester behaviour was that physical and verbal confrontation and the sabotage of logging equipment should be avoided. While this was rooted in ethical concerns, it was also predicated on the belief that such incidents would prejudice the public.[36]

Although non-violence was seen as promoting safer relations with the police and loggers, and protesters engaged in dialogue aimed at convincing these parties of the correctness of their views, the majority do not appear to have viewed "conversion", in the ONV sense, as a factor that would become decisive. "Openness" with opponents and police was limited to informal discussions and no detailed briefing of tactics took place. Rather, protesters wished to maximise disruption through the advantage of surprise.[37]

Cumulatively these political methods make the campaign a precursor of the approach that would be adopted in many of the environmental blockades that followed nationally and internationally: one which favoured loose rather than strict guidelines regarding normative behaviour. It is important to stress that concepts of nonviolence are almost always contested and evolving but were particularly so here due to the pioneering nature of the protest and the fact that its use of ODA was improvised rather than planned. Even the core commitment to avoiding physical violence at Terania Creek was occasionally tested with at least one incident of rock throwing prompting the intervention of other protesters.[38]

Despite having a protest camp in place, TNFAG still lacked a clear tactical plan at this point for how to respond to the impending arrival of logging trucks. When asked why this was this the case Nan Nicholson reflected:

> I felt that we were so right, and I was very innocent in this respect, that surely they would see sense and it would not come to a confrontation. We were also incredibly busy with trying to achieve a last-minute reprieve and inexperienced in protesting. As a result the question of what to do next was just pushed aside. Over time we became a lot more experienced and confident in blockading and were able to plan things very closely, but at that time it wasn't considered in any depth and so everything that followed was fairly spontaneous.[39]

The blockade

Once an agreement had been made to continue with some form of protest, however vague, preparations at Terania Creek rapidly took shape. Activists began reconnoitring the area using motorbikes and horses and communicated via walkie talkies. A "telephone tree", whereby people would be called and then in turn call others, was also set up.[40]

Generally, the obstructive tactics used at environmental blockades are aimed at first denying workers and equipment access to the site and then, where this fails, at interfering with ongoing work. Utilising the traditional logic of barricading, protesters at Terania Creek sought to achieve the former by placing cars, some with their tyres removed, on the main road.[41]

Two days after setting up the protest camp, the first ODA took place when a pair of forestry workers, claiming to be checking drains, entered Terania Creek Road in a truck. Faced with the choice of allowing them to pass or standing in their way, people took the latter option, thereby adding a tactic that was already widely known and commonly used in other contexts to the emerging environmental repertoire. Due to this spontaneous occupation the vehicle's progress was halted and the workers turned back.[42]

Although this was an important moment in that it marked the first use of disruptive protest in an Australian forest it received little commentary at the time, possibly because it did not represent a major attempt to begin logging operations and did not result in confrontation (merely a threat of violence from one of the loggers towards a camera person). Instead, the event which most protesters later recalled as foundational came on Thursday, 16 August, when a work crew was discovered attempting to sneak into the area along MacKays road. Their passage was quickly halted. Radio and telephone alerts brought hundreds to join those already sitting in the path of the vehicle.[43]

With the workers unable to advance or retreat, as vehicles had been hurriedly parked behind them, an agreement was struck whereby they and their Komatsu tractor were permitted to leave. A festive atmosphere reigned and, as throughout the entire protest and future events, music was used to promote group bonding and resilience.[44] Musician Lisa Yeates asserts:

> I think most people would agree that the protest would have been different without music, which was very powerful in such an emotionally charged environment. It was already a key part of this community and had the power to calm people down or ramp them up or make them laugh.[45]

Despite a lack of arrests and absence of violence the spectacle of hundreds of brightly garbed protesters blocking a bulldozer was one that captured media and public attention. With journalists on site, two TV stations sending helicopters from Brisbane, and TNFAG members and film makers Jenni Kendall and Paul Tait passing on footage of the protest to the ABC and other outlets, the event headed up news reports and the blockade would continue to feature as a major story over the coming weeks.[46]

Although some activists would later grant a primacy to ODA and signal their intention to use it from the onset of a campaign, its use at Terania Creek was in keeping with what will be seen to be a more common situation: that confrontational forms of protest are resorted to only after conventional approaches have failed. Challenged by political closure campaigners often face the choice of giving up or turning to unconventional and confrontational means. However, as Jasper has argued, understanding why some groups embrace disruptive tactics while others do not, as well as the how innovatively they create and use them, requires attention to "tactical tastes", which "like most cultural sensibilities, combine morality, emotion, and cognition".[47]

Scholars have observed that a variety of internal movement features can lead to higher rates of innovation. These include the existence of community-based institutions in which dialogue can take place, egalitarian decision making, and the advocacy of creativity as part of activist identity.[48] At Terania Creek the use of blockading and the wide array of new protest forms that followed can be attributed to these elements and more. Although people in the area were unaware of the recent tree-sitting campaign in New Zealand or the Wagerup occupation, many had participated in political actions concerning the Vietnam War, liberation movements, and trade unions, with some also having attended disruptive protests and pickets in connection with these campaigns. This background, the lack of a distinct plan, and a broad definition of acceptable behaviour would give protesters a wide remit to adapt and create tactics as they went along, producing a new protest repertoire in the process.[49]

Beyond such factors the nature of the New Settler movement and the alternative scenes it grew out of provided other resources that would be of vital importance, both during Terania Creek and blockades that followed. Chief among these was a cohort of people willing and able to undertake ODA within a community who viewed the use of protest positively. The frugal and communal quality of the region's lifestyles, and the ability of many of its members to gain support through Australia's welfare system, meant that there was a large pool of people who were not constrained by work and financial commitments from undertaking radical campaigns. The willingness of those with business, farming and other commitments to disrupt their daily routines and neglect duties underscored their commitment to the blockade.[50]

Some protesters also lacked familial responsibilities. Those who did have dependents were not prevented from participating however as children were either brought along or responsibilities communally shared or swapped between couples for a period in rotation.[51]

While a few among the New Settlers opposed the blockade, on the basis that it would intensify conflict with others in the region, a majority supported it and where they were not physically available donated food and other supplies. Further to this the community's focus on the arts provided another level of resources that could be drawn upon in terms of graphics, banners, and songs.[52]

As discussed in the Introduction, an important determinant of tactical innovation and selection is the role that activists assign to environmental blockading within the broader aims of their campaigns. The degree to which campaigners prioritise and balance goals, the importance they place on blockading within their overall strategy, and their views concerning the nature and role of the mainstream media all contribute to tactical direction. This is not least because they influence how much activists aim to apply actual, rather than perceived, disruption.

As a pioneering and largely ad hoc event, Terania Creek lacked a comprehensive strategy for how environmental blockading could achieve specific ends. Although gaining the public's attention and signalling opposition were core goals, the blockade had not begun with publicity foremost in mind. As a result, the prevention of work would remain a key concern. Tactics were mostly improvised according to what seemed to successfully obstruct, with the proviso being they should not overly alienate media and police and fit into the previously described definition of non-violence.[53]

Despite widespread publicity, blockading did not yet change the ALP's position. This was demonstrated by the arrival of more than 100 police officers in 38 vehicles on Friday, 17 August. In response protesters headed to MacKays road to once more supplement the dozens of vehicles in place with their bodies. Although tow trucks removed cars, and 17 people were arrested, it took 90 minutes to clear the road. A parallel incident, which demonstrates the wide variety of tactics employed within this ostensibly non-violent campaign, involved Hugh Nicholson pouring petrol along his property's boundary. Having threatened to set it alight he successfully forced the police to remove themselves and their vehicles from the family's land.[54]

Amongst others, McAdam has observed that unless success is achieved "even the most successful tactic is likely to be effectively countered by movement opponents if relied upon too long", leading to "an ongoing process of tactical interaction in which insurgents and opponents seek, in chess-like fashion, to offset the moves of the other".[55] In line with this dynamic protesters quickly changed tactics and continued to slow the progress of the tractor by walking in front of it, periodically stopping, but not long enough to be arrested for obstruction. The nature of the terrain and lack of powers allowing officers to clear the entire area favoured protesters as they were able to continually pass through the work site. The overall crowding of the forest meant that little work was completed and only a third of the track opened that day.[56]

Having called a mass rally five days earlier due to uncertainty as to how to proceed in the face of the immediate threat of logging, TNFAG now had a firm strategy in place, with the use of ODA at its core. Speaking to the *Australian* newspaper Hugh Nicholson emphasised the financial costs of the operation, stating "... by the time the State has paid for police to spend all that time moving us it could have easily afforded an environmental impact study on the area".[57]

Grasping the potential of the campaign to further their agenda of creating a series of new national parks, the major Sydney-based environmental organisations now offered their financial, political, and moral support.[58] As surprised as any at the effectiveness and level of organisation TNFAG had displayed, members of the Colong Committee later wrote:

> Terania became the symbol of rainforest preservation. More widespread public awareness and sympathy [was] achieved in two weeks than in five years by conventional means. For millions of people throughout Australia the event had dramatized the destruction of rainforest.[59]

Police and loggers left for the weekend and did not return until Wednesday, 22 August. During this time, the first act of property damage took place when the loggers' tractor, which had been left in the forest, had its wiring cut, ironically by a protester tasked with guarding it from such action. The nature of what constituted "sabotage", and whether it could be described as "violent", would be debated within the campaign and in many that followed.[60] In this case, concerned that the vandalism would give the Wran government a pretext to send in more police, as well as harm the campaign's image, members of TNFAG repaired the damage, called the machine's owner to apologise and placed more reliable sentries to watch over it.[61]

The police reviewed their strategies and upon their return demonstrated their capacity to adapt by rapidly removing vehicles parked in their way before forming a phalanx around the Komatsu. Forcing protesters aside, and carrying out 12 arrests, this allowed the tractor to continue clearing.[62]

Pushed to respond, protesters improvised new tactics the following day by climbing six or seven metres up trees to tie themselves onto branches or the trunk with rope. Although this tactic had previously been used in New Zealand, such use appears to have been unknown to those present. In this case, by mutual agreement and in the interests of safety, the sitter would usually return to the ground after being touched by a police climber. Other blockaders created further situations of manufactured vulnerability by wandering about and hiding in the bush. Although track clearing was completed and logging commenced, an operation that normally would have taken a day had been dragged out over a week.[63]

As is often the case in environmental blockades, the creation of safety hazards via tree-sitting and crossing through worksites did not always bring work to a halt. Attitudes towards the police varied amongst activists with some working hard to maintain good relations. Some police, particularly locals, responded cordially while others were accused of assaults as well as destroying tree-sitters' supplies and property.[64]

Many in the camp became despondent after workers decided to store logs on site rather than endure delays by daily trying to push through crowds to remove them. This new strategy allowed the rate of felling to increase, and a minority of protesters therefore took unsanctioned action: rendering the timber unusable by cutting it into unmillable lengths as well as spiking it and trees with nails. Such means were condemned by TNFAG, but effectively brought stockpiling to an end, thereby adding to the efficacy of less controversial tactics.[65]

Although the two can be complementary, there often exists a tension between prioritising effective obstruction, with its high potential for confrontation with police and opponents, and avoiding the negative publicity that can come with conflict. By controlling events and using tactics that create a spectacle without significantly disrupting their opponents, organisers may aim to project a benign image and keep the focus on their key messages, but they also run the risk of their actions being deemed inauthentic by activists and journalists. At Terania Creek, a minority swung to another pole in fully prioritising obstruction at all costs.[66]

This is the first time tree-spiking can be confirmed as having occurred in Australia. By hammering nails and steel spikes into the trees activists render any logging or milling expensive and highly dangerous as the metal can damage equipment or shear off and harm workers. As the goal is generally to save trees rather than to cause harm, proponents of the tactic have overwhelmingly argued it must be accompanied by warnings. Some interviewees claim that spiked trees were clearly marked at Terania Creek, while feller Ned Harvey recalled being taken by surprise when he cut into a tree and hit a spike. Arguably effective in this instance, the threat to opponents' well-being and the potential for negative public and media responses meant that the practice would not come to be widely employed in Australia, although it was popularised in the United States.[67]

While there is some consensus regarding the tactical effectiveness of tree spiking in preventing rapid felling, its broader impact at Terania Creek has been the subject of debate. Some have argued that tree spiking was the event that forced the state government to abandon its support for logging due to concerns that it marked an escalation that could result in serious violence. Others contend that the action was of minor significance and only one of many acts that contributed to the overall media attention they felt was decisive in changing decision makers' positions.[68]

Spiking was not used again, and protesters continued with other forms of ODA. Over the weekend they improvised new tactics aimed at hindering transport by digging ditches and diverting water into bog holes. Under strict ONV principles this would be forbidden, but with rains regularly damaging roads in the area such action was agreed to by the majority. At other times metal stakes, logs, and boulders were used to supplement the daily tactic of barricading the road with vehicles.[69]

After a break in logging due to showers, police and forestry workers returned in drizzling rain on Tuesday, 28 August. In response tree-sitters introduced further innovations that made it more difficult for the crew to manoeuvre around them. These included joining trees together with steel cables and stringing hammocks between others.[70]

Following another day in which logging was disrupted by tree-sitting, protesters opted for yet another tactic on Thursday, 30 August: doing nothing. Exhausted to the point where they believed they would not be able to carry out effective block-ading, the camp presented this break as a benefit to all. In return for agreeing to play an observer role, protesters negotiated with police to halt logging in one area as well as to allow them to carry out a short march within the coupe.[71]

As logging trucks left with their loads protesters threw wreaths in their path. Feelings regarding the future of the campaign appear to have been mixed with some reports indicating despair while another stated that "Spirits [were] high in anticipation of negotiations that have to come soon in favour of preservation of the forest".[72] This may have seemed wishful thinking to many, but the following day Standards decided to follow the protesters' lead and pulled their crew out of the forest for a rest of their own. The following week they extended this ahead of a meeting with the Premier.[73]

The NSW government shifts its position

During the previous weeks ongoing media coverage, mounting costs, and complaints of understaffed rural police stations had piled pressure on the state gov-ernment and Premier Wran now acted to defuse the issue. Acknowledging that "there is widespread backing for the conservationists" and that Terania Creek had "become a symbol for their fight to preserve the rainforests of New South Wales", he announced a formal halt to logging until cabinet could resolve the issue.[74]

Over the next month the government introduced amendments to the Forestry Act prohibiting the public to enter areas zoned off by the FC. These also made it an offence to obstruct or interfere with activities being carried out by anyone operating under a FC license, lease or permission, or to obstruct roads, damage trees, or change the flow of rivers and creeks. The combination of new restrictions with concessions was in keeping with strategies long employed by governments to neutralise successful challengers. In this case it would have a major effect on the tactics employed at future forest blockades but would not stop them from being mounted.[75]

The government also announced that it would extend the suspension of logging until an independent inquiry headed by Commissioner Simon Isaacs into Terania Creek was completed.[76] The 17 months of hearings would be fractious. Though its final report recommended continued logging, its findings were set aside. Logging at Terania Creek never resumed.[77]

During the time of the inquiry's deliberations the issue of rainforest logging was successfully diverted into conventional bureaucratic and political channels. This was accomplished not just by the suspension of logging at Terania Creek, but also by the government ordering a raft of EISs concerning forestry. Significant procedural changes introduced in October 1979 required increased public consultation and gave pro-conservation forces within the bureaucracy more influence.[78] The erosion of the FC's position and the expansion and deepening of the approval regime, as well as the direction of logging away from contentious areas, effectively met a number of environmentalist demands, thereby allowing the ALP to stall for time while it worked through internal matters.[79] The outcome of these debates, and the role of a second ODA campaign in them, will be returned to in the next chapter.

The impact of the Terania Creek blockade

The suspension of logging and changes in government policy and practice would clearly never have come about without the use of ODA at Terania Creek. Disruptive tactics slowed logging, imposed hefty economic costs, and through their novelty created a media sensation. In doing so the event not only demonstrates the initial efficacy and dynamics of the emerging repertoire, but also the role which locally based and focused campaigns can play more widely.

In a survey of various studies and events Rootes notes that "local campaigns are the most persistent and ubiquitous forms of environmental contention"[80] and discusses common outcomes of their activities, a number of which can be seen to apply to the events at Terania Creek. These include that they provide a typical entry point for new activists, potentially foster the development of specialist knowledge and innovations, and can reshape populations' conception of themselves and their region. In responding to political closure at a variety of levels, local groups often raise questions regarding the nature of democracy and challenge "the pro-development lobbies' conception of the public interest and of progress, [asserting] the necessity of prioritising use values over exchange values, well-being over economic development".[81]

Although widespread, Rootes notes that few localised campaigns "produce sustained mobilisations, fewer succeed in mobilising at a level beyond the local, and fewer still are effectively translated into national issues".[82] Because those with the power to change decisions are often not locally based themselves, he observes that a core factor in local groups achieving their goals requires them to shift their grievance "from a routine planning dispute into a high profile political issue", with this generally requiring "a coincidence between the interests of local campaigners and the campaign priorities of national organisations that is by no means automatic".[83] To gain such an alignment local groups often need to frame their issues as universalistic and in line with those promoted by trans-local organisations, with support more likely to occur if networks and connections are already in place.[84]

In the case of Terania Creek, local campaigners overcame a lack of cooperation from state and national-based bodies by employing ODA. In doing so their action tapped into broader issues that trans-local organisations had already raised and popularised them. It also notified the government that they had to pay more attention to local interests lest conflicts break out in similarly aggrieved communities. Further to this, the militancy of the blockade created a radical flank that forced trans-local organisations to adopt a local issue while also allowing them to position themselves as moderates.

Logging and attitudes towards the environment more generally would continue to be a flashpoint over coming decades. This was not least because the protest had caused many of its participants to deepen their appreciation of nature as well as created a cohort of radical environmentalists skilled in campaigning, willing to use ODA, and eager to share their skills.

More broadly, the campaign successfully challenged the FC's dominance over decisions concerning state forests and the basis upon which it made them. In doing so it reframed understandings of biodiverse places. The campaign to save Terania Creek aired and popularised emerging concepts regarding ecology and the interconnectedness of species and strengthened the position of those scientists, politicians, and bureaucrats who favoured them. It also, for the first time in Australia, brought a focus on rainforests as a particularly rare and endangered set of ecosystems. Once such framing had resonated with wider publics, others adopted it. Existing campaigns to save various ecosystems rapidly shifted focus and were unified on the basis that they included rainforest species.[85]

In promoting environmental blockading, the Terania Creek campaign would, as later chapters demonstrate, in time have an effect on international repertoires of protest. Its impact was initially felt on a national level as Australian campaigners applied and adapted the modular template to their own issues. Although the specific forms it had used would not always be replicated, it also established consensus decision making and "non-violence" as key organising principles.

Notes

1 Susan Ward and Kitty van Vuuren, "Belonging to the Rainbow Region: Place, Local Media, and the Construction of Civil and Moral Identities Strategic to Climate Change Adaptability," *Environmental Communication* 7, no. 1 (2013): 67.

2 Graham Irvine, "Creating Communities at the End of the Rainbow," in *Belonging in the Rainbow Region: Cultural Perspectives on the NSW North Coast*, ed. Helen Wilson (Lismore: Southern Cross University Press, 2003): 63–82; Lee Stickells, "Negotiating

Off-Grid," *Fabrications: The Journal of the Society of Architectural Historians, Australia and New Zealand* 25, no. 1 (2015): 109–11.

3 Nan Nicholson, *Terania Creek*, Unpublished manuscript, 1982, 6–9.

4 Renate Howe, "'Nobody but a Bunch of Mothers': Grassroots Activism and Women's Leadership in 1970s Melbourne," in *Seizing the Initiative: Australian Women Leaders in Politics, Workplaces and Communities*, ed. Rosemary Francis, Patricia Grimshaw, and Ann Standish (Melbourne: University of Melbourne, eScholarship Research Centre, 2012): 331–34.

5 Nigel Turvey, *Terania Creek: Rainforest Wars* (Brisbane: Glasshouse Books, 2006), 51–53; Ian Watson, *Fighting over the Forests* (North Sydney: Angus and Robertson, 1990), 83–87.

6 Nicholson, *Terania Creek*, 1, 5.

7 Joseph Glascott, "Conservationist Quits as Adviser to Govt," *Sydney Morning Herald*, 10 September 1979, 2.

8 Libby Connors and Drew Hutton, *A History of the Australian Environmental Movement* (Cambridge: Cambridge University Press, 1999), 117–27.

9 Colong Committee, *How the Rainforest Was Saved* (Sydney: Colong Committee, 1983), 2.

10 Craig McGregor, "Picnic at Terania Creek," *National Times*, 9 June 1979, 14–15.

11 James Somerville, *Saving the Rainforest: The NSW Campaign 1973–1984* (Narrabeen: James Somerville, 2005), 62–63.

12 Turvey, *Terania Creek*, 51–53; John Seed, Interviewed 26 January 2015.

13 McGregor, "Picnic at Terania Creek," 14–15.

14 Nicholson, *Terania Creek*, 8, 22–26, 34.

15 Peter Stephens, "Group Puts up Fight to Save Rainforest," *Australian*, 8 May 1979, 6.

16 Griff Foley, *Learning in Social Action* (London: Zed Books, 1999), 29; Lisa Yeates, Interviewed 29 January 2015.

17 Graeme Dunstan, "High on Trees," *Nimbin News*, 20 August 1979, 2–3.

18 Jasper, *The Art of Moral Protest*, 242.

19 MNS processes will be discussed in a following chapter. Yeates Interview.

20 Martin Branagan, "Art Alone Will Move Us: Nonviolence Developments in the Australian Eco-Pax Movement 1982-2003" (University of New England, 2006): 77–78.

21 Leggett Interview; Yeates Interview.

22 Cianchi, *Radical Environmentalism*, 161.

23 Ibid, 154–55.

24 Ibid, 55.

25 James Jasper, "Emotions and Social Movements: Twenty Years of Theory and Research," *Annual Review of Sociology* 37(2011): 8–9.

26 Dailan Pugh, Interviewed 8 June 2015; Patrick Anderson, Interviewed 9 June 2015; Ian Peter, Interviewed 8 June 2015.

27 Leggett Interview; Pugh Interview.

28 Stephen Brouwer, ed. *The Message of Terania* (South Lismore: Terania Media, 1979): 9.

29 Ron Chapman, " Fighting for the Forests: A History of the West Australian Forest Protest Movement, 1895-2001" (Murdoch University, 2008), 142–53; Barbara Epstein, *Political Protest and Cultural Revolution: Nonviolent Direct Action in the 1970s and 1980s* (Berkeley, Los Angeles, Oxford: University of California Press, 1991), 110–15.

30 Seed Interview.

31 Brouwer, *The Message of Terania*, 10; Nicholson, Interviewed 3 August 2015.

32 Yeates Interview; Foley, *Learning in Social Action*, 32.

33 Brouwer, *The Message of Terania*, 4.

34 Marty Branagan, *Global Warming, Militarism and Nonviolence: The Art of Active Resistance* (Basingstoke: Palgrave Macmillan, 2013), 32–34; Doyle, *Green Power: The Environment*

Movement in Australia (Sydney: UNSW Press, 2000), 49, 63; Coover, Virginia, et al. *Resource Manual for a Living Revolution* (Philadelphia: New Society Publishers, 1977).

35 Chapman, "Fighting for the Forests," 132–38.
36 Leggett Interview; Yeates Interview.
37 Bible, "Aquarius Rising," 47.
38 Leggett Interview.
39 Nan Nicholson, Interviewed 3 August 2015.
40 Nicholson, *Terania Creek*, 39–40.
41 Brouwer, *The Message of Terania*, 2–3.
42 "A Diary from the Festival for the Forest," 5.
43 Errol Simper, "Sit-in Blockade by Forest Lovers to Stop Loggers," *Australian*, 17 August 1979, 2; Dunstan, "High on Trees," 3.
44 Ibid.
45 Yeates Interview.
46 Dunstan, "High on Trees," 3; Colong Committee, *How the Rainforest Was Saved*, 21.
47 Jasper, *The Art of Moral Protest*, 238; Wang and Soule, "Tactical Innovation in Social Movements," 522–23.
48 Ibid.; 519–20; Polletta, "Free Spaces in Collective Action," 25–26.
49 Hayes, "Greenies and Government at Loggerheads," 775; Bible, "Aquarius Rising," 13–16.
50 Anderson Interview; Nicholson, *Terania Creek*, 15–17.
51 Sophia Hoeben, Interviewed 12 April 2015; Peter Interview.
52 John Watson, "Terania Mania," *Nimbin News*, 24 September 1979, 15–16; Brenda Liddiard, Interviewed 2 February 2015.
53 Yeates Interview; Leggett Interview; Nicholson Interview.
54 David Spain, "Report from Terania Basin," *Nimbin News*, 27 August 1979, 18–19.
55 Douglas McAdam, "Tactical Innovation and the Pace of Insurgency," *American Sociological Review* 48 no. 6 (1983), 736.
56 "Protest over Logging in Rainforest," *The Age*, 18 August 1979, 5.
57 Errol Simper, "17 Arrested as Police Break Forest 'Sit-In'," *Australian*, 18–19 August 1979, 3.
58 Bonyhady, *Places Worth Keeping: Conservationists, Politics and Law* (St Leonards: Allen & Unwin, 1993), 48–49.
59 Colong Committee, *How the Rainforest Was Saved*, 22.
60 Craig McGregor, "The Battle for Terania Creek," *National Times*, 28 August 1979, 7.
61 Turvey, *Terania Creek*, 60–62.
62 "Logging Protest: Twelve Arrested," *Sydney Morning Herald*, 23 August 1979, 3.
63 Turvey, *Terania Creek*, 64–68; Jeni Kendall and Paul Tait, "Give Trees a Chance," Gaia Films, 1980.
64 Spain, "Report from Terania Basin," 18; "Police Ride Shotgun as Logs Are Taken," *Daily Mail*, 30 August 1979, 3.
65 Nicholson, *Terania Creek*, 45.
66 Pugh Interview; Branagan, "Art Alone Will Move Us", 215–16.
67 Turvey, *Terania Creek*, 71; Dave Foreman and Bill Haywood, eds., *Ecodefense: A Field Guide to Monkeywrenching* (Tucson: Ned Ludd Books, 1987); Leggett Interview.
68 Bible, "Aquarius Rising," 48–49; Peter Meredith, *Myles and Milo* (St Leonards: Allen & Unwin, 1999), 267; Pugh Interview.
69 Joseph Glascott, "Protesters Form Barrier in Trees," *Sydney Morning Herald*, 29 August 1979, 12; Kendell and Tait, "Give Trees a Chance," 1980.
70 David Spain, "David Spain's Terania," *Nimbin News*, 3 September 1979, 10–11.

71 Joseph Glascott, "Truce Called in Forest Protest," *Sydney Morning Herald*, 31 August 1979, 4.

72 "The Continuing Story of up the Creek without a Study," 1979, 16; Spain, "David Spain's Terania," 10–11.

73 "Sawmillers Call Temporary Halt at Terania," *Sydney Morning Herald*, 4 September 1979, 1.

74 Catherine Harper, "Logging Halt Extended," *Sydney Morning Herald*, 5 September 1979, 1.

75 Peter Kennedy, "State Moves to Curb Forest Protest," *Sydney Morning Herald*, 8 September 1979, 3.

76 Catherine Harper, "Public Inquiry into Terania Logging," *Sydney Morning Herald*, 26 September 1979, 2.

77 Paul Ellercamp, "Wran Not Satisfied with Report on Logging," *Sydney Morning Herald*, 13 February 1982, 5.

78 Joseph Glascott, "New Forest Studies Ordered," *Sydney Morning Herald*, 13 October 1979, 3.

79 Bonyhady, *Places Worth Keeping*, 50.

80 Christopher Rootes, "Acting Locally: The Character, Contexts and Significance of Local Environmental Mobilisations", *Environmental Politics* 16, no. 3 (2007): 722.

81 Ibid; 732.

82 Christopher Rootes, "From Local Conflict to National Issue: When and How Environmental Campaigns Succeed in Transcending the Local", *Environmental Politics* 22, no. 1 (2013): 96.

83 Rootes, "Acting Locally", 735–735.

84 Rootes, "From Local Conflict to National Issue", 97–98, 108–111.

85 "Call for Logging Ban on All Rainforests," *Sydney Morning Herald*, 18 September 1979, 3; Pugh Interview.

2

"AN ATTITUDE, A WAY OF LIFE, A STATE OF HEART"

Organisational and tactical development within Australian environmental blockades, 1980–1984

Prior to Terania Creek similar protests using ODA to defend biodiverse areas had taken place in Brazil, India, New Zealand, and Finland. Although all played a role in popularising preservation within each country, none initiated protest cycles that significantly established, promoted, and developed the environmental blockading repertoire nationally or internationally.[1] This role was instead undertaken by Australian activists through a series of campaigns between 1980 and 1984.

This chapter will provide an analysis of the environmental blockades that followed Terania Creek. It demonstrates that the campaigns and publicity associated with these had a significant impact on Australian politics and attitudes towards the environment and embedded the action template as an enduring strategic option. It explores how the involvement of a variety of organisations and thousands of participants created a process of differentiation through which new tactics, organisational variations, and approaches to normative protester behaviour emerged. Associated with this it also traces how the success and threat of blockading drove authorities and businesses to spawn a repertoire of counter-strategies and tactics.

Middle Head, 1980: The template diffuses

After Terania Creek, the next environmental blockade was at Middle Head, an area on the mid-coast of NSW threatened by mineral sand mining. Although a local environment centre had unsuccessfully lobbied against the project, many in the area were caught by surprise when resource company Mineral Deposits arrived to begin clearing bush and dunes in March 1980. The decision to blockade was even more spontaneous than at Terania Creek, as upon seeing a bulldozer a group of surfers left the water to immediately halt work by climbing onto it, after which the company withdrew from the area. Concerned residents rapidly formed the Middle Head Sand Mining Action Group (MHSMAG). This was primarily made up of local

New Settlers for whom the precedent of Terania Creek provided the inspiration to set up a resident action group and protest camp as well as deploy ODA.[2]

As the area involved was one of significance to Indigenous people, members of the Gumbangerii community and MHSMAG worked in a loose coalition to undertake a range of protest actions over the next seven months.[3] Although a small number of prominent Bundjalung members had lent support to the Terania Creek protest this was the first preservation-oriented blockade in which Indigenous Australians played a major role.[4] Infuriated at a lack of consultation from the company and state government, members of the Gumbangerii and Dunghatti communities rapidly extracted an agreement to protect some, but not all sites of significance, with the result that opposition from the former continued. Feeding into increasing contestation and networking regarding land rights across NSW, members of other Indigenous communities visited and supported the blockade.[5] Although relationships between Indigenous and non-Indigenous protesters were described by John Seed as initially "aloof", he recalls that the shared experience of blockading broke down barriers.[6] Tensions concerning sovereignty and leadership would emerge in later Australian blockades, but appear to have been muted here with non-Indigenous blockaders later supporting the Gumbangerii to set up an Aboriginal Embassy outside state parliament.[7]

Following the initial actions, blockading would take place in two main phases. The first came in July when, having confirmed the state ALP would not rescind its license to mine, the company resumed clearing scrub to make way for a dredge pond. At this point up to 50 blockaders, primarily drawn from the local area, sealed entrances with their cars, used their bodies to block bulldozers and removed fencing.[8]

The last of these represented a new tactic within Australian environmental blockading, although the surreptitious removal of survey stakes and other minor property destruction had long been used by opponents of mining internationally and nationally. There were also local precedents with Garry Williams from the Gumbangerii Aboriginal Association speaking of Aboriginal people removing mining stakes at nearby Gumma since 1965 to impede construction.[9]

As at Terania, the use of ODA was reactive. Although it received some support from established environmental organisations, MHSMAG was made up of first-time campaigners who, given the unforeseen nature of the situation and the novelty of environmental blockading, lacked a defined strategy regarding its use. This resulted in a focus on obstructing work within improvised boundaries of normative protester behaviour similar to those at Terania Creek.[10]

With increased policing allowing clearing to take place, a lull set into the campaign until mining began and an influx of protesters from outside the area, including some who had been involved in Terania Creek, arrived from August onwards. Camp numbers from this point came to fluctuated between 50 and 100.[11]

Over the following weeks the protest camp became the focus of activity. It was expanded to include an office, alternative school, childcare tent, and a kitchen. In the furtherance of non-violent dialogue, this fed not only protesters, but also tourists and mine workers. Camp occupants bonded through morning circle meetings, workshops, singing and surfing, with the beach location proving a major draw.[12] Workshops and messages concerning New Age and Indigenous spirituality, with the accuracy of some of the latter disputed by Gumbangerri members, became a major feature of camp reports, press releases, and articles.[13]

The organisational, infrastructural, and lifestyle aspects of protest camps, and whether they are primarily seen as facilitating ODA or also as an end in themselves, reflect the degree to which members and blockading itself form part of differing collective identities. Despite being rarely focused upon in accounts of environmental blockading, the social and cultural aspects of protest camp life, along with the sense of adventure involved in ODA, provide an important incentive for many participants. Many protest camps associated with environmental blockading involve informal organisation, adherence to consensus decision making and provision, if not preference, for alternative diets, lifestyles, and appearances. Beyond offering a place to socialise and a base from which actions can be mounted, camps form key sites within what Melucci terms "submerged networks", the aggregations through which "activists make meaning and produce culture".[14] As such, the organisation and nature of life within protest camps means that they not only reflect social movement values, but are also a place in which they are developed. The processes and products of these "laboratories of insurrectionary imagination" are largely hidden from broader publics and often extend beyond the life of the event.[15]

In this case, the camp's ability to generate outcomes beyond the facilitation of ODA was demonstrated in a variety of ways. One of these was its instigation of the Green Alliance, an electoral and organising body aimed at more effectively coordinating environmental and social justice groups. This later fed into the formation of Greens parties in NSW.[16]

The involvement of radical performer Benny Zable and his setting up of a creative space in which banners and props were collectively made and theatre workshops carried out, added to the artistic colour of this and future protests. Middle Head also marked the debut of his iconic anti-corporate "Greedozer" character, who would appear at hundreds of protests in coming decades.[17]

Concerns on the part of some, such as Dudley Leggett, regarding incidents of property damage at Terania Creek meant that more restrictive guidelines were introduced regarding normative protester behaviour. Discussion intensified after fires were lit during blockading against bush clearing. Once mining began, further debate followed an action carried out by Graeme Dunstan, a leading countercultural figure active during Terania Creek, and an unidentified man. Boarding the mining dredge, the pair moved the operator out of the way and briefly turned off equipment while others untied ropes connecting the vessel to a walkway and rocked the pontoon by jumping on mooring. Actions such as these were condemned as "sabotage" by the majority of blockaders and marked a point after which acceptable tactics were confined to those solely involving placing bodies in harm's way.[18]

In a development that would have ramifications for future blockades, Middle Head also saw the introduction of a new tactic on behalf of the authorities. In a move that sapped the campaign's human resources, magistrates began applying bail conditions to those arrested, primarily for obstruction, which prevented their return to the mining site. Although a small number defied these, subsequently undertaking a hunger strike in jail, the threat of heavy fines and imprisonment provided an effective deterrent for the majority.[19]

Blockading continued sporadically until late October, but the occupation of the dredging pond and the placement of activists near mining operations did not have a major effect on the progress of work. Nevertheless, the delays ODA had already created, and the associated attention, caused Mineral Deposits to withdraw plans to level a neighbouring beach. It also invigorated an overall campaign that successfully phased out coastal mineral sandmining in NSW with the state government granting no new leases after 1980.[20]

Nightcap: A turn to confrontation

Following the end of the Terania Creek blockade in late 1979 the struggle over rainforest logging in NSW entered a stalemate. Resentful of government inaction, both the timber lobby and environmentalists moved to apply pressure during early 1982. For the FC and industry this came in the form of announcing they would log the Goonimbah State Forest around Mount Nardi, approximately 30 kilometres from Terania Creek. This was a provocative move as the forest had been named at a May meeting between the ALP and leading environmentalists as a "non-negotiable" area for preservation. It also included sections that bordered Nimbin's Tuntable Falls New Settler community. Emboldened by the Isaacs Inquiry's support for logging at Terania Creek, and frustrated that the Premier had vetoed its findings, the FC declined to carry out an Environmental Impact Study (EIS) and refused NSW National Parks and Wildlife Service officers, who favoured protection, access to the area.[21]

With logging set to take place later in the year, northern NSW residents formed the Nightcap Action Group (NAG) in early 1982. Wishing to avoid fighting the FC coupe by coupe, NAG called for the creation of a Nightcap National Park to cover over 4,000 hectares of forest, including Terania Creek and Mount Nardi. Via ties built on the success of Terania Creek, the proposal had the support of regional, national, and state-wide organisations.[22]

Unlike most environmental blockades, which respond to imminent threats, the first stage of the Nightcap protest was proposed as a pre-emptive action. The feeling among core local and Sydney-based activists was that their support base needed to be reactivated as far out from the looming confrontation at Mount Nardi as possible, and that ODA at a separate site at Grier's Scrub was the surest way to gain publicity and pressure the state government.[23]

As would also later occur in US radical conservation circles, growing experience initially led NAG members to strategically, rather than reactively, choose a target, and to view ODA as a means of achieving broader reform rather than primarily protect a single area. Underscoring the importance of emotions and moral shocks, the heartfelt attachment that the forest engendered soon lent the blockade added meaning. John Seed recalls that upon arrival blockaders "discovered this magnificent Flooded Gum forest that had somehow been sheltered on the side of the range". Furious at the possibility of facing ODA, loggers destroyed sections of the forest not required for milling to demonstrate their power in the situation and punish the protesters for being there. This however bolstered NAG's commitment, with Seed arguing "Seeing the area and what they were prepared to do, made us incredibly passionate to save it, even though we really didn't have the numbers."[24]

Beginning ODA in late July, the campaign's tactics would represent a shift away from Middle Head's emphasis on restrictive notions of acceptable behaviour. Maintaining a public commitment to "non-violence", activists defined this minimally and showed a willingness to explore what NAG member Lisa Yeates describes as "the grey areas".[25] Initiating physical assaults and engaging in major sabotage was considered out of bounds by the majority, but verbal goading of loggers and minor property damage incurred through graffiti and barricading generally condoned. Internally there appears to have been an understanding that in some cases camp members were also justified in using violence in self-defence if attacked by timber workers.[26] When instances of significant sabotage of opponent's equipment and physical violence happened during the protest, they were officially rejected or denied by NAG. For some this had an ethical basis, but for others it derived from a strategic view, as outlined in the campaign's handbook, that the "state always wins when it comes to violence" and that such actions could allow protesters to be portrayed as "the Red Brigades or Charlie Manson".[27]

The Nightcap handbook appears to have been an innovation in Australian environmental blockading. Produced after action at Grier's Scrub had commenced, it included articles concerning the history of NSW campaigns and the ecological importance of the area as well as information relating to consensus processes and suggestions for dealing with police. Information regarding why and how the publication was produced and by whom is scant. Sophia Hoeben recalls various groups in Canberra and elsewhere discussing producing something similar for earlier protests, but outside of pamphlets outlining legal rights it appears that it may have been the first of its kind produced for any protest in Australia. It is likely that the handbook was based on those previously produced for UK and US anti-nuclear protests.[28]

Publications combining background material with organisers' positions on decision making and appropriate behaviour, and practical information on matters such as legal rights, accommodation, and first aid have become a standard element of Australian protest camps and blockades. The degree to which they, and other declarations of organisers' official positions and ideas, influence subsequent behaviour varies greatly. In this case while some interviewees do not recall seeing the publication, the positions it outlined appear to be in line with those carried out by the majority.[29]

Although the campaign had begun with a broader focus, the emotional connection to the forest meant that many involved in the campaign came to place an equal, if not greater, emphasis on obstructing logging as on gaining positive media coverage. Familiarity with the issues involved and the loss of novelty associated with blockading meant that media interest was not as high as at Terania Creek, in any case.[30]

Changes in protester norms can also be related to more aggressive behaviour from workers and police. Loggers would be accused of menacing protesters with brush hooks, axe handles, and other weapons. Under the combative leadership of former Army commando Harold Fredericks, the police were more determined to prevail than at previous protests, with one officer allegedly telling protesters "You might have won Terania, but you're not going to get this one."[31]

Although activists often question the wisdom of overly focusing on short-term victories, such tussles, which may be as simple as causing an opponent to lose their temper or to withdraw for a day, can take on major significance for all involved. Events during environmental blockades would often bear out Jasper's observation that such "satisfactions of action, from the joy of fusion to the assertion of dignity — become a motivation every bit as important as a movement's stated goals".[32]

The shift in tactical approach also reflected a change in the make-up of those participating. Bren Claridge, the Nicholsons, and others from the nearby Channon district who had been core decision makers at Terania Creek remained involved. However, the campaign was mainly run by those residing in the Nimbin area, including a number who were new to blockading. Some of these were drawn from a group who were less peaceable than their predecessors. Activist Ian Cohen recalls in his memoirs that camp life was "hectic, with a sprinkling of unbalanced behaviour".[33] The rowdy nature of some involved, and the rejection of demands by Dudley Leggett that tactics reflect those used at Middle Head, meant that he and others who had previously been involved in blockades did not participate in this one. Frustration at the inability of the Terania Creek campaign to procure a longer term solution, and that the Middle Head one had failed to save the beach, likely diminished the persuasiveness of arguments that stricter norms be applied here too.[34]

Blockading at Grier's Scrub began on 21 July when a large group of protesters were denied entrance to the forest after FC staff hastily erected a gate across the access road. The use of regulations enabling the FC to prevent the public, including journalists, from entering large areas of the forest during logging operations encouraged activists to improvise new tactics. On this occasion, in a means previously used during anti-war, student and union protests, NAG left the area to occupy the Murwillumbah District Forestry Office.[35]

Faced with being easily spotted and arrested if they appeared in larger groups, activists began to infiltrate the forest to create situations of manufactured vulnerability by climbing or standing under trees. They also, in the words of musicians Mook and Shanto, sought to disrupt work by "nagging forestry workers and going up to [those] who had chainsaws and making them chase you through the bush …"[36] Seeking to further minimise arrests and distract and wear down workers, as well as give the impression that their numbers were larger than they were, small groups of activists would hide and call out from different points. With police and media rarely on site, confrontations took place with each side accusing the other of starting scuffles and carrying out physical assaults.[37]

The imposition of bail conditions preventing protesters from returning to blockade sites throughout the protest prevented a minority of protesters from continuing their involvement in ODA. Demonstrating the importance of geography, this and the FC boundaries, failed to exercise the impact they had at Middle Head as the much larger area, multiple sites of entry and a general lack of police presence made them largely unenforceable.[38]

By early August, NAG conceded that Grier's Scrub would be lost but determined to continue to interfere with operations and gain any publicity they could, activists pushed on with daily incursions. Having successfully mobilised a core of 50 participants, with many others visiting, the blockade was expanded over the next month to include three protest camps. In response the pro-forestry Lismore council introduced a new counter-tactic in unsuccessfully attempting to have one camp deemed an illegal development.[39]

On 4 August, NAG introduced tactics focused on denying police and timber workers access to the forest and over 100 protesters sat on the road ahead of a series of rocks and parked vehicles. With a logging truck unable to turn around easily on the narrow thoroughfare, and protesters surrounding a tow truck, it took from 6 a.m. until 3 p.m. for police to clear the entrance. When they arrived at the gate that had been placed to mark the boundary of forestry operations, they discovered that that the mechanism of its lock, and others placed upon it, has been rendered inoperable by superglue. This was the first time that such a tactic was employed during an Australian environmental blockade. With minor property damage condoned by the campaign, workers discovered shortly after, in another innovation, that protesters had used chains and padlocks to lock down gears and doors on their equipment. Slogans such as "Workers' Sweat = Mill Owner's Profits" were also painted on log piles, a bulldozer, and a fuel tank.[40]

As a result of this action, NAG's barricading tactics and isolated incidents of more serious sabotage, including sand being poured into a bulldozer's hydraulics, contractors began to stay overnight in the forest as well as post guards. This in turn saw protesters extend their campaign of harassment by disrupting workers' sleep with hectoring speeches and the atonal playing of instruments.[41]

Protesters continued to hinder loggers entering and exiting the forest throughout the month. Introducing new variations on barricading, one action involved six vehicles, some with their axles chained together, being placed in the road. Once these were towed away protesters briefly stalled the entry of two logging vehicles by removing and hiding a cattle grid, thereby leaving a large hole. According to Cohen these acts not only sought to obstruct opponents, but to also stretch state resources by ensuring that "only a fully equipped police contingent giving around-the-clock support would allow logging to occur".[42]

From early September NAG pursued a new strategy in calling for a boycott of Tooth's breweries, the David Jones and Georges department stores, and all other firms owned by the Adelaide Steamship Company. This was prompted by the company's 37 per cent interest in Standard Sawmills, the Lismore business carrying out logging.[43] Campaigns targeting investors and companies linked to opponents had long been used by unions and anti-war campaigns, and would become increasingly commonplace and sophisticated in environmental activism in decades to come. Their use in conjunction with EB was innovative at this time.

Little logging took place during most of September. This was partially due to heavy rain, but also because, as John Seed recalls, "They thought that if they waited, we'd get sick of it and go home. However, people liked being in the bush and someone from Tuntable had donated kilos of dope so everyone was happy to hang around." To avoid the potential of a camp being raided Seed wrote to the Minister of Planning and Environment stating he had received information indicating that the police were planning to "plant" drugs on protesters. Whether connected or not, no drug-related arrests took place during the protest.[44]

The impasse ended on the eve of the Commonwealth Games, when up to a third of the camp had left to join Indigenous and civil rights protests in Brisbane. Warned of the authorities' intentions, NAG mobilised around 200 supporters by the time 50 police and 40 loggers arrived at Mount Nardi on Wednesday, 29 September. Building on previous tactics, campaigners constructed their most elaborate and spectacular barriers to date, interspersing human blockades with a series of barricades. From 4.30 a.m. onwards police broke up groups occupying the road, towed away vehicles, replaced four cattle grids, and removed a truck and logs which had been set alight. Having cleared the road, they constructed a new gate, set up a second roadblock, and established their own camp, informing the media they planned to maintain a presence for up to four months. One group of timber workers constructed a compound surrounded by barbed wire and searchlights while a second, who had entered the forest undetected, began logging.[45]

Other than during a break on the first and second day of blockading – when police and protesters joined together to locate Ian Cohen who, to the chagrin of his fellow blockaders, had pioneered a new tactic by pretending to be lost in the bush – much of NAG's efforts centred on preventing the movement of trucks. Although the narrow road allowed them a tactical advantage, the number of police meant that such actions met with mixed success. Exasperation appears to have led to a series of incidents in which a minority of protesters were accused of hurling rocks, throwing coffee in the face of a truck driver, spiking trees, and placing nail-studded boards on the thoroughfare. During an action in which a small group sat in the path of a convoy, one protester was run over and had his ankle broken.[46]

As has been observed in many social movement campaigns, innovation often occurs at points of defeat or when existing tactics are perceived to be inadequate.[47] At times, these improvisations are embraced, at others they promote division. During the Nightcap campaign internal conflict was contained, but the aggressive intentions and actions of some protesters were noted by members of the Tasmanian Wilderness Society (TWS) who, unbeknown to NAG, had briefly joined the protest. Having also observed increasingly negative reporting by journalists, who outside of the region were ignoring the protest unless it involved violence, TWS activists drew the conclusion that the media-oriented blockade they were planning would need to be closely controlled.[48]

With over 100 protesters arrested, many of them in the previous three days, the Nightcap blockade came to an end on Friday, 1 October, when the Land and Environment Court (LEC) placed an injunction on further logging. This came in response to an application by protester Di Kivi and NAG lawyer Murray Wilcox QC requesting operations cease until an EIS was carried out. The injunction was subsequently extended and became permanent on 22 October when the court ruled, in line with regulations which had come into force under the Environment and Planning Act following Terania Creek, that an EIS "was a necessary prerequisite of logging where there is likely to be a significant effect on the environment".[49]

This was the first time in Australia that an environmental blockade had been successfully combined ODA with legal action, with the former designed to gain publicity and delay work, and the latter advanced to obtain a more lasting halt to operations. The coupling of such strategies would later become common in preservationist campaigns. In this case it also added the political pressure required to force the state government to enact legislative change as on Tuesday, 26 October, Premier Wran announced that rainforest logging would be immediately phased out. Subsequent legislation added 118,128 ha of forest to national parks or reserves, including the creation of the Nightcap National Park taking in Terania Creek and the Goonimbah State Forest.[50] Although some conservationists were unhappy that other areas of rainforest remained open to logging, most celebrated the ALP's decision.[51]

Out of the three NSW blockades undertaken to this time, a cohort of experienced blockaders had, in Benny Zable's words, "come of age". As a "travelling activist team" this group, known as the Nomadic Action Group (NAG), would go on to provide logistics, music, banners, and props, and generally spearhead tactical innovation, at later anti-nuclear and environmental blockades.[52] For their part NSW's mainstream environmental organisations' satisfaction with the government's legislation led them to favour strategies based on cooperation with the ALP and away from ODA.[53] When combined with the removal of immediate threats to forests in the areas where New Settlers lived, this would see a cessation of environmental blockading in NSW for seven years.

The Franklin Dam blockade

The next Australian environmental blockade was the first held outside of NSW. It was also to be the largest and longest event of its kind to the time, involving more than 2,500 protesters and 1,340 arrests, gaining widespread national and international media coverage. As such it did the most since Terania Creek to nationally promote the template of action.

The campaign also introduced new priorities, forms of organisation, and patterns of decision making. These came in part due to the event being initiated, resourced, and largely directed by a mainstream environmental organisation, TWS, using a pre-formulated orthodox non-violence (ONV) approach that prioritised publicity over obstruction. This was in contrast to the organisation of previous blockades, which had been initiated by informally organised, New Settler-based local action groups applying ideas concerning organisation and normative protester behaviour that they had largely developed themselves.

TWS had been formed in 1979 after a long series of conflicts between environmentalists and the powerful Tasmanian government owned Hydro-Electric Commission (HEC). Australia's modern conservation movement had largely emerged out of a five-year campaign against the flooding of Lake Pedder and its surrounding forests by the HEC that began in 1967 and ended with the flooding of the lake in 1972. When the HEC moved to build a new dam that would flood sections of the Franklin and Gordon rivers in the southwest of the state, TWS ran a media and lobbying campaign that made the project a national issue. While opposition to the dam was strong outside the state, it remained popular within Tasmania. The communities directly affected were split, and blockaders regularly harassed by police and abused and occasionally assaulted by dam supporters.[54]

Although all TWS members were given opportunities to take part in branch activities, the organisation was a largely hierarchal one with positions informally determined by specialisation, personal relationships, and length of service. From 1980 to 1982 the organisation debated the use of an environmental blockade and a working group was tasked to study precedents such as Terania Creek, the Wagerup occupation, and overseas protests.[55]

Following the failure of various political solutions proposed by the state ALP, the conservative Liberal Party came to power in May 1982, determined to commence the project. Faced with political closure at a state level, TWS undertook unsuccessful legal action, began lobbying the conservative federal government to intervene, and convinced federal opposition parties to reject the project.[56]

Although some within TWS were wary of the possibility of ODA sharpening conflict and generating negative publicity, representatives met with ONV[57] advocates to discuss planning for potential direct action. Approaches and ideas developed by US organisation Movement for a New Society (MNS) in the 1970s, primarily in relation to anti-nuclear campaigns, had been indirectly diffused to Australia via handbooks, films, and articles. Direct diffusion came from MNS activists who toured Australia as well as from Australians and New Zealanders who underwent training in the United States. By the early 1980s small training collectives had formed in various cities. The Franklin campaign provided the impetus and platform for these to form into an active network with a magazine, both named Groundswell.[58]

Impressed by the model, TWS would introduce campaign guidelines which adhered to MNS principles of openness, democracy, and "non-violence," with activists required to brief police, avoid all language and physical actions that could be deemed "aggressive", utilise consensus decision making and agree not to engage in any form of property damage. To prepare them for direct action and ensure they were aware of these boundaries, all attending the protest would have to first engage in training workshops, involving ONV theory, role plays, and tactical discussions.[59]

The beginning of road building, and market research demonstrating ODA would not alienate the public, bolstered pro-protest feeling within TWS with the result that in July 1982 it publicly announced it would hold a blockade. Although, as with earlier campaigns, the turn to ODA was only taken after more conventional means of influence had failed, it was far from a last-minute bid: TWS provided its opponents with months of notice. Following extensive preparations, they decided to begin blockading on 14 December, the date upon which the Southwest region would formally gain World Heritage status.[60]

Located around six hours boat ride from the nearest main town, the dam construction area was hilly and densely forested. Due to the remote nature of the site, and in keeping with MNS precedents, TWS's vision was that during mandatory non-violence training people would form into affinity groups. These small clusters of protesters would make decisions and be responsible for planning and carrying out the specifics of actions that the wider campaign had agreed upon. Although protesters would form into affinity groups, TWS's original plan for each to be responsible for its own shelter and supplies and head upriver on a rotational basis was not adopted.[61]

Arriving in November, activists from NAG argued that a permanent protest camp and kitchen should be established. This was based on the large numbers of people involved, the distance between drop off points and worksites, and the need for some protesters to remain upriver to act as guides and maintain communications. Following debate that indicated TWS distrusted NAG, in part due to reports of aggressive and undisciplined behaviour during the Nightcap campaign, it was agreed that a camp would be put in place. A number of others were later set up to stage actions and carry out reconnaissance as well as ease pressure on the main camp.[62]

TWS set up an Information Centre, to handle media relations and logistics, and a base camp, from which protesters would travel upriver, in Strahan. The camp was initially on council land in the town, but following physical threats and a petition from opponents, later moved to nearby private land dubbed "Greenie Acres". The Information Centre, like other parts of the campaign, was understaffed and run primarily by women, who were forced to learn on the job.[63]

Communications between the camp, Strahan and Hobart were made difficult by mountainous terrain that precluded direct radio contact in an era when satellite phones were unavailable. The parlous state of communications increased the isolation of those upriver and added to distrust and misunderstandings between different parties. This issue was partially alleviated, but never resolved, late in the campaign when activists based upriver became "River Communicators" travelling back and forth to brief the Information Centre.[64]

Adding to regulations it had already introduced in October restricting access to areas under HEC control, the Tasmanian government changed the state's laws on 24 November, making trespass an arrestable offence, subject to fines of up to A$100 or six months in jail.[65] The imposition of bail conditions during the protest denied arrestees the ability to return to the Franklin. It would result in a rapid turnover in participants, but the penalties failed to deter involvement.[66]

The quarantining of large areas for HEC operations, as well as the remote and inhospitable location of the worksites, made it difficult for large numbers to remain near them for extended periods. Given that access to the upriver camp was not prohibited in the initial stages of the blockade, a pattern of protest soon emerged. Having registered with TWS, blockaders would travel to Strahan where those who had not already taken part in non-violence training would do so before being cleared to travel upriver. They would then be ferried in tour operator Reg Morrison's boat the *J-Lee-M* to the base camp. From there the majority were taken by guides to locations where they would be arrested and returned downriver by police to be processed. Those who accepted bail conditions returned home. The 450 who refused to do so were imprisoned on remand, spending up to 27 days in jail.[67]

Actions involving groups of people occupying workplaces, businesses, and other sites of contention before passively submitting to arrest had a long history in Australia and overseas. During the Franklin campaign most land-based blockading targeting helipads, tracks, and other work sites would follow this pattern.[68]

Along with other social movement theorists, Wang and Piazza have argued that groups seeking to appeal to "heterogeneous supporters" are often discouraged from using violent or heavily disruptive tactics "because doing so might undermine existing and potential support for their claims".[69] TWS's overall strategy echoed this view as it was predicated on mobilising public opinion to force the federal government to intervene and override state government decisions. Although the tactics it favoured were less likely to cause effective obstruction than others, they were perceived to be media-friendly, easy to participate in, and unlikely to generate confrontation with police and workers.[70]

Most ODA tactics incorporate logics traditionally associated with civil disobedience. Not only do they potentially allow protesters to demonstrate their opposition by deliberately contravening the law on the basis of ethical beliefs, but also enable them to "bear witness".[71]

The majority of the land-based actions at the Franklin prioritised these themes and caused little obstruction. A large number of protesters also took part in water-based actions that, while resulting in a high number of arrests, generally required protesters to actively evade the police in order to remain in place before the arrival of HEC craft, as well as to minimise the confiscation of resources. The tailoring of tactics to avoid controversy, the carrying out of actions at times that suited journalistic deadlines, and a preference for publicity stunts and events involving celebrities all marked a major shift from previous blockades.[72]

Most who attended the protest remained for less than a fortnight. Among those who played a longer-term role in the protests three broad and informal factions emerged: TWS, the non-violence trainers (NVT), and the upriver activists (UA). As has been previously noted, strategic and tactical differences stem from a complex mixture of cognitive, cultural, and emotional factors. In this case all parties were publicly committed to "non-violence" and "consensus", but what each understood these to mean, and what they sought to gain through ODA, differed. Each faction would come to be defined by their political and tactical approach, rough geographical location, lifestyle preferences, and role in the campaign. Their interplay would not only have a major impact on the tactics and direction of this protest, but each would also become prototypical approaches to ODA in Australia over the coming decades; in consequence, a detailed analysis of each is necessary.

TWS, which combined the long-term strategic focus, resources, and professionalism of a formal environmental organisation with the tactical approach of ONV, dominated the protest. Its activists were based in the organisation's offices in Hobart and the Information Centre in Strahan and drew on support from branches across the country. TWS was responsible for and controlled campaign finances and logistics as well as media liaison, legal support, and publicity. It directed the overall political strategy of the campaign, negotiated with politicians, and did much of the recruiting for the blockade.[73]

TWS's decision to limit tactics to those defined as non-violent under ONV was in part ethical, but also strategic in that they suited its media-oriented strategy. The state government's continual predictions of violence and description of protesters as "guerrillas" and "extremists" strengthened TWS's resolve to avoid any action that might trigger confrontation. It also believed, in keeping with ONV theory, that as long as protesters acted in a "non-violent" manner, repressive acts by the authorities had the potential to lend the blockade moral authority and public support.[74]

Although the organisation would come to embrace consensus decision making over time, at this point TWS mainly confined its use to the particulars of running the camps and carrying out actions. TWS director Bob Brown – who came to embody the campaign in the media's eyes – told author James McQueen that consensus decision making was "the only satisfactory way of unifying a changing group of blockaders over a period of time; but only because overall objectives were clear and visible". Decisions concerning these objectives were made by the TWS leadership.[75]

In the light of concerns regarding poor publicity the use of consensus decision making, and training was also seen as a control mechanism. These aspects were particularly emphasised by those within TWS who rejected aspects of ONV philosophy as impractical and felt uncomfortable with its cultural milieu and focus on emotional sharing. TWS media liaison worker Chris Harris argued that the chief redeeming feature of consensus was that "it led to the adoption of the lowest common denominator and thus probably prevented any outrageous or dangerous actions and decisions being undertaken".[76]

In keeping with ONV principles TWS engaged in regular liaison with police. As with most environmental blockades, it appears to have seen dialogue with police more as a way of maintaining safe and cordial relations than as a means of conversion. Although future protests would see the organisation reveal more comprehensive information to its opponents, during the Franklin blockades the police were generally told only of daily upriver arrivals and departures as well as the locations of planned actions; further details were omitted. While this was in part deliberate, the absence of TWS activists upriver and communication issues constrained the organisation's ability to be completely "open".[77]

Lacking experience in direct action and unable to fulfil all the requirements of the campaign through its own membership, TWS effectively outsourced non-violence training to activists associated with ONV, Quaker, and feminist groups. The trainers at the Franklin were far from a monolithic group and included TWS members, non-aligned protesters, and members of the Groundswell network. The latter would to some degree come to form a separate bloc or tendency within the protest.[78]

Responsible for maintaining the base camp in Strahan, the trainers carried out mandatory one and later two to three-day sessions with arriving protesters. Applying procedures largely developed by MNS, Groundswell activists advocated complete openness with opponents. Believing that everyone involved should have control over all aspects of blockading, so long as they remained within the ONV frame-work, they applied formalised and lengthy consensus decision-making processes involving the use of an agenda, clear proposals, and detailed procedures for checking decisions.[79]

Runciman et al. argue that training influenced new arrivals not to take aggressive action, but claim it was often inappropriate and focused overly on theory with little information about practicalities and tactics. Overworked trainers either lacked the time or inclination to take part in ODA upriver and due to ideological and personal differences, as well as problems with communication, rarely interacted with UA.[80]

Unlike TWS, who primarily viewed ODA as an extension of their lobbying efforts, the core of NVT did not prioritise working within the established political system, instead arguing that the only way to effectively protect the environment was to convert opponents and the wider public to the need for radical change via non-violent means. A desire to explain the exact ethics and ideology behind ONV, and to ensure consensus about every decision, generally guided their activity rather than concerns regarding media messaging, obstruction or even victory in this par-ticular campaign.[81]

Key members of NVT chafed at the fact that consensus decisions made by protesters at the Strahan camp were regularly overruled or ignored by TWS and UA. There was also frustration that time pressures, the need to keep protesters moving upriver, and the work involved in coordinating other tasks meant that programmes and processes were at times reduced in length. One unnamed trainer commented that "TWS has not been consciously organized using nonviolence principles; rather the technique of nonviolent direct action has been lifted from the range of discip-lines associated with nonviolent social change." This echoed many trainers' belief that ONV's effectiveness could not be fully realised without deep commitment to the values they espoused.[82]

While some protesters perceived training and formal consensus decision making as an imposition and a form of "middle-class control", others were converted to ONV principles.[83] The formation and use of affinity groups – common in over-seas anti-nuclear, peace and feminist campaigns, but formally introduced into an

Australian environmental blockade for the first time here – proved effective and popular.[84]

The Upriver Activists were the third broad faction to emerge. Responsibility for maintaining the remote forest camps located near where work was taking place and for coordinating actions fell to those with the most experience and interest in such activities. This grouping had the loosest definition of "non-violence" with activist Alice Hungerford arguing in a letter that it "is not a set of rules to be rigidly followed … it is an attitude, a way of life, a state of heart".[85] With NAG members such as Benny Zable, Brenda Liddiard, and Lisa Yeates present it was also responsible for much of the music and performance during the protest.[86]

Having unsuccessfully sought to begin ODA ahead of TWS's official start date for blockading, UA were the most focused on obstructing work, not least because they were witnessing the daily destruction of the forest. They too acknowledged the importance of media coverage but felt that many of the actions that were carried out were formulaic. They argued these could have been more newsworthy, as well as effective in disrupting work and imposing financial costs, had they minimised arrests from the beginning, been planned once newcomers had experienced conditions upriver, and not been constrained by ONV limits.[87]

UA had a commitment to egalitarianism, but in keeping with previous Australian environmental blockades, upriver decision-making processes were less formal and involved than those used by NVT. As with TWS and NVT, there nevertheless appear to have been innumerable meetings covering almost every aspect of the campaign. As elsewhere the pace of action and unexpected events often meant that decisions were unilaterally made or overturned by prominent individuals.[88]

UA engaged in some level of police liaison, but believed detailed briefings were disadvantageous. With the intention of diverting police resources, and preventing protesters from being arrested before they made it to the site of actions, activists regularly eluded police and broadcast false details via channels they knew were monitored. To avoid their skills and knowledge being lost to the campaign, scouts, guides, and others avoided arrest. Although UA generally deemed sabotage as counterproductive there were incidents of property damage carried out by those within the group.[89]

The range of tactics UA could employ were constrained by practical considerations concerning their remote location, large police numbers, and the need to maintain logistical support from TWS. They also had to accommodate the views and desires of the mass of short-term protesters flowing in who were either less radical than themselves or had been influenced by TWS and NVT. Despite this they arguably increased the tactical creativity of the campaign, with activist Phil Cushing observing that the "Nightcap people brought … a realistic understanding of what was what, and especially of how to be sneaky."[90]

All three groups were united by an overriding desire to see the dam discontinued. Although major disagreements would occur at the end of the blockade, their activities during the campaign corresponded enough to keep the cycle of protesters and ODA flowing. While a lack of changeover in group composition entrenched personnel and positions, the geographical isolation and autonomy of each group may have lessened open conflict. The fact that ODA was instigated and continued in the absence of an agreed upon strategy concerning its role led at times to conflict and inertia, but also arguably produced a creative tension that generated momentum and innovation at others.[91]

In addition to these main groupings members of Tasmania's Aboriginal community also played a role. Many felt that their views were side-lined, and expressed anger that TWS and others had failed to acknowledge their sovereignty or consult with them before entering the area. Despite this, a small number engaged in ODA and the Tasmanian Aboriginal Centre campaigned against the dam.[92]

The blockade was carried out in three stages, with the first beginning on Tuesday, 14 December. The goal of early actions, and many that followed, was to bring media workers upriver where the stunning nature of the area and its destruction would be highlighted in the context of arrests associated with ODA.[93]

Despite most action taking place on land, water-based actions came to represent the protest. These mainly involved people in canoes, inflatable boats, and other vessels, lining up in an attempt to halt the passage of HEC vessels. Precedents existed for such water blockades in the floating picket lines used over decades across Australia by the Waterside Workers Federation, as well in anti-nuclear actions in Melbourne, New Zealand, the United States, and United Kingdom. This was the first time they had been carried out as part of EB. With the authorities paying little heed to safety regulations, the tactic rarely delayed opponents: police vessels would advance ahead of those belonging to the HEC, ramming and dispersing blockaders before officers carried out arrests. What these actions lacked in obstructive efficacy they made up for in their photogenic and symbolic quality.[94]

Of the 417 protesters who registered for action, 202 were arrested and 167 imprisoned during four days of action before a Christmas moratorium on blockading began. Following precedents set by civil rights and free speech protesters overseas, some anticipated that mass arrests would clog up the justice system and generate further costs and pressure for the Tasmanian government.[95] Throughout the blockade volunteer legal advisors supported detainees as did a Hobart-based support group, although difficulties associated with accessing prisoners often hampered their efforts. Attempts were made to maintain affinity groups within the jails and regular meetings, games, and other activities were engaged in to maintain morale. Despite the women's section of Risdon prison being forced to place beds in its corridors during the first phase of the blockade, issues related to extra prisoners did not become a decisive factor in the campaign, partly because the government responded by placing a general one-week limit on detention.[96]

As anticipated by TWS, the blockade immediately generated extensive media coverage, nationally and internationally. Starting in the run up to Christmas was fortuitous, as there were fewer political stories and issues to compete with. An analysis of Hobart's *Mercury* and Melbourne's *Age* by Hutchins and Lester indicates that the frequency of stories in relation to the dam tripled with the beginning of the blockade, with the former newspaper featuring the protest on its front cover almost every day and the latter running front page stories 12 times in December and 14 in January. Footage of one arrest, involving British television presenter and botanist David Bellamy alongside 53 others, was broadcast in more than 30 countries.[97] As will be seen in subsequent chapters such coverage brought the campaign, and the concept of blockading, to the attention of environmental activists globally, playing a role in inspiring them to undertake similar action.

A tension often exists within media-oriented campaigns between organisers' desire to control the agenda and the need to provide enough "authenticity" and conflict to ensure newsworthiness. At times journalists complained that TWS was overly managing proceedings and that the blockade resembled a marketing campaign. The routinization of protest and the media's desire for novelty often sees coverage drop off after a campaign's beginning, but in this case the repetitive nature of tactics added to journalists' cynicism. Hutchins and Lester argue that "The key to the initial impact and success of the blockade, carefully coordinated tactics that few journalists had witnessed before, gradually became a liability, dismissed as artificial and aimed simply at attracting coverage."[98] Nevertheless, although attention slowed from mid-January onwards, it remained extensive: the major broadsheet, the *Canberra Times*, for example, published 58 pieces on the Franklin issue during February and March 1983.[99]

During the second phase of the blockade, which ran until 21 February, an average of 50 protesters arrived daily until the resumption of university, and the resultant loss of student protesters, led numbers to drop off.[100] Due to pressure from those upriver who wanted more disruption of work as clearing intensified, as well as the need to maintain media interest and respond to increasingly successful counter-moves by the HEC and police, tactics during this period began to shift. The deception of police via diversions and the setting up of "phantom blockades", actions that would never be carried out, increased. Protesters also more commonly avoided arrest. When occupying worksites some would hide in the bush when police arrived, only to reoccupy the area when they left. Decisions regarding the location of actions were also taken with a view to stretching police resources over the widest possible area.[101]

The shift to increased obstruction was echoed in the decision by blockaders to begin targeting the transportation of equipment and materials through Strahan by sitting in front of trucks and blockading the HEC's town compound. An associated tactical innovation involved Strahan-based activists attaching their arms and torsos with high-tensile chain to barges, gates, and equipment. Those upriver soon followed suit by doing the same with machinery and trees. This required fewer blockaders to face arrest and had a much greater potential for disruption as the only way to remove chains was with an oxyacetylene torch. Using such devices had a long history among protest movements but had not been used within an Australian environmental blockade before.[102]

Despite its efficacy and novelty further innovation in this direction did not occur during the Franklin campaign. This may have been because increased policing soon prevented activists from getting access to targets in Strahan; perhaps, too, "locking-on" was too dangerous to undertake upriver without a media presence to ensure police and HEC workers progressed safely.[103] Organisational dynamics possibly also played a role with Runciman et al. arguing that the blockade lacked both a clear and unified strategy and a method of decision making that might have allowed the groups within it to make major changes. They conclude, "Because no single group was able to achieve overall strategic control, and because the tensions and differences between the three powerful groups remained unresolved, the Blockade was unable to respond easily to changing circumstances."[104]

During this period authorities changed their strategy with police becoming more confrontational and magistrates attempting to impose stronger bail conditions.[105] This harder line was illustrated when, in actions that involved over 70 arrests, police staged a major operation in order to bring a bulldozer upriver. Between 1 a.m. and 6 a.m. on Wednesday, 12 January, windows at the TWS Information Centre were smashed by rocks and communication cables cut, radio communications jammed, and public telephones in Strahan and Queenstown switched off. A large contingent of police then blocked the road out of Greenie Acres, arresting protesters who attempted to leave or engage in ODA.[106]

The following day, aiming to waylay the barge carrying the bulldozer, activists attempted new variations on river-based tactics. In an action that had been vetoed during meetings, but was carried out anyway, the police vessel *Freycinet* was briefly stalled at a narrow point in the river by the release of 30 logs. This delay allowed a scuba diver, whose position was marked by a diver's flag, to enter the river. However, when the *Freycinet* arrived its driver ignored regulations banning motorboats from coming within 120 metres of a diver and rode over the top of him. People jumping into the water to block river traffic were similarly handled and protester vessels pushed aside. Having received news of police tactics, activists hastily abandoned an attempt to lower a person in front of the HEC convoy via a harness.[107]

By the end of January, the campaign had entered a period of inertia. The federal Liberal government offered to compensate the Tasmanian government for the project's cancellation. This was rejected and, in keeping with a long-standing conservative commitment to "states' rights", the Commonwealth declined to use its constitutional powers to override the Tasmanian government's decision.[108] New opportunities arose with the federal government's decision to call an early election on 3 February. This was widely considered a gift to the campaign because with the Federal ALP (as well as the smaller Democrats party) opposed to the project, the future of the dam became a key election issue.[109]

With 70 branches set up, membership having doubled in recent times, and more than $1 million flowing into the organisation by mid-February, TWS shifted quickly into electoral mode holding huge rallies around the country and mobilising thousands of volunteers to canvas on behalf of anti-dam parties.[110] Some within TWS, concerned that UA would engage in embarrassing actions, argued for an end to blockading. However, it was decided that the symbolic and media value of ODA was too great to abandon. On 8 February, TWS announced that a third phase of the protest, featuring intensified blockading, would commence in two weeks and culminate in mass action on Green Day, or G-Day, on 1 March, four days before the election.[111]

In response, the Tasmanian National Parks and Wildlife Service was instructed by the state government to prohibit camping in the entire Gordon River State Reserve. On 23 February police evicted the upriver base camp. While the relocation of blockaders to smaller dispersed camps hindered communication and preparation for ODA it ultimately aided TWS's strategy by bringing journalists back to the story and site. Publicity increased with the arrival of G-Day: 231 protesters were arrested at ten different locations, protests was carried out in Launceston and Hobart, and TWS's flag was even raised above the HEC building.[112]

Although TWS had ceded much of the day-to-day tactical decision-making to others out of necessity, it preserved overall control. On 3 March a decision was made to end the blockade, close the Information Centre, and remove upriver resources by election day. This was partially due to a perception that the role of blockading was complete, but also stemmed from concerns that conflict would intensify to dangerous levels in Strahan once, as was looking likely, the Federal ALP was elected and the project cancelled. No consultation had been carried out with activists located upriver and the shock of the decision has seen animosities linger up to the present. Following the refusal of some UA members to leave, a compromise was reached whereby activists maintained an upriver vigil to symbolically protest and document continued environmental destruction.[113]

This decision proved worthwhile because work on the dam continued at an ever-greater pace until July. The Federal ALP came to power on 5 March and passed regulations on 31 March prohibiting HEC work in a World Heritage area. This, as well as later legislation and injunctions preventing work, were ignored by the Tasmanian government. Clearing only ended after the High Court affirmed the Commonwealth's ability to use its "external powers" to override a state government decision.[114]

In overcoming huge resistance from the Tasmanian government, and the various bodies at its disposal, the campaign claimed a major victory. The size, profile, and length of the blockade and the involvement of TWS and NVT, also accelerated a process of differentiation amongst ODA-oriented environmental activists in Australia. By its conclusion three currents had emerged, each favouring different versions of normative behaviour, strategic focus, and organisation. Much debate would be engaged in following the Franklin and during future campaigns all parties would emphasize the different lessons they had drawn from the experience.

For TWS and other formal organisations, the key conclusion was not that ODA should be at the core of campaigning, but that specific policy outcomes could be achieved through alliances with the Federal ALP as well as the leveraging of electoral preferences. As a result of internal debates and developments, TWS would initially withdraw from direct involvement in ODA, returning to involvement in blockades a few years later. Building on the Franklin example, this would primarily be through means that allowed for control by core campaigners while employing training and tactics that stressed publicity over obstruction.[115]

For their part, activists from both the NVT and UA factions criticised TWS, arguing that the focus on defeating one project, and involvement in mainstream political processes, did not guarantee long-term protection for Southwest Tasmania. In articles published during and after the blockade in *Groundswell*, NVT activists claimed that a more comprehensive adoption of ONV principles, and a broader focus, could have been a major step towards building a movement capable of tackling the economic and social issues in society that had led to the dam. They also felt that a closer focus on civil rights issues and a willingness to break bail conditions and re-enter HEC land could have more effectively challenged, if not defeated, the state government's ability to control protest via trespass laws and other punitive measures.[116]

In line with these criticisms, and having raised the profile of its ideas, those associated with the NVT group continued to provide ONV training and workshops around the country. Some would take part in Tasmanian forest blockades organised by TWS in the second half of the decade, but their focus for much of the 1980s would be on the burgeoning anti-nuclear movement.[117]

For the activists involved in the upriver vigil there were further reservations concerning the campaign, as they thought the ending of ODA had been undemocratic and premature, allowing the HEC to destroy a greater area of forest after the election than before.[118] Unlike TWS and NVT they would play a major role in EB during 1984 and 1985, providing camp infrastructure and pushing for actions that emphasised disruption.

The First Daintree blockade

Campaigns aimed at preserving Queensland's Wet Tropics – which covers close to 9,000 km^2 of the state's northern coastline and contains 30 per cent of Australia's species of marsupials, 65 per cent of ferns, 40 per cent of birds, and 60 per cent of birds and butterflies – were a major focus of Australian environmentalists during the 1980s.[119] During the 1980 Second World Wilderness Congress in Cairns Queensland Premier Joh Bjelke-Petersen describing the area as a "living museum of plants and animals".[120] However, due to his conservative state government's even stronger support for development, it rejected the international gathering's recommendation that it "create a single, very large National Park… that would encompass the entire coastal, valley and mountain regions from Cooktown to the Daintree River".[121]

The August 1983 announcement by the Douglas Shire Council of its intention to complete a 33 km four-wheel drive track linking Cape Tribulation to Bloomfield became the issue that fully focused national attention on the region's rainforests. With the Franklin recently saved and the state government fully committed to the project, environmentalists lobbied the Federal ALP to stop it from going ahead. However, with the election over, conservationists' political influence had declined and their relations with the ALP cooled. The federal government was focused on issues regarding macro-economic and industrial relations reform and reluctant to take on the obstinate Queensland government. With his position weak within the government, Federal Minister for the Environment Barry Cohen demurred from intervening decisively. An agreement he brokered with the shire council to hold off work until consultation was completed was subsequently broken on 30 November when bulldozers crossed the Daintree River, thereby triggering Queensland's first sustained environmental blockade.[122]

Mike Berwick, who conducted police liaison and later became local Mayor, described those involved in both stages of the blockade, the first of which occurred in late 1983 and the second in mid-1984, as mainly "local people and the local cops and the local bulldozer driver, who all knew each other".[123] Many protesters were in their twenties and thirties and belonged to regional New Settler and alternative communities that had sprung up in the 1970s. There were also some older and long-term residents involved, including those in industries such as fishing, which stood to be harmed by the soil run off associated with construction. Some local Indigenous people and the North Queensland Land Council campaigned against the road, but those to whom the affected area belonged, the Wujalwujal community, did not. Despite the fact that it threatened cultural sites of significance they prioritised long-standing relations with pro-development forces and welcomed the increased access the road could provide.[124]

During the blockade, a handful of full time, largely voluntary activists within the Cairns and Far North Environment Centre (CAFNEC) provided logistical and media coordination and conducted lobbying, with their Cairns' office acting as a stop-off point for those travelling to the protest. Founded in 1981, CAFNEC drew from the region's alternative community and held a blockade in November that year against logging in the Mount Windsor Tableland. This was organised by local conservationist Rupert Russell, who had been screening Kendall and Tait's film about the Terania Creek campaign, *Give Trees a Chance,* across the region. The group camped in Mt Windsor, handing out flyers concerning the issue to timber workers and other passers-by before beginning a daily picket involving up to 36 people. These were mostly local activists, but also included NAG's Benny Zable. After a truck forced its way through a group standing in the road, blockaders laid down, at which point 13 were arrested. This effectively brought the protest to an end, but in the week it had lasted it had demonstrated the ability of ODA to disrupt work and bring public attention to regional rainforest issues.[125]

The Port Douglas based Wilderness Action Group (WAG), which had been founded in 1983 to carry out lobbying and education campaigns regarding local environmental issues, was primarily responsible for running the blockade at Cape Tribulation. Having already made rough plans they rapidly responded to the beginning of construction. Following now established precedents they set up a protest camp including a kitchen, information stall, toilets, and other basic infrastructure on privately owned land. Smaller camps would be established in the rainforest as required.[126]

Ian Cohen believes the remote location and the antipathy of Queensland's state government and police force kept away many interstate protesters. New Settler properties had been subject to raids and evictions and hundreds of marchers had been arrested and assaulted since the Bjelke-Peterson administration effectively banned public demonstrations in 1977. The presence of local officers in the first phase of the Daintree blockade meant that relationships were generally cordial, but its second phase would see the use of violent repression.[127]

An initial meeting of blockaders indicated that only two out of eighty had previously been involved in a similar protest. Largely unfamiliar with the details of what had occurred during interstate blockades and reacting to events as they unfolded, the campaign lacked an overall strategic plan for blockading. As a result, organisation, tactics, and ideas concerning normative behaviour would evolve as it progressed. Some non-mandatory training workshops would eventually be carried out, but these focused on practical information and role-plays and, as with the rest of the campaign, were not based on ONV theory or processes.[128]

NAG members had maintained their connections following the Franklin, playing a militant and contentious role at the recent Roxby Downs anti-uranium protest. A small number travelled from Tasmania and NSW upon hearing of the blockade. Although there were differences over issues such as how protesters presented themselves to the public via their appearance, the need for experienced people, and the numbers to properly carry out ODA, meant that local activists generally welcomed the involvement of NAG members and were open to their ideas. Much smaller and more personally interconnected than the Franklin campaign, the Daintree blockades, despite a broadly similar separation of locations and roles, did not see major divisions immediately manifest between office-based political campaigners and forest-based blockaders.[129]

Although following in the mould of the pre-Franklin blockades, ideas concerning normative behaviour were nevertheless stricter than those adopted and implemented during the Nightcap campaign. As with all blockades, behavioural norms reflected the existing standards of those involved, as well as their perceptions regarding practical and political considerations. Foremost among these here was the question of how to build local and wider support in a highly conservative state. Arguing that "bad publicity is the main weapon used against us" WAG assured its opponents that it would be peaceable and that "no interference with machinery will take place".[130]

With the aim of promoting good relations, daily police liaison was carried out and the authorities were sometimes informed of WAG's intention to blockade before work started. Many actions however were set up in secret in order to maintain surprise and maximise obstruction. Other than one night, when protesters kept police awake with singing and trumpet playing, harassment tactics against workers and authorities were not employed. Despite increasing police violence during the second stage of the blockade, protesters generally avoided verbal and physical aggression or resistance on their part.[131]

Unlike at the Franklin, the blockade had begun as a largely unplanned reaction to imminent destruction. This, plus the impending wet season – which all were aware would bring a complete halt to work – meant that it was as much focused on obstruction as publicity. The lack of large numbers willing or able to be arrested, combined with NAG input, led to an emphasis on methods that would delay clearing while minimising arrests.[132]

Blockading initially focused on two areas: the southern end of the track at Cape Tribulation, which attracted the majority of protesters, and the even more remote northern end at Bloomfield, which was only accessible by foot or boat. Any advantage the Douglas Shire Council had hoped to gain by beginning roadwork unannounced was quickly lost due to poor preparation. Issues relating to confusion over the track's route and the boundaries of an adjacent national park, and disputes with private landowners and National Parks and Wildlife Service officers, regularly delayed construction.[133]

Around 500 people would take part in the blockade with each day seeing 30–100 directly involved. During the first days of action, protesters employed tactics from previous blockades, including parking vehicles and blocking equipment with their bodies. Work was halted at both ends of the project, but only after protesters came close to being seriously injured.[134] Arrests were carried out during this time, but attempts to use roadblocks to prevent people from entering the southern end of the protest were circumvented by boats, four wheel drives and a ferry.[135]

On Friday, 2 December, a declaration from the National Parks and Wildlife Service banned members of the public from coming within 300 metres of a bulldozer.[136] Although this made it easier for blockaders to be taken away, it did not prohibit them from entering the forest altogether and further specified that work had to be halted on safety grounds if they entered the exclusion zone. With workers, police, and equipment withdrawn daily at 4.30 p.m., blockaders also had the advantage of being able to set up elaborate actions overnight.[137]

Although protesters engaged in intermittent blockading at Bloomfield, Cape Tribulation was the focus of ODA. Inspired by actions carried out during the Franklin campaign, one NAG member had brought a high-tensile steel chain with him. Combining two previous tactics, he and other protesters climbed trees with some chaining themselves to branches. This innovation was initially successful in preventing clearing, but after two days police stopped enforcing safety rules. Having allowed a bulldozer to push between occupied trees, and realising it was easier to cut padlocks than chain, they climbed up and removed sitters.[138]

Following further interruptions to work, due to a combination of ODA, mud and poor weather, authorities introduced a second exclusion order closing the National Park to all but police and council workers.[139] In response NAG members came up with an entirely new tactic that took their opponents by surprise on Monday, 12 December. Arriving early in the morning, police and workers found a man tied to a wooden cross. Symbolising the "crucifixion [sic] of an ancient life-force, the rainforest",[140] this device was unlikely to delay work for long, but the line of five people buried up to their waists and necks behind him could not be so easily removed. Despite cramping, and stretches of involuntary panic for those fully buried, this simple innovation allowed protesters to once more maximise obstruction with the use of small numbers. With soil packed tightly around them, and shovels difficult to use without harming those buried, police employed a fruit tin to slowly dig out one protester before a bulldozer was driven precariously between those remaining in the ground.[141]

Realising that their daily retreat was allowing protesters too much of an advantage, police and workers eventually began to camp onsite. Heavy rains and landslides continued to interfere with clearing and made it difficult for police to drag away protesters, some of whom covered themselves with mud. Despite being authorised to use "reasonable force" to remove protesters, police halted work on 14 December after an incident in which a bulldozer driver pushed through groups of protesters and nearly killed a local resident by lifting him into the air.[142] Hampered by poor weather workers did not return until 19 December and soon left after they found five people buried closely together up to their necks astride a barrier made of logs, posts, signs, and banners to which others had chained themselves. Regardless of Douglas Shire Council claims that the dispute was "all over bar the Greenie's shouting" and the track only in need of "finishing touches", it was far from complete. With the wet season fully underway, clearing would not resume until August.[143]

The Errinundra Plateau Campaign, 1984

During the break between blockades in the Daintree, NAG members initiated the first environmental blockade in Victoria. Despite the existence of ongoing campaigns, which had already extracted minor concessions, NAG members largely failed to consult with Victorian groups before launching their own protests in early January 1984. NAG's initial plan was to engage in a series of "hit and run" actions at sensitive sites across the state. This innovative strategy, which would have employed brief ODAs to gain media attention while signifying the ability of activists to potentially engage in more persistent protest, soon foundered. Highlighting the ability of close interaction with the natural environment to shift priorities and devalue more strategic considerations, most of the group became emotionally attached to old growth forest at East Gippsland's Errinundra Plateau. Having committed to saving the area they remained on site for close to a month.[144]

Due to Christmas holidays and poor weather logging crews were largely absent from the forest, but the presence of the protest camp, as well as some initial blockading with vehicles and bodies, secured much media attention. Some local environmentalists supported the protest, but the state's major organisations distanced themselves, as they believed it endangered progress they had made in their own negotiations.[145]

The threat of a blockade drew the enmity of local timber-workers and their families. This was exemplified by events that occurred during a public meeting the protesters called in the small town of Bonang on 12 January. Radical environmentalists, vastly outnumbered by their opponents, were heckled, and punched and their vehicles pelted with beer bottles.[146]

Just as social movements share tactics, so do those who oppose them. Following the success of Terania Creek, activism and networking by logging supporters slowly emerged around Australia. Modelling their tactics and strategy on those employed in the United States, groups such as the Grafton based Ladies for Environmental Awareness in Forests employed conventional social movement tactics of collecting petitions, lobbying politicians, running education campaigns, and holding public meetings and rallies. These were not successful in preventing significant reform in NSW, but a popular pro-dam movement emerged in Tasmania, with government support, in 1983. The Errinundra protest appears to have triggered similar efforts in Victoria, as the timber industry and unions moved to mobilise thousands of supporters in counter rallies.[147]

Social movement scholar Herbert Haines coined the term "radical flank" to describe how the actions and positions of militants affect those of other actors. In some cases they can strengthen the position of moderates by placing pressure on opponents and making demands and actions previously considered excessive appear reasonable. Militant activity can also discredit those involved in a campaign and give rise to harmful counter-movements.[148] The Errinundra campaign compelled the state ALP to respond but resulted in negative outcomes. The government complained that protesters had not flagged their maximalist demands before taking action and found it difficult to engage in discussions with those who lacked "identifiable leadership or an executive which could present and represent the views of the Group as a whole".[149]

On 17 January 1984, the state government guaranteed logging on the Errinundra Plateau for a further three years. A week later it followed the example of other states in introducing rules that made it "illegal for anyone to go within 200 metres of timber harvesting to obstruct, impede or hinder operations".[150] When logging crews arrived on 2 February protester numbers had dwindled to 30. An equal number of police officers prevented them from blocking equipment with their bodies and the blockade soon collapsed. Following a split among those involved, a small number of protesters continued with their original plan and disrupted logging in the state's Otway region, drawing further criticism from other environmentalists.[151]

Although its initial strategy had been innovative, in the light of its failures, no further blockades would be carried out in Victorian forests until the 1990s. According to NAG member Lisa Yeates, who had not been involved, the main lesson radical environmentalists drew was that "You can't go in gung-ho without consulting and working under the auspices of the local people who have been campaigning there for years and who know the conditions."[152]

The second phase of the Daintree blockade

Back up at the Daintree rains had seen much of the work done on the track washed away. Despite documents – later released following a Freedom of Information (FOI) application – demonstrating that the Queensland Main Roads Department had advised against the road on the grounds of need, expense, and environmental concerns, the local council and Queensland government remained firm.[153]

The issue had gained prominence, but the federal government continued to avoid intervention, leading WAG to plan a second blockade. National and state networks and organisations increased their involvement, with the ACF declaring 1984 the "Year of the Daintree". The ACF unsuccessfully attempted to persuade local activists to call off the blockade, which they believed was jeopardizing attempts to broker a broader deal on forests with the Federal ALP.[154]

With the second blockade looming, prominent members of TWS, which had progressed from a Tasmanian into a national organisation, attempted to persuade CAFNEC and WAG to isolate NAG and impose ONV rules and training along the lines of those used at the Franklin blockade. WAG members shared some of TWS's concerns regarding the negative media attention NAG members' alternative appearance and unpredictable behaviour could bring, but did not follow the advice, likely because the existing model and relationships had been successful. TWS increased its campaigning efforts, but played no significant role in the blockade itself.[155] For their part NAG members deepened their involvement, bringing in extra equipment and more than 40 protesters as well as running direct action workshops and preparing a campaign handbook.[156]

By 6 August 1984 a large contingent of police had established their own camp three kilometres from the road's southern entrance at Cape Tribulation. With a continuing focus on obstruction and plenty of time to prepare, blockaders created the most complex devices aimed at extending manufactured vulnerability yet seen in Australia. A number were buried in the ground, up to six feet deep, with their legs chained to concrete slabs and logs. Others had a series of "fiddlesticks" (logs spiked with nails to prevent them being cut with chainsaws) stacked in a criss-cross pattern above them or were chained aboveground by hand or foot to the structures. More familiarly, boulders and timber were placed in the road as was NAG's bus, which had had its wheels and battery removed.[157]

With police and council workers waiting two days to move in, those in the ground were dug out each night. Being buried and unable to move, for fear of making the dirt compact itself more tightly, some reported experiencing panic followed by feelings of deep calm and transcendence. As Graeme Innes later recalled in an article appearing in the *Earth First Journal*, printed in the United States:

> Last minute preparations complete, we entered our holes to await the arrival of the police. I felt a strange serenity. There was no fear in waiting, rather a calm understanding that this was the right action and stood above the law of the land.[158]

When police and council workers arrived on 8 August, unfamiliarity with these new tactics meant a day was taken up removing up to 100 blockaders. Attempts to dig blockaders out with a backhoe were held up after a man, in the manner of actions that had happened at the Franklin, attached himself to the top of the machine with high-tensile chain. Once this obstacle was removed trenches were dug on either side of those buried before police tunnelled in with shovels and cut through chains with bolt-cutters. The operation required the involvement of up to three officers at a time; one pair of bolt-cutters broke during proceedings. Thereafter a crane was used to precariously remove the logs one by one before the final protesters were dug out. Although council workers were cavalier in their attitude, police intervened at times when it appeared serious injuries were imminent. This may have been partially motivated by the substantial media presence, as the story gained national and international coverage.[159]

Disappointed that their actions at the entrance to the forest had failed to delay their opponents for longer, protesters focused on a new line of defence, a series of tree-sits. These were placed in rows located at points through which it would be difficult for bulldozers to traverse without endangering sitters. Building on previous practices, and allowing protesters to remain in place for extended periods, small platforms were erected for the first time in an Australian blockade. Ropes, nets, and hammocks also connected trees to prevent bulldozers moving past or working around them. Although not yet described by the term, such "tree villages" would become a major part of forest and anti-blockades in the United States and the United Kingdom during the 1990s.[160]

The presence of multiple tree-sitters led police and council workers to withdraw from the area during the afternoon of 10 August. With only 1.5 km of the road completed and action now taking place where few journalists could observe events (due to bans on entering the national park), authorities moved to employ increasingly repressive means.[161]

On the evening of 11 August, up to 100 blockaders held a candle-lit vigil, at which point police assaulted them and deployed dogs, eight suffered bites.[162] After one man was admitted to hospital, and graphic recordings of protesters screaming released, police claimed that the protesters had "orchestrated" the event.[163] Amidst continuing arrests and a major increase in fines for hindering police and disobeying a police order, the state government depicted protesters as "hordes of people on the dole" and announced it was considering legislation to provide police with extra powers.[164]

Despite the dog patrols, protesters continued to resupply those occupying trees, using pepper to obscure their scent. Police taunts and rock throwing, as well as the flaunting of safety rules by chainsaw operators and bulldozer drivers, failed to remove sitters. What did eventually force many of them down were threats to destroy a much larger section of forest to build a path around them. Regardless, one NAG member remained in place for six days. Asked by a reporter how he felt when he finally was arrested, he stated, "I'm out of my tree, man."[165]

Meanwhile at the northern end of the proposed track, protesters had maintained a vigil since July. Deep pits, dubbed "elephant traps", were dug to prevent bulldozers moving or crossing the Bloomfield River at low tide and a yacht kept on hand that could be beached in their path as required. With the road coming through from the southern end, these tactics had not been tested, but in mid-August a device was erected which replicated the manufactured vulnerability of a tree-sit. According to Ian Cohen this put a small platform in the "fork of a seven-metre pole" wedged in a tree trunk with an "old boat with the bottom torn out placed on the roadway, a hole excavated, and the installation lifted upright".[166]

Tripods and other devices which place protesters precariously at heights over tracks and roads would become a regular feature of Australian and overseas protests from the late 1980s onwards. The efficacy of this particular innovation was not tested however as protesters were confronted by members of the Wujalwujal community who demanded they remove the device. Threatened with it being pulled down by a front-end loader, and with work only occurring at the other end of the track, the protesters complied.[167]

Blockade numbers began to dwindle and with them the ability to effectively obstruct road works. Media attention was also flagging and divisions between CAFNEC and blockaders had begun to open after the former refused to focus on police violence, for fear of shifting attention from rainforest destruction. Concerned that events were getting out of their control, local activists called an end to the blockade on 19 August. NAG members initially opposed this but following another week of intermittent blockading they withdrew and some of group headed to an anti-uranium blockade in South Australia.[168]

The track would open in early October, but its launch was an embarrassing affair. Even before the wet season had fully arrived a 50-vehicle convoy, including those carrying the Douglas Shire Council mayor and Queensland's Environment Minister, became bogged with eight or nine cars stuck late into the night. Within months sections of the road had collapsed; despite remaining inaccessible to most vehicles, promised upgrades never eventuated.[169]

Although successful in gaining publicity, the limitations of tactics exploiting manufactured vulnerability in the face of sustained state repression were clearly demonstrated during the second Daintree blockade. Nevertheless, the blockade had drawn major attention to threats to the region's ecology. From here on in the movement to protect the Wet Tropics was dominated by full time, paid activists from the peak environmental organisations. During the 1987 federal election these formed a new alliance with the federal ALP that, following the government's re-election, led to the granting of World Heritage status and greater protection for over 900,000 hectares of forests in North Queensland.[170]

The end of a protest wave in Australian conservation activism

By 1984 a series of campaigns had fully established environmental blockading as a strategic option for Australian environmentalists. A cohort of experienced blockaders had emerged and with them distinct and differing approaches concerning the role ODA could and should play in achieving campaign goals. Connected to these were differing codes of normative protester behaviour regarding attitudes and actions towards opponents, acceptable tactics, how much information should be revealed to those outside of campaigns, what constituted acceptable property damage, and other standards.

Interweaved with these developments – all of which were influenced by factors such as changing counter-tactics, rules and laws affecting protest, political opportunities, the attitude of authorities towards safety, and geographical location – was the emergence of an expanded body of ODA tactics. All of the campaigns employed soft blockades. As at Terania Creek this mainly involved people occupying entry points and work sites as well as climbing onto equipment. Due to the geographical location and nature of work involved, the Middle Head and Franklin Rivers campaign saw soft blockading extended to water locations with protesters swimming and using vessels. Soft blockading was increasingly countered through the willingness of some state governments to introduce regulations and bail conditions prohibiting protesters from entering certain areas and to expend sizeable police resources to remove them.

For some campaigners this was not an issue as they primarily saw ODA as a method of expressing opposition, bearing witness, and gaining media attention. As such, large numbers of arrests could be efficacious and innovation came to be focused on how these incidents could be leveraged to pressure decision makers.

Those who also prioritised disruption at the point of destruction were the main drivers of tactical innovation. In order to minimise arrests, and conserve protester resources, some activists took to infiltrating sites in small numbers before carrying out soft blockades. In some cases the exhaustion of police and opponents' resources became a key determinant of the placement and timing of actions as well as the use of diversions. To this end, as well as due to interpersonal enmities, some blockaders also began using psychological means to harass opponents and police.

The dynamics involved with maximising obstruction while minimising arrests drove activists to come up with new forms of barricades and new tactics of enhanced vulnerability. Vehicles, boulders, and logs were commonly placed on roads and ditches were dug during protests. Other means of physically preventing access were added including removing cattle grids, placing and setting alight car bodies, and immobilising gate locks with super glue.

During the Franklin campaign, activists began chaining themselves to equipment and trees. Two blockades in the Daintree rainforest did the most to expand the range of enhanced vulnerability tactics. Ladders and ropes were used to go higher into trees. Activists increased their use of hammocks and introduced platforms and chains to help secure themselves for longer. Tactics such as burying people in the ground, locking their feet to concrete slabs and logs, chaining bodies to barricades, and placing trees in "fiddlestick" arrangements were innovated. Although a lull in EB followed the Daintree, the efficacy of such tactics would be further extended during the next major wave of Australian environmental protest in the late 1980s through the introduction of tripods and lock-ons.

The shift in the Daintree campaign from locally based campaigning and blockading to deals brokered between ALP and organisational elites exemplified changes which swept the Australian environment movement from the mid-1980s onwards. The ACF and the larger state based and regional organisations increasingly eschewed grassroots mobilisation for direct lobbying, submission writing, and incorporation into government processes via membership of consultative committees.[171]

Although it too had shifted towards practices that largely centralised control in the hands of its national leadership, TWS still contained activists committed to egalitarian decision making and grassroots campaigning as well as those who saw the strategic value of ODA. Stemming from this, and the inability of a large and popular movement to otherwise curb logging, it would continue to support carefully controlled, media-oriented blockades in Tasmania that adhered to ONV principles and included training and police briefings. Intermittent ODA carried out through the 1980s incorporated tactics developed elsewhere and, inspired by developments taking place in the United States, included the longest and highest tree-sits yet seen in Australia, one 16 days in length. They did not otherwise significantly add to the EB tactical repertoire or represent a new current amongst Australian environmentalists. In consequence, they will not be analysed here.[172]

During this period those environmentalists outside of Tasmania who favoured ODA found themselves largely side-lined. Radicals' energies mainly went into the peace movement with many taking part in blockades of warships, uranium mines, and military bases. Here they would continue to innovate tactics, such as bolting their necks and supergluing their bodies to gates at Roxby Downs in 1984, tactics which would be adopted and extended upon in later environmental blockades.[173] Other radicals continued their work on projects associated with reafforestation and environmental sustainability while some, such as NAG member John Seed, became focused on the global dimensions of rainforest destruction.[174]

Notes

1 Rik Scarce, *Eco-Warriors*, 157–58; Roger Wilson, *From Manapouri to Aramoana*, 120–23.
2 Robert Rosen, Interviewed 29 January 2015; Kent Trussell, "Save Middle Head Beach," *Maggies Farm*, April 1980, 8.
3 Cohen, *Green Fire*, 31–39.
4 Foley, *Learning in Social Action*, 29.
5 Robin Osborne, "Confrontation at the Beach-Head," *National Times*, 13–19 July 1980, 5; "Mining: 24 Hours Warning," *Nambucca Guardian News*, 27 June 1980, 27.
6 Seed Interview.
7 "Middle Head Protest Widens," *Macleay Argus*, 9 September 1980, 1–2. Robyn Meehan, "On the Beach," *Maggies Farm*, October 1980, 12–13.
8 "The Mining Front... Fence Goes up; Man Arrested," *Nambucca Guardian News*, 9 July 1980, 2.
9 " Middle Head Protest Widens," 1–2.
10 Rosen Interview; Leggett Interview.
11 "The Fanfare and the Tumult Gone," *Macleay Argus*, 9 August 1980, 1; "Press Release 22 September" (MHSMAG, 1980), 1–2.
12 Cohen, *Green Fire*, 33–34; Peter Geddes, "Untitled," *Maggies Farm*, September 1980, 2–3.

13 "Middle Head Protest Widens," 9 September 1980, 1–2; MHSMAG, "Report," *Conservation Camp Newsletter Middle Head*, 28 September 1980, 1–3.

14 Alberto Melucci, *Nomads of the Present: Social Movements and Individual Needs in Contemporary Society* (London: Hutchison Radius, 1989), 35.

15 Feigenbaum, Frenzel, and McCurdy, *Protest Camps*, 116.

16 "New Group in Mining Protest," *Nambucca Guardian News*, 8 October 1980, 4; Dudley Leggett, "Green Alliance," *Sunshine News*, October 1980, 13–14.

17 Benny Zable, Interviewed 25 March 2015.

18 Pip Wilson, "Impressions of Middle Head," *Maggies Farm*, October 1980, 9; Graeme Dunstan, "Lessons in Losing," *Maggies Farm*, October 1980, 11.

19 Joseph Glascott, "43 Arrests at Sand Site," *Sydney Morning Herald*, 23 September 1980, 4.

20 NSW Department of Education. Australian Environmental Activism Timeline, Available [Online]: www.teachingheritage.nsw.edu.au/section03/timeenviron.php (Accessed 2 June 2014); "Workers Hear the Bad News: Sand Miner Quits Macleay," *Macleay Argus*, 13 December 1980, 1,5.

21 "Editorial: Logging War Heats Up... Still the Government Is Silent," *Northern Star*, 1 October 1982, 2.

22 "Environment Groups to Support Protesters," *Northern Star*, 22 July 1982; Seed Interview.

23 Ian Cohen, *Green Fire*, 41; Yeates Interview.

24 Seed Interview.

25 Yeates Interview.

26 Cohen, *Green Fire*, 43–52; Seed Interview.

27 *Nightcap Handbook* (Nimbin: Nightcap Action Group, 1982), 3.

28 Sophia Hoeben, Interviewed 12 April 2015; Zable Interview.

29 Ibid; Yeates Interview.

30 Ibid; Seed Interview.

31 Jeni Kendell and Eddie Buivids, *Earth First* (Sydney: ABC Enterprises, 1987), 66; Turvey, *Terania Creek*, 134.

32 Jasper, "Emotions and Social Movements: Twenty Years of Theory and Research," *Annual Review of Sociology* 37 (2011): 285–303.

33 Cohen, *Green Fire*, 52–53.

34 Leggett Interview; Ian Watson, *Fighting over the Forests*, 94–95.

35 "Forestry Offices Occupied by 200 Conservationists: No Arrests Made," *Northern Star*, 22 July 1982, 1.

36 Andy Parks, "Environmental Protest Songs of North East NSW 1979–1999" (Honours thesis, Southern Cross University, 1999): 3.

37 Turvey, *Terania Creek*, 135.

38 Hoeben Interview.

39 "Date Set to Hear Mt Nardi Dispute," *Northern Star*, 21 September 1982, 2.

40 "Work Disrupted by Forest Protesters," *Northern Star*, 5 August 1982, 3; "Press Release: 4 August," (NAG, 1982), 1.

41 Yeates Interview.

42 Cohen, *Green Fire*, 53.

43 "Anti-Loggers Call for Ban," *Northern Star*, 10 September 1982, 1.

44 Seed Interview.

45 "Logging War Flares," *Northern Star*, 30 September 1982, 1,5; Cohen, *Green Fire*, 54–56.

46 "Four More Arrests at Mt Nardi," *Northern Star*, 1 October 1982, 2; Cohen, *Green Fire*, 57.

47 Wang and Soule, "Tactical Innovation in Social Movements," 519.

48 Yeates Interview; Kendell and Buivids, "Earth First", 70,86.

49 "Nightcap," *Colong Committee Newsletter*, no. 75 (1982): 5; "Anti-Logging Protest," *Canberra Times*, 2 October 1982, 3.

50 "NSW Cabinet Saves Rainforests from the Axe," *Canberra Times*, 27 October 1982, 3,55.

51 Meredith, *Myles and Milo*, 281.

52 Zable Interview.

53 Jeff Angel, "Letter from TEC Executive Director to John Seed Concerning Blockading," Undated, 1988, 1–2.

54 G. Holloway, *The Wilderness Society: The Transformation of a Social Movement Organization* (Hobart: Department of Sociology, University of Tasmania, 1986), 1–7.

55 Pam Waud and Robin Tindale, eds., *The Franklin Blockade* (Hobart, TWS, 1983): 18–19.

56 Roger Green, *Battle for the Franklin* (Sydney: Fontana/Australian Conservation Foundation, 1984), 255–56.

57 These activists did not generally use the term "Orthodox Non-violence," preferring "Non-Violent Action", "NVA", or simply "non-violence". However, because almost all of the activists involved in the Franklin campaign and others described themselves using such terms I will use ONV to identify those adhering to this particular form.

58 Ron Chapman, " Fighting for the Forests", 132–39; Kathy Brouillette and Michael Lockwood, "Guide for Trainers," *Groundswell: Newsletter for a New Society* 1, no. 1 (1982): 4–5.

59 Ibid.

60 Waud and Tindale, *The Franklin Blockade*, 17–18.

61 Claire Runciman et al., *Effective Action for Social Change* (East Ringwood: ACF Books, 1986), 12.

62 Yeates Interview; Alice Hungerford, ed. *Upriver: Untold Stories of the Franklin River Activists* (Maleny: UpRiver Mob, 2013): 52–53, 140–43.

63 Ibid, 198.

64 Yeates Interview; Waud and Tindale, *The Franklin Blockade*, 10–11.

65 Ibid, 10.

66 Martin Branagan, "Art Alone Will Move Us," 122.

67 Hungerford, *Upriver*, 76–80, 114–17, 35–36.

68 Ibid, 67–69, 76–80.

69 Dan Wang and Alessandro Piazza, "The Use of Disruptive Tactics in Protest as a Trade-Off: The Role of Social Movement Claims," *Social Forces* 94, no. 4 (2016): 1676.

70 Waud and Tindale, *The Franklin Blockade*, 28–31.

71 Margaret Dicanio, *Encyclopedia of American Activism: 1960 to the Present*. (Lincoln: iUniverse Inc., 2005), 208.

72 James McQueen, *The Franklin: Not Just a River* (Ringwood: Penguin, 1983), 36; Peter Thompson, *Bob Brown of the Franklin River* (Sydney: George Allen & Unwin, 1984), 161.

73 Runciman et al., *Effective Action for Social Change*, 5–6.

74 Brett Hutchins and Libby Lester, "Environmental Protest and Tap-Dancing with the Media in the Information Age," *Media, Culture & Society* 28, no. 3 (2006): 438.

75 McQueen, *The Franklin*, 58.

76 Waud and Tindale, *The Franklin Blockade*, 44–45.

77 Ibid, 48; Branagan, "Art Alone Will Move Us," 141.

78 "Cover," *Groundswell* 1, no.2 (1983): 1–2; Karl-Erik Paasonen, Interviewed 7 October 2015.

79 Branagan, "Art Alone Will Move Us," 123–27.

80 Runciman et al., *Effective Action for Social Change*, v,6,19,29–30.

81 Ibid, 17–22.

82 "Blockade," *Groundswell* 1, no. 2 (1983): 5.

83 Martin Branagan, "'We Shall Never Be Moved': Australian Developments in Nonviolence," *Journal of Australian Studies*, no. 80 (2004): 146–49.

84 Hungerford, *Upriver,* 31, 37.

85 "TWS Blockade Feedback," *Groundswell* 1, no. 4 (1983): 3.

86 Yeates Interview.

87 Hungerford, *Upriver,* 11, 40–41.

88 Ibid, 44; Runciman et al., *Effective Action for Social Change,* 7, 34–35.

89 Seed Interview; Hungerford, *Upriver,* xii, 5, 76–80, 85.

90 Ibid, 225.

91 Michael Lockwood, "Nonviolence and the Southwest," *Groundswell* 1, no. 2 (1983): 6–7; Runciman et al., *Effective Action for Social Change,* 12–13.

92 Michael Mansell, "Comrades or Trespassers on Aboriginal Land?," in *The Rest of the World Is Watching,* ed. Cassandra Pybus and Richard Flanagan (Chippendale: Pan Macmillan, 1990): 101–6.

93 Waud and Tindale, *The Franklin Blockade,* 11.

94 Hungerford, *Upriver,* 116–19.

95 Ibid, 74–75.

96 Tony Faithfull, "In Prison," *Groundswell* 1, no. 2 (1983): 9; Runciman et al., *Effective Action for Social Change,* 57–62.

97 Hutchins and Lester, "Environmental Protest and Tap-Dancing with the Media in the Information Age," 441.

98 Ibid, 444.

99 This is in contrast to the 89 pieces that ran during December and January.

100 Waud and Tindale, *The Franklin Blockade,* 11.

101 Ibid, 53, 74; "Protest Plan Vigil in Wilderness," *Canberra Times,* 19 December 1982, 3.

102 Waud and Tindale, *The Franklin Blockade,* 3; "Former Minister Amongst 42 Arrests," *Canberra Times,* 11 January 1983.

103 Cam Walker, Interviewed 28 September 2015.

104 Runciman et al., *Effective Action for Social Change,* 8.

105 Waud and Tindale, *The Franklin Blockade,* 12, 53, 74.

106 "Police Thwart Dam Protest," *Canberra Times,* 13 January 1983, 3.

107 Waud and Tindale, *The Franklin Blockade,* 74–79, 86–87; Hungerford, *Upriver,* 84, 108–10, 205.

108 McQueen, *The Franklin,* 29–30.

109 Ross Dunn, "No Dams Campaign: It Depends on the Election," *Sydney Morning Herald,* 14 February 1983, 2.

110 "Franklin-Lower Gordon Rivers Campaign," *Wilderness News* (1983): 2; "Wilderness Society Turnover $1m," *Canberra Times,* 12 February 1983, 7.

111 Waud and Tindale, *The Franklin Blockade,* 12.

112 Ibid, 103–14.

113 Hungerford, *Upriver,* 193–95, 233–39.

114 "On the River," *Wilderness News* (1983): 2; "Workers' Day Off as HEC Told Project to Stop," *Canberra Times,* 2 July 1983, 23.

115 Holloway, *The Wilderness Society,* 23.

116 Hungerford, Upriver, 280; Kathy Brouillette et al., "TWS Tactics: An NVA Scenario," *Groundswell* 1, no. 3 (1983): 3–5.

117 Peter Jones, Bryan Law, and Margaret Pestorius. The Story of the Australian Nonviolence Network, Available [Online]: www.nonviolence.org.au/downloads/ann_story.pdf (Accessed 19 March 2010).

118 Cohen, *Green Fire*, 79–80.
119 Natasha Bita, "'Greenies' Lost Battle but Won the War," *Australian*, 10 June 2014, 2.
120 "Police Clash with Forest Protesters," *Canberra Times*, 13 August 1984, 3.
121 Quoted in Mary Burg. How the Wet Tropics Was Won, Available [Online]: http://cafnec.org.au/about-cafnec/how-the-wet-tropics-was-won/ (Accessed 26 August 2014).
122 Wilderness Action Group (WAG), *The Trials of Tribulation* (Port Douglas: Wilderness Action Group, 1984), 6; Connors and Hutton, *A History of the Australian Environmental Movement*, 166–74.
123 Wilderness Action Group, *The Trials of Tribulation*, 33.
124 Christopher Anderson, "Aborigines and Conservationism: The Daintree-Bloomfield Road," *Australian Journal of Social Issues* 24, no. 3 (1989): 218–19.
125 Bill Wilkie, *The Daintree Blockade* (Mossman: Four Mile Books, 2017), 65–71. Zable Interview.
126 WAG, *The Trials of Tribulation*, 5, 24.
127 Raymond Evans, *A History of Queensland* (Cambridge: Cambridge University Press, 1997), 222–26; Cohen, *Green Fire*, 83–86.
128 Bill Wilkie, Interviewed 3 September 2014; Yeates Interview.
129 Ibid; Yeates Interview.
130 WAG, *The Trials of Tribulation*, 27–28.
131 Wilkie, Interviewed 3 September 2014.
132 WAG, *The Trials of Tribulation*, 7; Cohen, *Green Fire*, 83.
133 Adrian McGregor, "8 Arrests as Dozer Team Cuts Road," *Courier Mail*, 3 December 1983, 3; WAG, *The Trials of Tribulation*, 6–7.
134 Adrian McGregor, "Treetop Protest Stops Dozers," *Sunday Mail*, 4 December 1983, 5; "Troubled Road at the Cape," *Courier Mail*, 3 December 1983, 1.
135 WAG, *The Trials of Tribulation*, 12–15.
136 McGregor, "8 Arrests as Dozer Team Cuts Road," 3 December 1983, 3.
137 Cohen, Green Fire, 87.
138 Ibid; McGregor, "Treetop Protest Stops Dozers," *National Times*, 4 December 1983, 5; Wilkie, *The Daintree Blockade*, 142–53.
139 WAG, *The Trials of Tribulation*, 35–37.
140 Ibid., 40.
141 Bill Ord, "Forest Protester Tied to Cross," *Courier Mail*, 13 December 1983, 3; Wilkie, *The Daintree Blockade*, 154–64.
142 Bill Ord, "Tenni: Forest Homework Fails," *Courier Mail*, 15 December 1983, 10; Cohen, *Green Fire*, 87–88.
143 Bill Ord, "Cape Road Fight 'All Over'," *Courier Mail*, 14 December 1983, 3; WAG, *The Trials of Tribulation*, 57.
144 "Protest in Victorian Rainforest," *Canberra Times*, 12 January 1984, 7; Cohen, *Green Fire*, 96–103.
145 ; Peter Morgan, "Contested Native Forests: A Theoretical and Empirical Study" (PhD thesis, RMIT, 1997), Available [Online]: www.mcmullan.net/pmorgan/docs/phd/TCH7.htm (Accessed 14 August 2014).
146 "Violence after Gippsland Meeting: Inquiry Announced into Victoria's Timber Industry," *Canberra Times*, 14 January 1984, 3; Walker Interview.
147 "Orbost Angry with Conservationists," *Canberra Times*, 6 February 1984, 7; "Loggers Fight Has Just Begun," *The Sun*, 4 February 1984, 7; Nigel Turvey, *Terania Creek*, 122–25.
148 Herbert Haines, "Black Radicalization and the Funding of Civil Rights", *Social Problems*, 32 no.1 (1984): 31–33.

149 R Joiner, *Errinundra Plateau: Resolution of Conflict* (Melbourne: Government Printer, 1984), 2.

150 David Humphries, "Government Halts Police Swoop on Logging Protest." *The Age*, 30 January 1984, 3.

151 "Conservationists Arrested at Errinundra," *Canberra Times*, 4 February 1984, 7; Walker Interview.

152 Yeates Interview.

153 Michael Rae, "Daintree Road," *Wilderness News* (1985): 3.

154 Timothy Doyle, *Green Power: The Environmental Movement in Australia* (Sydney: UNSW Press, 2000), 52.

155 Doyle, *Environmental Movements in Minority and Majority Worlds* (New Brunswick: Rutger University Press, 2005), 88; Wilkie, *The Daintree Blockade*.

156 Yeates Interview.

157 "Daintree Protesters Dig In," *Canberra Times*, 7 August 1984, 3; Wilkie, Interview; "Daintree Update," *Nimbin News* (1984): 3.

158 Quoted in Jeni Kendell and Eddie Buivids, *Earth First* (Sydney: ABC Enterprises, 1987), 138–39.

159 Wilkie, *The Daintree Blockade*, 217–31.

160 "Daintree Lull Ends: Police Clash with Daintree Protesters," *Canberra Times*, 13 August 1984, 3; Cohen, *Green Fire*, 93.

161 Wilkie, *The Daintree Blockade*, 260–71.

162 Ibid; "Daintree Update 2," *Nimbin News* (1984): 3–5.

163 "Rainforest Row Turns to Violence," *Courier Mail*, 13 August 1984, 1.

164 "Cabinet Probe on Protesters," *Courier Mail*, 14 August 1984, 10.

165 Wilkie, *The Daintree Blockade*, 272–82.

166 Cohen, *Green Fire*, 88–89; "Greenies Retreat to Fight Another Day," *Courier Mail*, 17 August 1984, 11.

167 Cohen, *Green Fire*, 90; "Aboriginals Stop Anti-Road Group," *Courier Mail*, 18 August 1984, 13.

168 "Rainforest Blockades Are Dropped," *Courier Mail*, 29 August 1984, 9; Cohen, *Green Fire*, 93–94; Wilkie, *The Daintree Blockade*, 284–94.

169 "Buses Stuck on Daintree Road," *Courier Mail*, 8 October 1984, 1, 12; Burg, "How the Wet Tropics Was Won."

170 Doyle, *Environmental Movements in Minority and Majority Worlds*, 38–50.

171 Connors and Hutton, *A History of the Australian Environmental Movement*, 233–38.

172 G Holloway, *The Wilderness Society*, 35–38, 42–43; Dave Heatley, "The Early 80s: A Rising Momentum," in *For the Forests: A History of the Tasmanian Forest Campaigns*, ed. Helen Gee (Hobart: The Wilderness Society, 2001): 209–12.

173 Jeff Angel, "Letter from TEC Executive Director to John Seed Concerning Blockading," 1–2; Branagan, "Art Alone Will Move Us", 182.

174 Seed Interview; Hoeben Interview.

3

"THE FIRST VOLLEY IN THE NONVIOLENT WILDERNESS WAR"

Environmental blockading spreads to the United States, 1983–1986

The previous two chapters have shown how the environmental blockading template became established in Australia and a repertoire of contention formed. By providing a detailed account of a single wave they demonstrated how the use and innovation of tactics were influenced by factors such as the goals of ODA, the terrain involved and police and opponents' reactions and counterstrategies. Campaign organisation, protest personnel and resources, emotions, and definitions of normative protester behaviour were also analysed as further contributing elements.

This chapter provides a history of how environmental blockading was in part diffused from Australia to activists in the United States and how these developed their own tactics and organisational approaches within a series of events that made old growth logging a national issue. In focusing on transformative events the chapter contributes to a history of the repertoire and further illustrates aspects of tactical influences and dynamics more generally.

The political context of resource extraction in the United States during the 1980s

As in other countries a popular environmental movement emerged in the United States during the 1960s and 1970s. This was divided by various issues and approaches with a major distinction emerging between environmental justice movements and wilderness/biodiversity/conservation movements'.[1] The former are said to focus on the effect of polluting industries on communities, particularly those whose economic and social disadvantage place them at workplace and geographical risk from toxic materials. The latter, with which environmental blockading has been primarily associated, is mainly concerned with preventing the destruction of biodiverse ecosystems and the species associated with them.[2]

As in Australia, overlapping local, state, and federal governments and bureaucracies have governed extractive industries with logging generating the majority of grievances that led to blockades. The various layers of the US legal system and legis-lation passed in the 1970s have allowed conservationists greater opportunities than elsewhere to challenge authorities' decisions and demand investigations regarding environmental impacts as well as review and overturn those previously carried out.[3]

Unlike Australia, where logging primarily took place on publicly owned and controlled lands, much logging in the United States also occurred on private property. Changes in the political economy of the industry during the 1980s saw increasing corporatisation and concentration of ownership. These led to marked increases in rates of felling and extensive job losses.[4]

US conservationism had come to be dominated by roughly ten formal peak organisations by the late 1970s. Known as the "Big Ten" they included the Wilderness Society (which has no connection to the Australian organisation of the same name), Sierra Club, and National Audobon Society. Standard practice amongst these was to engage in fundraising and recruitment aimed at financially supporting a small number of professional activists. These specialists, often based in national and state capitals, were responsible for setting the direction of each organisation and carrying out research and campaigning. Activity was largely focused on lobbying politicians, and later corporations, as well as forging and maintaining relationships with them.[5]

Processes of political incorporation, which occurred in Australia during the mid-1980s, began earlier in the United States. Following major legislative successes in the early 1970s increased involvement of professional activists in govern-ment processes led to a curtailing of movement demands and a demobilisation of grassroots support. A tendency towards minimal claims, compromise, and a reli-ance on insider connections with the Democratic Party weakened the movement by the end of the decade. This was exemplified by the outcomes of negotiations with US Department of Agriculture (USDA) in the late 1970s and early 1980s regarding the fate of publicly owned forests given potential protection by the 1964 Wilderness Act. Despite the second Roadless Area Review and Evaluation (known as RARE II) being headed up by a former executive for the Wilderness Society, the process resulted in decisions that gave immediate protection to only 15 million of the 80 million undeveloped acres of land under Forest Service (FS) control. A significant proportion of this covered alpine and other areas generally not sub-ject to exploitation, with 36 million acres of the rest being immediately opened to logging, mining and agricultural activity. The findings were partially overturned by litigation on a state-by-state basis. Nevertheless, the initial outcome reflected the failure of existing strategies to counter initiatives from the corporate sector and pro-development policy-makers.[6]

These issues deepened as conservation peak bodies became politically isolated following the election of President Reagan in 1980. The Republican administration's opposition to regulatory protections and aggressive exploitation of publicly owned lands sparked new interest in conservation issues. A spike in grievances delivered mainstream organisations new members and resources. In turn their inability to shift repertoires and effectively counter their opponents drove the rise of a new movement based on confrontation and direct action rather than compromise and institutionalised appeal and reform.[7]

Earth First!

The formation of the Earth First! (EF!) network in 1980 came as a definitive rejection of the culture and praxis of mainstream conservationism. As the first group to demand a complete end to the logging and development of old growth forests, as well as the restoration of ecosystems by closing existing roads and developments, it would have a deep and lasting impact. Based on the slogan "No compromise in defence of Mother Earth" EF!'s creative and militant use of protest involved an estimated 10,000 to 15,000 activists by the 1990s.[8] Most significantly for this study, the network and many of its members would initiate and participate in dozens of environmental blockades from 1983 onwards.

Beyond believing that a new organisation based on maximalist demands could act as a "radical flank" for the rest of the movement, and better defend biodiverse areas through confrontational protest, EF!'s founders were inspired by concepts of biocentrism and deep ecology. The premise of these, that all "life on Earth has an intrinsic value and that human behaviour should and must change drastically",[9] challenged the development-based ethos of US society and its economy as well as the reformism of many environmentalists. Although many involved were not long-term residents of the areas they defended, their understanding of the biological importance and science of ecosystems and profound emotional connection to nature encouraged militancy in a similar way as it had for Australians.[10]

The apocalyptic nature of the milieu has been overemphasised by some commentators, but revolutionary ideas were commonly expressed in public statements and the network's flagship publication, the *Earth First! Journal (EFJ)*. Although it is difficult to ascertain the degree to which those taking part in EF! campaigns were influenced by such beliefs, key figures and factions favouring misanthropic Malthusian or left-leaning social justice-based ideologies would compete for influence throughout the 1980s.[11]

As with much of the US conservation movement, EF! was primarily made up of Anglo-Americans aged between 18 and 40, with a smattering of older activists. The network's blunt, direct action-based approach reflected its rejection of the polite liberalism of environmentalism in the late 1970s. EF! initially projected a masculine, "buckaroo" image that reflected its founders' taste for fusing hard drinking and country music with radical action.[12]

Within a short period, people from backgrounds extending from "animal rights vegetarians to careful followers of Gandhi, backwoods buckaroos to thoughtful philosophers, bitter misanthropes to true humanitarians" joined the network.[13] Despite such diversity a preference for rowdy confrontation, satire and self-deprecating humour continued to characterise its internal culture. The network's embrace of ODA, refusal to obey police instructions, pay fines, or abide by judicial rules similarly displayed a lack of respect and belief in existing processes, institutions, and conventions.[14]

Despite sharing a common identity and chapters in several locations, Earth First! groups operated as a decentralised network. Reflecting its founders' anarchist and libertarian leanings, as well as their rejection of the mainstream organisations, each EF! group was autonomous and responsible for its own activities. The network's resources were miniscule compared to those of mainstream groups with only a few members, mainly early organisers and those working on the *EFJ*, paid a small stipend. Many of those travelling to or working on campaigns relied on informal and part-time work with accommodation sourced through camping and sleeping on friends and supporters' floors and couches. This reflected, and in turn encouraged, an orientation towards non-institutional tactics as these could be mounted with minimal funding, when combined with the skills, materials, and office space available within the milieu and through connections to student activists and colleges.[15]

Coordination, discussion, and information sharing came via the roughly bi-monthly *EFJ* as well as through gatherings such as the annual "Round River Rendezvous" (RRR) and smaller regional events that combined workshops and campfire discussions with partying and sing-a-longs.[16] Starting in 1981 "roadshows", described as an "old-fashioned medicine show with a new message", saw campaigners and leading members regularly tour the United States.[17] Eschewing the typical lecture or public meeting format these were highly theatrical, combining music, speeches, and what EF! founder Dave Foreman described as "humour [and] passion" to "describe visionary wilderness proposals", found chapters, and build campaigns.[18]

EF!'s search for new and radical means of protecting nature, along with its grassroots, combative and decentralised form, drew on various precedents. Although opinions regarding class and other social justice issues varied within EF! there was an imaginative link to the syndicalist union the Industrial Workers of The World (IWW). In the late 1980s actual links to the organisation would be built with some EFers carrying dual membership, but initially it was knowledge of the union's activities and approach during its peak from the 1900s and 1920s that was most influential. In part modelling its tactics and approach on the IWW, EF! similarly mixed sabotage and direct action with grassroots organising and extra-parliamentary activism. Culturally the IWW's attitude, as expressed in its bitingly satirical songs and cartoons, influenced EF's self-deprecation and disrespectful approach to opponents. EF!'s sloganeering stickers were referred to by the IWW term of "Silent Agitators" and the sabotage manual *Ecodefense* issued under the name of early union leader "Bill Haywood".[19]

A more direct precedent came from early 1970s groups and individuals' application of sabotage and pranks to issues related to pollution, mining, and development in Arizona, Florida, and other states. Most famously an anonymous individual known as "The Fox" specialised in harassing corporate polluters in the Kane County area of Illinois by blocking waste pipes and chimneys, dumping sewage in their offices and publicising environmentally destructive activities via letters, signs, and labels attached to company products.[20]

Environmental Action, an organisation which had coordinated the inaugural Earth Day in 1970, a founding event which signalled the concern of millions for ecological issues, launched a National Ecotage Contest in 1970, inviting entrants

"To think up ideas that the ordinary citizen might use to bring attention to and pressure against corporate polluters."[21] The group claimed to have coined the term 'ecotage'.[22] Receiving hundreds of submissions outlining how to utilise means from letter writing and pranks through to billboard removal and the disabling of machinery, the organisation compiled a book titled *Ecotage*, which was released as a mass market paperback by commercial publishers Pyramid in 1972. Buoyed by the response, the book's authors called for a movement whose strength would be "that it is not formally organised" and therefore "could not be stopped by the elimination of key leaders".[23]

Interest in this approach was confirmed by the book's publication and the popularity of Edward Abbey's 1975 novel *The Monkeywrench Gang*, in which a small team of misfits carry out ecotage across America's Southwest before destroying the Glen Canyon Dam.[24] However, it would not be until the formation of EF! − a group that took much inspiration from Abbey's book − that an informally organised environmental movement based on provocative and extra-parliamentary tactics emerged.

EF!'s outlook and organisational form made for a highly experimental political culture. Its founders explicitly encouraged members to apply and invent a range of confrontational tactics and strategies previously eschewed by conservationists. As will be seen, the existence and brokerage of such a network facilitated faster and wider diffusion regarding ideas and tactics than had been possible in Australia.[25]

The network's repertoire of contention initially focused on rallies and publicity stunts such as faking a crack in the Glen Canyon Dam in 1981 and disrupting appearances by the anti-environmentalist Secretary of the Interior James Watts. Members submitted detailed proposals to governments and bureaucracies demanding maximum environmental protection, filed legal appeals, and called for boycotts of companies connected to deforestation.[26]

Despite the majority of its members engaging in other activities, EF! soon became synonymous with the practice of sabotage. Dubbed "monkeywrenching" by milieu members, information about how to disable and damage opponents' equipment appeared in the *EFJ* and members compiled a detailed handbook in 1985. In publicly advocating practices such as tree spiking as a means of halting work, leading EF! members deliberately courted controversy. The association initially boosted public interest and created moral shocks via media coverage that exposed audiences to the destruction caused by clear-felling. Highlighting the way in which disruptive tactics "can both enhance and suppress protest effectiveness"[27], it also formed the basis of attacks by politicians, companies, and mainstream environmentalists and was used to justify monitoring and infiltration by state, federal, and private security agencies.[28]

The turn to blockading and its connections with Australian and other activism

Although the United States had a long tradition of obstructive direct action by labour, civil rights, and anti-nuclear movements, EF did not make plans to use it until two years after the network was founded.[29] Having removed survey stakes on various occasions EFers, held a day long rally of hundreds at Little Granite Creek in Wyoming's Gros Ventre range as part of its 4 July RRR. Threats made at this and other events to begin a blockade strengthened local efforts and helped force Wyoming's State Governor to move against the project.[30]

Actual ODA began in New Mexico in November 1982 when around 10 to 15 local and travelling EF! members engaged in two occupations against illegal road construction and oil drilling in the Salt Creek wilderness area. They blocked vehicles with their bodies and a small encampment. These actions gained national media and spurred an injunction that prevented further drilling.[31]

Another action, against strip-mining in New Mexico's Bisti Badlands, in 1982 involved 50 Earth Firsters (EFers), Native American activists and other locals in trespassing on private lands. After further campaigning, including EFers dressed as clowns disrupting a public hearing, the Bureau of Land Management relisted the region as a Wilderness Study Area and close to 42,000 acres were put under protection in 1984.[32]

Moderate conservationists had long advised against the use of litigation on the basis that it could upset their political allies and present opponents with opportunities to pass adverse legislation. In the case of Salt Creek their fears were borne out as Congress modified existing laws to validate leases existing before October 1982, such as this one, and allow for drilling in designated wilderness areas. Following the company's return to the site, two EF members defied a Department of the Interior ban on anyone other than employees and wildlife officials from entering the area and blockaded work on 3 February 1983. An *EFJ* editorial subsequently claimed the protests had demonstrated that "non-violent direct action has a tremendous role in galvanizing popular public support and delaying destruction".[33]

These early actions were enthusiastically supported by key activists and given extensive coverage in the *EFJ*. The network raised money to help cover the running and legal costs of future blockades as well as encourage EF members to travel to protests as part of the sardonically titled Save Wilderness At Any Time (SWAAT) team. Neither project was lasting, but with 50 groups and contacts listed in its contact list by March 1983 and further roadshows and coverage in mainstream, conservation and leftist media, the network was set for its first major blockades in 1983.[34]

The turn to blockading had in part been inspired by actions opposing the enlargement of California's New Melones reservoir from 1979 to 1982 during which activists sought to prevent the flooding of the Stanislaus River canyon by hiding in the overflow zone and chaining themselves to trees and rocks as well as blocking roads with their bodies to prevent clearing. These actions generated publicity and led to intermittent moratoria and limits on flooding, but the project was eventually completed. While the activism involved was covered in state and national media, as well as the *EFJ*, there appears to have been little direct connection between EF! and the activists involved. It therefore serves as a precursor to, rather than the beginning of, the wave of American EB which the network fostered from the early 1980s.[35]

EF! was also influenced by the Australian campaigns of the late 1970s and early 1980s. A prime source of diffusion was NAG activist John Seed. He had first come in contact with EF! in 1981 when poet Gary Snyder gave him a copy of its journal while on tour in Australia. Impressed with the group's grassroots, militant and direct-action-oriented approach, he began corresponding with core members.[36]

Seed's letters, articles, and columns for the *Earth First! Journal* (*EFJ*) began with a report in September 1982 describing blockading at Grier's Scrub.[37] Although EF! had already threatened to carry out a blockade in July, this material led core members to advocate more ambitious action with Foreman writing "Our brothers and sisters in Australia have set a powerful example for us."[38] The *EFJ*'s enthusiasm for Australian blockading increased with the Franklin campaign. A March 1983 front page story declared:

> Clearly, the world leadership in wilderness preservation has passed to Australia. While the environmental establishment in the United States preaches moderation and practices meekness, the "Greenies" of Down Under are taking courageous/exemplary action to protect their wilderness and are sending the world a message.[39]

Further news regarding the Franklin, Daintree, and other campaigns would be carried in *EFJ* and referenced in articles concerning US actions.[40] Songs composed by Australian activists, shared directly by Seed and via cassette tapes, would be performed at American events and blockades.[41] As co-founder of the Rainforest Information Centre Seed also joined 1984 and 1985 US EF! roadshows that visited numerous cities.[42] These fed into EF! members forming the Rainforest Action Network (RAN) and disseminated information about the environmental blockading template and repertoire through discussion of Australian campaigns and screenings of films about them. In turn Seed's experience of US EF! had an impact on Australian practices.[43]

A pattern of protest emerges

The combination of a high concentration of old growth forests subject to rapidly increasing exploitation with pre-existing progressive student, activist and alternative communities meant that much of EF!'s blockading in the decade following its foundation would be centred on logging in the United States's Pacific Northwest (PNW) states of Oregon and Washington, as well as California.[44] Although Australian examples of blockading provided a key inspiration for American activists, the scale and form of blockades they employed during the 1980s would differ greatly. Until the 1990s, American activists generally employed intermittent actions involving 5–30 people that were specifically planned to last a day or two. Where open-ended blockading was entered into, it rarely lasted more than a few days as arrests removed all the available activists or a decision by the courts or opponents brought an end to logging. When the tactic of tree-sitting emerged, it allowed actions to last for weeks, but these involved small numbers of people disrupting a limited area. Blockading was primarily based around small affinity groups entering threatened areas to carry out actions rather than operating out of a long-term base camp at or near the site.[45]

The key reason driving this pattern was that activists simply lacked the support, resources, and numbers to safely and successfully launch sustained blockades in remote locations. As Mike Roselle recalled in a 2013 interview, "With the loggers it could be very intimidating and even with the law enforcement it could be dangerous because there were no witnesses... most of the actions we did it was just a dozen or two dozen people."[46]

Although interest in PNW/California was higher than anywhere else in the country, environmental concerns regarding logging had not yet reached or resonated with significant sections of the public. As a result, the FS narrative that old-growth forests were a "rotting" and "decadent" resource that needed to be exploited, removed, and replaced remained dominant. National environmental organisations were unlikely to engage in litigation, let alone ODA. EF! itself was still emerging and could not yet mobilise significant numbers.[47]

The other major source of ODA-oriented protesters at this point could have come, as in northern NSW, from the communities living in areas subject to logging. The cheapest properties in the PNW/California often bordered government-controlled FS and Bureau of Land Management (BLM) land and it was these that had been purchased by the American equivalent of New Settlers since the late 1960s. Witnessing clear-felling first hand and largely unconnected to the timber industry, many "back-to-the-landers" supported a reduction in cutting. While this generated a level of activism that would grow as the decade progressed, widespread support for logging from others in the area, particularly outside California, mitigated involvement in controversial practices such as ODA.[48]

Although support for existing timber management practices was mixed, anger towards those advocating reductions in the level of logging ran high in rural communities where the timber industry remained the key employer. During the 1970s environmentalists had successfully mobilised broad coalitions of local residents to oppose the aerial spraying of forests with pesticides, but issues regarding logging itself proved fraught and divisive. Isolated numerically, socially and politically, and in some cases already suffering regular harassment from authorities due to their association with marijuana growing, few back-to-the-landers were yet ready to invite the opprobrium of their neighbours and the immediate physical and legal risks associated with ODA.[49]

Operating within these constraints EFers were nevertheless determined to start using environmental blockading, both to immediately oppose the destruction of ecosystems and in the hope of triggering a radical movement capable of undertaking more ambitious action in the future. They found just such an opportunity in 1983.

Blockading begins: Bald Mountain, 1983

A dense, rugged, and biodiverse region incorporating 160,000 acres of forest, the Kalmiopsis roadless area forms part of the Siskiyou coastal mountain range, which extends along an 160 km arc from California to Oregon. A section in southwest Oregon was under marked threat in the 1980s as the FS had embarked on an ambitious program of road building and timber sales allowing logging companies access to public lands. An additional reason for opposition came on the basis that the new roads would separate the forest's northern section. As legislation required areas to be contiguous, this would effectively reduce the area available for future protection under existing wilderness provisions.[50]

The Oregon Wilderness Coalition (OWC), with reluctant support from the Sierra Club, had appealed the construction of Bald Mountain Road in 1982, claiming that a comprehensive EIS was required. OWC had been founded in the 1970s and established a network of local environmental groups across the state. Initially small, voluntary, and operating on a tiny budget, the organisation would go on to popularise an alternative to the dominant national organisations' lobbying strategies through grassroots organising and the aggressive use of appeals and litigation.[51] Limited in scope by the refusal of the Sierra Club to address the deficiencies of RARE II, and with OWC lacking experience and funds, the appeal failed and as a result both national and local organisations were resigned to the loss of the region.[52]

EF! was alerted to threats to the area by an anonymous letter, purportedly from a FS soil biologist. Travelling to Oregon in 1983, members suggested that local environmentalists try a blockade. Some had previous experience with ODA concerning pesticide spraying, nuclear power, and peace issues, but the majority were exhausted and unwilling to carry on. With the nearest police station distant from the worksite, the risk of violence from workers was also perceived to be high. Leadership of the campaign thereby largely passed to EF members such as Roselle and Ric Bailey. These moved to the region, basing themselves at a supporter's home in Takima, 40 miles from the construction site.[53]

Planning for another blockade in the Californian section of the Siskiyous was also taking place during the same period. The issue of a proposed road to link the towns of Gasquet and Orleans and open up the region to a further 200 miles of road construction saw core EF! members meet with Native American activists and members of the local communal and lesbian activist scene in Arcata. Disagreements over the issue of normative protester behaviour caused the latter groups and EF! to agree to "follow the example of our sisters and brothers in Australia and stop the Freddie bulldozers and chainsaws with their bodies" as part of "a peaceful blockade of road construction".[54] Explicitly non-violent actions including training and affinity groups were planned for lower elevation areas while the Native American group chose to hold their own, potentially more aggressive, actions on sacred land further up the mountain.[55]

Rallies and protests were held at FS offices and in the forest, EF! founders Foreman and Bart Koehler toured the state, and affinity groups from various cities and towns in California formed. No ODA would be carried out however as court action, based on Native American religious rights, initially delayed construction before a federal court ruling on 25 May 1983 banned road building within a 27 mile radius of Siskiyou High Country.[56]

With snow clearing and road building under way at Bald Mountain, EF! finally got their chance to carry out a more extensive series of blockades in what Foreman told the media was "an experiment, a test case".[57] The first of a series of one day actions kicked off on 26 April when EFers Roselle and Kevin Everhart and two local men, Steve Marsden and Pedro Tama, walked out of the woods to link arms and stand in front of a bulldozer. Prior to this the FS and construction crew had been told there would be action and that it would not involve violence or sabotage. The timing and type of blockading was otherwise kept secret. The driver attempted to move the men by rolling dirt and rocks in their direction, but eventually walked away leaving them to hang an EF! banner on the vehicle and talk with media representatives. Police arrived three hours later to arrest the group for Disorderly Conduct.[58]

Having received media coverage throughout the state, more blockading came on 5 May when eight protesters, six of them handcuffed and chained to the bulldozer, shut down clearing for half a day. Five days later another group of seven, including two who chained themselves to a bridge, stopped work for an hour. Most of those arrested during blockades were subsequently made subject to bail conditions that, similar to those applied in Australia, banned them from re-entering FS lands. With vehicle entrances often shut by locked gates, blockaders and their support teams generally had to hike to action sites on foot, avoiding workers and others along the way.[59]

A fourth blockade on 12 May involved EFers Foreman and David Willis positioning their bodies behind a felled tree in the road. After police removed the tree, a truck arrived, narrowly missing Willis in his wheelchair before stopping and then pushing Foreman at increasing speed uphill. EF!'s best-known figure then went under the truck and was dragged along the stony road. Illustrating the ideological diversity within EF! at the time a famous exchange followed in which supervisor Les Moore screamed at Foreman "You dirty communist bastard. Why don't you go back to Russia!", to which the battered EFer infamously replied "But Les, I'm a registered Republican."[60]

In keeping with EF!'s culture the campaign was informally organised with people taking on roles as required. Some involved in the campaign hailed from peace groups and were critical of the "combative attitude of EF! [and their] use of the green fist and monkeywrench as symbols".[61] Drawing on personal experience and pre-existing beliefs, these activists advocated that blockading be based on non-violent guidelines developed during the Civil Rights era and refined by the US anti-nuclear protests of the 1970s. Such principles were readily adopted in EF! circles. Karen Coulter, who had been active amongst anti-nuclear groups before joining EF! in 1984, recalls:

> A lot of the EF! founders had no experience with direct action at all and hadn't done non-violence training. Given the long history of non-violent

civil disobedience in the US with the civil rights, women's suffrage, anti-war and other movements, this model was something that everybody had ingrained in them to some degree anyway. I don't think there was really any other overt model in the early days.[62]

EF! members at Bald Mountain shared an understanding that ODA would be safer and more effective in terms of publicity if activists avoided using aggression. They agreed to adopt standard non-violence principles, with some qualifications. Non-violence workshops were carried out by trainers, mainly from Peace House in Ashland; formal consensus processes used to decide on actions; and agreements adopted in which blockaders promised to avoid physical and verbal aggression and submit to arrest. However, the majority's focus on obstruction meant that a move to notify the FS, workers and police before actions was rejected. This was on the basis that such briefings would allow police to arrest activists too quickly.[63] Such a divergence from strict MNS/ONV-style principles was not unprecedented, as various anti-nuclear activists had already adopted approaches to non-violence allowing for secrecy and spontaneity.[64]

Some advocates and analysts of nonviolence practices have characterised those who diverge in this way as "strategic" or "pragmatic" rather than "ethical", "conscientious" or "principled" practitioners of non-violence as their tactical palette is said to stem from a focus on conflict and coercion over conversion.[65] Such a dichotomy, rooted as it is in a particular definition of non-violence and ethical behaviour, is not only highly loaded, but also of minimal use in analysing how conceptions of nonviolence shape tactics and standards of normative protester behaviour within the majority of avowedly "nonviolent" campaigns.

As in Australia, those taking part in US forest-based blockades held a variety of motivations for identifying as "non-violent". For some the choice was purely strategic as while they were potentially in favour of violent political struggle they did not believe it was currently effective to use such means in the context of public campaigns. Others perceived their choice to employ nonviolence – in terms of not harming others, acting aggressively, and carrying weapons – as stemming as much from an ethical rejection of what they defined as "violence" as from strategic concerns. Beyond this, activists held a variety of positions on issues such as whether practices of carrying out sabotage away from public campaigns, digging up roads, wearing masks, and practicing secrecy constituted "violent" or "nonviolent" behaviour. Undoubtedly many did wish to use nonviolent means to defeat rather than convert their opponents, but this did not make their particular version of non-violence any less rooted in principles or ethics.[66]

Although general principles of behaviour were agreed to, during the Bald Mountain campaign specific decisions regarding each action were left to individual affinity groups. These were drawn from people based in various cities and towns across the PNW who firstly travelled to the campaign office and then on to the work site. Each group decided on matters such as the date and time of actions, the tactics to be used, whether to notify the media beforehand, and whether to plead to charges individually or as a group. Those willing to be arrested were supported by others carrying out tasks such as reconnaissance, transport, and photography. This division of labour and form of decision making would be employed at most US blockades during the 1980s, although in some cases organisers and others not present at the site would also have considerable input into the details of actions.[67]

Such means of organisation plus the adoption of a form of the standard US non-violence code allowing for secrecy would also become the typical way in which blockades would be carried out in the forests and in other EF! actions for the rest of the decade. Mikal Jakubal, who joined EF! in 1985, recalls that unless activists were rapidly responding to a situation:

> The way we would typically do things would be to have a weekend protest gathering. The Saturday would usually be informational hikes, Saturday night music, camp out, discussions and on Sunday workshops and nonviolence training. Out of that would come an affinity group or two that would then plan an action following the '80s nonviolence code of "We will use no violence, verbal or physical. We will carry no weapons. We'll be honest, open and friendly, etc." Typically we wouldn't share all the details of where and when because we didn't want to be intercepted down at the pavement, we wanted to get to the action site.[68]

Karen Coulter, who ran workshops at several protests and gatherings, describes nonviolence training as being:

> Very much focused on practical issues. When I worked with the anti-nuclear movement, sessions often went as long as a day, but when I became involved with EF! I was as ready as anybody not to have it last that long. I think the longest I ever did with EF! was eight hours, but usually it was more like two and a half to three hours and one was only 45 minutes. Often you were reacting to something and with all the other preparation required just didn't have much time.
>
> The sessions I developed with my then partner Assante would start with ethical and practical reasons for acting non-violently. We wouldn't go into ideology, but we'd talk about the nonviolence code, what it usually covers and some history. We'd look at the legal process and what the decision points are from talking to an officer through to a jury trial with all the different split offs and options along the way.
>
> We'd then alternate between presentations using stories from our experiences – funny ones, tragic ones, scary ones – which people really liked, and role-plays. These were designed to help people defuse situations and get to the point where they could prioritise issues and find consensus in order to respond really quickly to changing situations. When you're dealing with things like angry loggers and police threatening people with felony charges you often don't have much time.[69]

A survey of relevant literature and more than 390 environmental blockades indicates that specific tactics and strategies come to be used in campaigns and locales via a combination of existing knowledge, innovation, and diffusion. As noted in the Introduction, diffusion always involves some degree of translation and tactics are constantly adapted to suit new conditions and political, geographic, and other dynamics. Tactics that appear in more than one setting can also be the outcome of parallel innovation. This is more likely to take place when the technique is relatively simple and open to being easily innovated under similar geographic and campaign conditions. For instance, despite no known connection, campaigns in New Zealand in 1978 and NSW in 1979 both employed tactics in which people climbed trees to prevent them being cut down. As tactics become technically complex, they are more likely to require conscious modelling, and therefore diffusion and brokerage.

At Bald Mountain, indirect and weak direct diffusion via correspondence and alternative media coverage led EFers to apply and adapt the environmental blockading template from Australia to a new locale and situation. With EF! lacking detailed knowledge of Australian practices or an organisational model and approach to normative behaviour of their own, the blockade employed and adapted organisational and ethical principles that had been regularly applied to other issues within the United States. The decision to engage in sporadic blockading was influenced by local conditions, and tactics were drawn indirectly and directly from experience and information concerning recent US, and to a lesser degree Australian, nonviolent direct action.

Three more actions followed from late May to early July resulting in a total of 44 arrests, primarily for criminal mischief, disorderly conduct, and criminal trespass. Workers regularly endangered protesters with some being perilously buried in soil and mud and one man having his foot run over. Police were generally not present and refused to make arrests when such actions were brought to their attention.[70]

According to campaigner Ric Bailey, blockading tactics were chosen according to what was likely to cause "maximum equipment shutdown time".[71] Drawing on local precedents established during an anti-pesticides campaign, protesters blocked vehicles in three waves in order to stretch police resources and thereby maximize obstruction. Each group managed to keep their presence hidden, forcing officers to travel roughly 40 miles at a time to deal with disruption.[72]

With the blockades stalling work, and the associated publicity enabling litigants to raise required funds, EF! in conjunction with members of ONW, reconstituted as the Oregon Natural Resources Council (ONRC), filed a citizen-initiated lawsuit. Free of the constraints imposed by the Sierra Club their case successfully argued that construction was illegal without an EIS. A halt to work on 1 July and associated media attention regarding threats to the Kalmiopsis helped pressure the Oregon legislature to subsequently pass a new wilderness bill. While expanding protection for the region, Bald Mountain was left out and the bill included provisions nullifying the legal decision that had saved it. Following further protest, federal legislation prevented any extension of the road in 1988, but logging in the area continued, as did environmental blockading.[73]

Although some EFers dismissed blockading and nonviolent tactics as insuffi-
cient to prevent long-term environmental destruction, many within the network
were ecstatic with the campaign's immediate outcome. As Roselle later recalled,
"We came out to Oregon on a mission to get arrested saving old growth... We
knew this was the first volley in the nonviolent wilderness war."[74] A RRR held at
Bald Mountain in July 1983 went from being a planned launching point for further ODA to a celebration involving 300 participants.[75] Between this and the *EFJ*
word soon spread and the blockade entered EF! lore with activists still singing songs
applauding it 22 years later.[76]

Debates regarding the efficacy, ethics, and parameters of "non-violence", which
had carried on for decades within US peace and anti-nuclear circles were revisited
in the wake of Bald Mountain. While acknowledging that the campaign guidelines
had ensured a positive public image as "we did not instigate any confrontations
while some of the construction workers did", Bailey claimed in a retrospective
piece for the *EFJ* that the approach had restricted "diversity and creativity", with
actions tending to "serve the code itself, and not always the causes which the code
was designed to serve".[77] In further debate carried out in the *EFJ*, he and others
suggested the network either dispense with such rules or devise their own to better
reflect the network's ethos, encourage greater tactical initiative and flexibility, and
allow for self-defensive violence.[78]

In environmental blockades that followed, conceptions of acceptably "non-
violent" tactics varied, most particularly in regard to definitions of what constituted
sabotage. However, a desire to minimise physical conflict on safety and publicity
grounds meant that, regardless of one or two incidents of protesters striking back,
none openly extended their guidelines to allow for "self-defence".[79]

Unlike in Australia, where the radical environmental movement generally
disassociated itself from the practice, the role of monkeywrenching was widely
and openly debated. Due to its decentralised nature, EF! did not take an "official"
position, but significant support for major sabotage was present. Key proponents
such as Foreman argued that property damage was "nonviolent" as "it is not aimed
at harming any life" and required safety warnings in order to effectively stop
logging. Despite defining ecotage in this way Foreman also argued, on the basis
that blockaders could be blamed for sabotage and provide a target for media and
worker hostility, that:

> Non-violent direct action and monkeywrenching are like milk and beer.
> They are very good individually but godawful when mixed. When a NV
> action is taken in defence of wilderness, we must strongly discourage any
> kind of ecotage. The slightest monkeywrenching will muddy the water ...[80]

For much of the 1980s and 1990s, and since, such reasoning has generally, but not
wholly, ensured that a basic division between the use of repertoires concerning
monkeywrenching and those associated with environmental blockading has been
maintained in the United States.[81] The issue of monkeywrenching, and tree-spiking

in particular, would not cause a major split in EF! until later in the decade, but from the network's inception some opposed its use anywhere. In an article for the *EFJ*, Bald Mountain protester Peter Swanson argued that activists would only be respected and safe if they were consistent in all activities:

> When we engage in civil disobedience, the actions of the people we face are influenced by their perceptions of us. Right now, EF! is seen as both Dr. Jekyll and Mr. Hyde: peaceful protestors acting out of moral imperatives by day, wild-eyed revolutionaries wielding sticks of TNT by night.[82]

Further blockades, tree-sitting, and the development of a counter-repertoire, 1983–1986

The political and economic context outlined at the beginning of this chapter continued to drive forest exploitation throughout the 1980s, generating new grievances and campaigns. As a result, the strategy of combining environmental blockading with litigation slowly grew in popularity in PNW and California.[83]

Responding to an attempt by loggers to undermine a lawsuit brought by the Environmental Protection and Information Centre (EPIC), a group who would often work in concert with blockaders in coming years, the next blockade began on 7 October 1983 in the Sinkyone's Californian Sally Bell Grove. Fifty activists ran through the area, hugged trees, and lay in the path of felling. Little work was carried out before logging company Georgia-Pacific agreed to halt work and by the next day an injunction was in place.[84]

With various legal manoeuvres underway the company moved to restart logging on 25 October. Employing a new counter-tactic the company hired 50 Sherriff's deputies and security guards to patrol its privately owned land. Determined to "stop the logging and not just make a statement" blockaders split into small groups and infiltrated the forest. Although most avoided arrest by hiding and running through dense scrub, while sounding air horns to signal their presence, they were unable to prevent work as fellers ignored safety concerns, pinning one woman under a tree.[85]

Similar tactics did little to stop work the next day and in response protesters introduced new, more covert tactics. While the definition of what constituted unobjectionable levels of sabotage would shift over the coming years, minor vandalism appears to have been already defined as suitably "nonviolent" in this case. By jamming an entrance lock with a toothpick and hiding logging equipment, activists slowed proceedings on 26 October for long enough for a new injunction to bring logging to a halt. The area was subsequently spared through a court finding which set a precedent for EPIC to repeatedly force "the California Department of Forestry to consider cumulative impacts to water quality, wildlife and other resources when reviewing timber harvest plans".[86]

Buoyed by this success, EFers joined an existing local campaign in 1984 to threaten ODA before logging commenced in the Middle Santiam forest. Located in west Oregon, activists claimed it contained "the second largest biomass in the world" including rare species and 300-foot-tall trees.[87] The campaign came in the context of a Wilderness bill before the Senate, which had been criticised for incorporating only 20 per cent of the state's 4.5 million acres of unprotected roadless forest. Included within this was merely 7,500 out of 80,000 acres of the Middle Santium campaigners had sought for preservation. With multiple attempts at litigation failing to block timber sales, blockading against logging in the area would go on for three years and involve a range of tactics, expanding the global repertoire and igniting much debate.[88]

As at Bald Mountain tensions between strict non-violence advocates and others were evident from the start. These would recur less during future EF! protests as many of the former withdrew from campaigns and a rough consensus regarding normative protester behaviour emerged. On this occasion the debate primarily concerned symbolism rather than specific tactical or organisational issues. EF! member George Draffan later recalled that those identifying as pacifists "didn't like the attitude ... the talk of direct action to the point of violence, and they didn't like the fist logo".[89] Shared commitment was nevertheless sufficient at this point to see joint action carried out with the non-violence code from Bald Mountain applied and opponents not given detailed briefings on actions. It was agreed that this would be done under the banner of the Cathedral Forest Action Group (CFAG) in order to distance the campaign from EF's association with monkeywrenching.

During the first action on 5 May 1984, protesters held up the blasting of a hillside for road construction by sitting on boxes of dynamite before five were arrested. Over the next three months a model of intermittent one day ODA was employed as CFAG members occupied FS offices, worksites, and bridges on five further occasions. Decisions were made by consensus and actions mostly carried out by affinity groups, but also included larger assemblies.[90]

During recent protests, police had usually issued citations and immediately released arrestees, but authorities now hardened their approach. On 17 July activists, including one who attached diving weights to his waist to make himself harder to move, successfully blockaded timber trucks for five hours. In response police cleared the area before twisting arms and dragging protesters away by their hair and beards. Officers denied claims of brutality only to find that images of their actions, surreptitiously taken by photographer Leo Hund, had been provided to newspapers and television stations.[91]

With a judge granting Willamette Industries a Temporary Restraining Order (TRO) against blockading, protesters were taken to jail and presented with restrictive bail conditions. In addition to this, an action on 23 July led to three protesters, including Hund, being arrested on felony charges for refusing to help deputies load arrestees into a patrol car. These charges related to the 1927 Riot Act and came despite the presence of FS employees who could have fulfilled the role.[92]

Demonstrating the way in which personal enmities can shape protest policing strategies, Hund later claimed he was told by one officer that his colleagues had spent a week combing legal texts to find a way to arrest him on more serious charges. The District Attorney's office initially shared enthusiasm for the novel approach but following negative media coverage dropped the charges the day before they went to court. With the support of the American Civil Liberties Union (ACLU), Hund mounted a legal challenge that resulted in a financial settlement and a ruling finding the Riot Act provisions unconstitutional.[93]

Some of the 48 arrested during the 1984 campaign engaged in further protest by employing tactics commonly used in civil disobedience campaigns. This included refusing to provide names or information, insisting they face court as a group, and engaging in hunger strikes and other forms of non-cooperation. With the Linn County jail already facing legal challenges concerning overcrowding, some judges released protesters without bail or with conditions allowing them to return to the forest so long as they agreed to remain law abiding. Others laid charges of contempt of court, sparking further protest and resulting stays of up to eight days in jail.[94]

Further drawing on decades of social movement practice, and setting a precedent for future environmental blockades, a number of protesters chose to face trial in September and October rather than plead "no contest" to charges of "disorderly conduct". Utilising the classic defence of "choice of evils" members of CFAG and EF! claimed they had acted in order to "prevent an imminent and irreversible harm to the public domain" and that "the crime committed was lesser than the crime [they] sought to prevent".[95] They further argued that their actions had not been disorderly as, in the words of EFer and non-violence trainer Mary Beth Nearing, "Many hours of preparation go into each action and every participant is obliged to adhere to an agreed upon code of conduct."[96] The overall attempt to turn the trial from one concerning activist behaviour to that of FS policy did not result in an acquittal. It seemingly won over the jury with their foreperson stating, "We greatly admire your cause and your convictions, but we had to find you guilty."[97]

The company carrying out logging also took its own steps to contain protest activity. Utilising a tactic previously used against protesters in other movements, Willamette Industries sued for $17,000 in losses in December 1984 naming 30 protesters, CFAG, EF! and others as defendants, and requesting a permanent injunction against all interference with its operations. The trial would not conclude until after the following summer's blockading, which continued undeterred, and resulted in four individuals and CFAG being ordered to pay $13,500.[98]

The use of such litigation would increase markedly over the rest of the decade as numerous companies undertook what became known as SLAPP (Strategic Lawsuits Against Public Participation) suits. Hundreds of activists and organisations would be sued on the basis of "defamation, conspiracy, nuisance, invasion of privacy or interference with business/economic expectancy".[99] Through such tactics companies sought legal findings that would directly prevent campaigners from undertaking activities ranging from education and lobbying to ODA as well as to indirectly curtail protest by draining time and resources. EF! activists were rarely forced to pay any money to their opponents as few of them had the property or resources to do so. Although few activists were willing to publicly admit it the counter-tactic, whose potential penalties far outweighed the more regular costs associated with fines and imprisonment, were daunting for some.[100]

Although the Middle Santium campaign was successful in drawing some journalists to remote logging sites it failed to have a major impact on logging. Frustration at this and the continued unwillingness of mainstream groups to pursue further legal action led a small group to spike a stand of trees at Pyramid Creek in January 1985. The action was carried out by Roselle and others under the name of the Bonnie Abbzug Feminist Garden Club, a moniker drawn from a character in *The Monkeywrench Gang*. The FS and company were informed, spiked trees clearly marked, and a press release sent out. No arrests ensued and Roselle later admitted the action "hardly slowed them down", except for the fact that a large number of officers had to be deployed as the FS "didn't have a policy, and they didn't know what was going on".[101] The action attracted negative media coverage and enraged some within CFAG and Oregon EF! who publicly denounced it.[102]

In 1985 the two main forest stands targeted by CFAG and EF! in the area were at Pyramid Creek and Squaw Creek, with the latter renamed the Millennium Grove for the fact many of its Douglas Fir trees had been growing since the Fall of Rome. With the ACLU appeal preventing further use of the Riot Act and a federal court decision limiting the number of people who could be held at the Linn County jail, authorities changed tactics. Although protesters would once more be cited and released without bail, the official closure of a 1200-acre area surrounding logging by the FS on 8 April and the use of arrests and fines for trespassing made it much more difficult to access the site.[103]

Tensions between EFers and CFAG members continued over issues such as the wearing of the EF! fist logo on T-shirts during actions and whether the placement of EF! stickers on FS signs constituted "sabotage" and went against the spirit of nonviolence by inviting conflict. A formal split never occurred, but EFers set up a separate meeting space/dwelling and began carrying out actions in their own name while continuing to participate in those under the CFAG designation. The key reason for this according to Mike Jakubal, who joined EF! that year, was frustration over the slow pace of decision making as, "Once we started Earth House consensus was as simple as 'I've got an idea, what do you think? Everyone's up for it, let's go!'"[104]

During campaigning a major tactical variation was introduced into the international repertoire. Jakubal, a rock climber and the first person to undertake a US tree-sit, recalls:

> During the preparation weekend there was a discussion about the previous year's actions and how they were removing people easily and making our actions short-lived and symbolic. [CFAG member and strict nonviolence advocate] Brian Heath said "I wish there was a way to get up in the trees." With old growth trees there are no lower branches, no way to just scamper up them. Friends and I had climbed trees with ice climbing gear for fun in the summer and I had also climbed lots of big walls and had all the gear, so replied "I can do that."[105]

Jakubal (who adopted the non de plume "Doug Fir" after the species he was occupying) and a support crew of five entered the forest at Pyramid Creek on the evening of 17 May. Despite the forest closure and the fact that a well-publicised CFAG gathering had just occurred there were no FS security present in the area. Jakubal recalls: "We found this surprising, but the Forest Service hadn't really worked out what they were doing yet." Summing up the dynamic interplay of tactic and counter-tactic in the United States during the period he contends:

> As is typical with the environmental, anarchist, lefty milieu, because we lacked a metric to gauge strategic rather than tactical success we often used the same tactics over and over. The more we repeated them, the more the other side were able to learn from us.
>
> In general I was surprised at how slow [the FS, police, and other authorities'] institutional learning could be. During the 1980s they were very slow to share information with each other. They were easy to out-manoeuvre, particularly when a tactic spread to a new area or involved a different agency. Nevertheless, they caught up eventually.[106]

At this point EFers were convinced of blockading's efficacy but had not yet developed clear approaches regarding its strategic use. Nevertheless, the network's action-oriented culture and the desire of those involved in campaigns to protest and obstruct work at the point of destruction continued to drive tactical innovation. Pioneering a new method of tree-sitting, which allowed activists to ascend to points much higher than before, Jakubal drove spikes (which would be removed later to avoid accusations of sabotage) into a tree and using ropes and a simple webbing ladder climbed 70 feet. Utilising a small climbers' aluminium-framed platform, or portaledge, to sit on, he then secured himself to the tree. Well supplied, and having dropped a banner reading "Don't Cut Us Down", he planned to stay for up to four days before being relieved by another sitter.[107]

As it turned out, once loggers and the FS arrived and escorted other protesters and journalists out of the area, felling continued around him with branches brushing his platform. Although he had brought bags for the purpose, Jakubal climbed down later that night to go to the toilet. FS officers came out of hiding and arrested him, later dumping him in a logging town in the early hours of the morning. The final tree in the stand was cut down the next day.[108]

Despite the tactic's obstructive value proving limited on this occasion, it gained local and national media attention and many within the EF! milieu recognised its broader potential. Diffusion rapidly occurred along what were then typical EF! lines via travelling activists, roadshows, personal conversations, and demonstrations at regional and other EF! gatherings. Coverage of actions and new tactics regularly appeared in the *EFJ* and a range of magazines and other publications produced by movement members. Many also read and received coverage in other radical and progressive publications. According to Karen Coulter mainstream media was rarely a channel for tactical ideas within the movement as, "We didn't really pay attention to it unless we knew we were going to be in the news for something we'd done."[109]

A workshop led by Jakubal and others from the CFAG protest was held in June 1985 at a Washington State EF! gathering and discussions regarding how tree-sitting could be made more effective featured in an *EFJ* article shortly after. This suggested multiple sitters "be scattered through the unit, especially if they can be positioned in such a way as to bollix up the projected felling pattern". The piece underlined the importance of providing emotional and other support to sitters, as well as giving the media and FS "the impression that your crew is up there to stay".[110]

On Sunday, 23 June 1985, six climbers put these ideas into practice. Participants hailed from across the country and formed part of the EF! Nomadic Action Group (EFNAG), named after the Australian assemblage. Where previous attempts had failed to cohere, this recently formed, small and loose grouping would travel across the country to support and initiate campaigns and carry out ODA over the coming period.[111]

Scaling trees within the Millennium Grove with food, books, musical instruments, and platforms constructed from plywood in tow, the party ascended to heights of up to 100 feet. With the area not subject to a FS ban, CFAG held a press conference in the forest resulting in national and international media coverage.[112]

Mainstream media coverage was viewed as strategically important by most EFers, as evidenced by the fact that most action reports in the *EFJ* mentioned the degree to which their story had been covered by others. Having been told during a meeting with representatives from the large national organisations that it was not worth trying to save old growth or the Spotted Owl, as the media would not cover the story, EF! actions illustrated the value of ODA during 1985 as the network's actions were regularly covered in mainstream newspapers. In a 1990 interview Roselle claimed that unlike moderate environmentalists:

> We understand media. They want pictures, they want action. We ignored the environmental reporters and went to the newshounds … We got 40 people out to Millennium Grove … climbed the trees and draped them with banners that said "Give A Hoot- save the Spotted Owl".[113]

While the desire for media coverage was important, for many it was not the prime factor for undertaking ODA or in selecting locations and tactics. Karen Coulter argues:

> It's always been about pretty much wanting to immediately stop what was happening and to end the destruction. Later on, the media became more of a concern, but we were often reacting to a situation. Our actions were prioritised by the significance of the [forest involved in a] timber sale rather than how a particular place might play in the media, especially in the early days.
>
> We did want the media there as sometimes, but certainly not always, they could provide some protection. The actions, and the attention they drew,

also affected court decisions by demonstrating that there was strong public opposition. In retrospect I think the attention that blockading brought made certain concepts such as "old growth" into household words.[114]

Mike Jakubal argues that the focus on immediate issues detracted from the movement's effectiveness:

> The problem on our side was that we didn't have a long term strategic focus. We were very focused on outwitting the authorities on the day of the action and winning each particular battle and didn't think too much about how that fit into the overall picture or how NVDA or a particular fight might serve a total strategy for saving forests. Even when people did consider this stuff they were more intuitive than deliberate and kept winging it. At times it worked and overall we changed the narrative around ancient forests, but things were very hit and miss. At key points we dropped the ball and there are ancient trees and roadless areas that no longer exist, that we may have been able to save.[115]

Other than a small break to visit a café, talk to the media, and deal with legal matters, Huber was the only one to remain on site for the entirety of the action, although others occupied platforms for shorter periods. An innovation, apparently developed in parallel to similar Australian tactics used at lower heights, was added and the extent of the area under protection expanded through the use of grappling hooks and ropes to connect neighbouring trees.[116]

With the help of supporters camped nearby, eight of whom were arrested during supply runs carried out nightly at random intervals, Huber managed to stay aloft for 28 days on a platform made from a door. By 20 July, authorities improvised a solution and a basket crane was brought in. After manoeuvring it into position two Linn County deputies attempted to drag Huber from his perch. After two hours of climbing and struggle, which in Huber's words "experimented with the outer fringes of nonviolence", he was finally removed.[117]

A flyer promoting the action had described tree-sitting as an "inter-species" tactic as "the hominid participants will prevent the death of a tree" while "the tree species participation will keep the humans safe from arrest and incarceration/citation."[118] The deep emotional and spiritual connection many tree-sitters would come to report was in evidence here. Huber returned to the site after being released from jail to take a graft from branches remaining around its stump. This was later taken to a forestry lab before being replanted among second-growth firs southwest of Corvallis, Oregon.[119]

The tactic soon became popular across the United States with hundreds of people employing climbing gear in the decades to come in order to set up tree-sitting platforms at ever greater heights. Although its media profile, and the excitement involved in the experience itself, ensured that the tactic persisted, its obstructive value varied. As Karen Coulter explains:

In small timber sales of 40 to 200 acres it can be effective because people can rope into other trees. If they're near a town they can draw on a supportive populace and the media will come out because it's close. With large sales in rural areas like eastern Oregon, Idaho, and Montana it hasn't really been used because we're too isolated and the media just doesn't show up. It doesn't hold up the loggers because there are too many access points and they can easily just log all around it.[120]

Within a year the tactic had also diffused to Australia where then novice TWS member Alec Marr, who would eventually become the organisation's long-running Executive Director, took to a tree at Farmhouse Creek, Tasmania, for 16 days from 25 February 1986 to obstruct logging.[121] Diffusion did not come via John Seed, who had toured the United States in 1985, or the *EFJ*, but via indirect channels as former TWS director turned state MP Bob Brown mentioned he had read about people using platforms for tree-sits in Oregon. With TWS long ostracised from radical environmentalists, information about previous Australian tree-sits does not appear to have been passed on and no one from the NAG milieu was involved. Lacking any details beyond Brown's comments, the group improvised a climbing method and patched together a platform from planks of timber and Masonite that proved far from comfortable. The sit delayed work for a fortnight and the blockade gained increased media attention, initially for its novelty and then for violence, as timber workers attacked protesters while police stood by. An attempt was also made to cut Marr's tree down with him in it and Brown was shot at two days later. None of this dissuaded Australian activists as having witnessed its obstructive and media potential they would return to the tactic regularly in decades to come.[122]

The forest blockading template spreads across the United States and U-locks enter the repertoire, 1986

Although the majority of EF! activity continued to focus on rallies and other forms of protest, forest-based ODA was carried out in four states during 1986. The first action to be held in Washington involved six people who blockaded a logging truck in the Wenatchee forest in late June by sitting in the road.[123]

At the end of March 1986, a large number of loggers entered Millennium Grove with a plan to curb protest by rapidly felling all remaining trees. FS officers and police employed new tactics by placing roadblocks and around the clock security in the forest and surrounding areas. With little notice local activists were only able to occupy one tree for less than a day.[124]

With logging complete EFers responded the following day with an April Fool's Day occupation of Willamette Industries' Portland office. Demonstrations at premises related to forestry and other extractive industries were already a regular part of the EF! repertoire. Based in towns and cities, such targets posed fewer safety risks and were more accessible than remote work sites, allowing EFers to easily carry out solidarity and protest actions.[125]

This particular office action bears mentioning here as it introduced a tactic that soon spread to biodiverse settings. Having previously poured sawdust from Millennium Grove over the Regional Forester's desk while the officer was seated behind it, Mike Jakubal occupied the office's entrance by placing a bicycle U-lock around his head. The lock was attached to thick logging cable wound through entry posts, and it took police and a maintenance worker 30 minutes to work out a solution. As the lock and cable could not be easily cut, the door handles were removed and the activist sent on his way with the lock and cable still encircling his neck.[126] As he recalls:

> The police walked me out the back entrance and just kicked me out on the street. They'd threatened to arrest me, but they didn't know what else to do. They couldn't take me to jail because I might hang myself or hurt someone with this stuff around my neck.
>
> This illustrated a paradox that has served me and other activists well over the years. If you don't care about the consequences, you get away with more things. You actually get arrested less and are more effective than if you take lots of precautions. If you have to do a lot of logistics and reconnaissance and maintain security and look outs and all these things then (a) It often won't get off the ground because it all becomes too difficult to organise and (b) You're more vulnerable because it takes more time and you're more exposed and more likely to tip off the opposition. If you go in with a devil-may-care attitude and do something unexpected then they're caught off guard.[127]

Bicycle U-locks had been first developed in the early 1970s and by the end of the decade the "Kryptonite" brand had become widely available. Activists had long chained themselves to gates, equipment, and other objects. Like cyclists they embraced these new locks for their ability to withstand most bolt-cutters as well as the fact that their close fit around a protester's neck or feet increased the difficulty of safe removal.[128]

As has been previously noted, tactics and approaches employed by activists associated with the US peace, civil rights and anti-nuclear movement had a major influence on EF's initial use of "non-violent" tactics. As in Australia the close association between environmentalists and anti-nuclear activists meant that ideas and tactics passed freely between the movements. This was the case here as anti-nuclear activists had first employed U-locks to obstruct opponents, both by locking gates together and by locking body parts to them. Jakubal had heard about the tactic two years earlier from a friend Eric who, along with his family, had been involved in anti-nuclear campaigns.[129]

Internal diffusion led to the device first being applied at the point of environmental destruction in October 1986. During the disruption of clearing and burning of old growth forest at Four Notch, Texas EFer Bugis Cargis leapt onto a tree crusher and attached his neck to it with a U-lock. Texan activists had long opposed such operations and previously employed ODA during 1985 when activists climbed and chained themselves to trees to prevent clearing in Boggy Creek, Austin.[130] On this occasion the lock proved too tough for bolt-cutters and so the protester remained atop the vehicle for more than 24 hours until a locksmith arrived to drill out the mechanism. With 15 press vehicles and more journalists than protesters present, the one-day blockade received much TV and newspaper coverage and a seven-day moratorium was imposed, while the Environmental Protection Agency investigated the situation.[131]

With the use of U-locks becoming a regular feature of environmental blockades, it rapidly diffused to other movements. Most notably US anti-abortionists employed the device at dozens of blockades of medical clinics from 1987 onwards, extending its effectiveness by locking themselves to blocks and concrete-filled barrels as well as under and within strategically placed cars. Others interlaced U-locks to connect protesters' necks together. By 1990 *Newsweek* reported that the Kryptonite company was fielding "increased calls from police departments asking how to thwart the locks" with their advice being to "get a locksmith to drill out the cylinders or employ gigantic Jaws of Life cutters used for auto wrecks".[132]

Given the timeline it is likely that anti-abortionists learnt of the tactic indirectly via mainstream media coverage of EF! actions. Despite the two movements extending the effectiveness of the tactic in similar ways, there is no mention of anti-abortion actions in the EF! milieu's media. Despite some EFers taking part in clinic defence, interviews indicate that few, if any, were aware of the tactic's use by their opponents as they did not personally witness or hear of it. As a result, innovations appear to have taken place in parallel and their spread to have been the result of internal rather than inter-movement diffusion.[133]

American developments, 1980–1986

From 1980 onwards a radical movement emerged in the United States that rejected the "insider" strategies of dominant national conservation organisations in favour of using confrontational means to pursue protection for biodiverse areas. Amongst these was the deployment of environmental blockading. By the end of 1986 members of Earth First! had established EB as a response to ecological threats in parts of the United States. As the next chapter demonstrates this left them poised to extend its use over the next five years geographically as well as in terms of the number, length, and size of actions.

Australian examples played an important role in inspiring EF! members to undertake their own environmental blockades. Diffusion regarding tactics and approaches to normative protester behaviour however was minimal, with American activists innovating their own, in part based on local precedents.

As in Australia, exclusion orders increased the tendency for US activists to focus soft blockade tactics on closing access points as well as to develop tactics that required fewer activists to infiltrate sites. Compared with Australia, settlement patterns and a greater risk of assault and repression from workers and authorities meant that EF! campaigns had fewer resources and were less able to draw on support from local communities. As a result, action was more intermittent, and blockaders were initially unable to launch blockades that lasted more than a few days.

Due to these factors, and others, such as the network's loose approach to normative protester behaviour and priority on maximising disruption to work at the point of destruction, EF! activists rapidly innovated tactics involving enhanced vulnerability, adding major additions to the repertoire.

Sabotage played a much larger role in American environmental activism than in Australia or Canada. Due to the heavy penalties involved, and the likelihood of poor publicity, the damaging of property, beyond that incurred during barricading, rarely occurred during environmental blockades. As the following chapter illustrates, attempts to project sabotage and environmental blockading as separate repertoires were largely unsuccessful. Resulting bad publicity and other pressures meant that internal debate regarding sabotage would increase over the coming years.

Notes

1 Bob Edwards, "With Liberty and Environmental Justice for All," in *Ecological Resistance Movements*, ed. Bron Taylor (Albany: SUNY Press, 1995): 27–35.

2 Ibid.

3 Kathie Durbin, *Tree Huggers: Victory, Defeat & Renewal in the Northwest Ancient Forest Campaign* (Seattle: Mountaineers, 1996), 32–33; Douglas Bevington, *The Rebirth of Environmentalism: Grassroots Activism from the Spotted Owl to the Polar Bear* (Washington DC: Island Press, 2009), 3–4.

4 Lawrence Rakestraw and Mary Rakestraw, History of the Willamette National Forest, (Eugene: USDA Forest Service Pacific Northwest Region, 1991), www.foresthistory.org/ASPNET/Publications/region/6/willamette/chap6.htm; Braggs, "Earth First!," 22–25.

5 Robert Mitchell, Angela Mertig, and Riley Dunlap, "Twenty Years of Environmental Mobilization: Trends among National Environmental Organizations," ed. Riley Dunlap and Angela Mertig, *American Environmentalism* (New York: Taylor and Francis, 1992). 11–25.

6 Christopher Manes, *Green Rage: Radical Environmentalism and the Unmaking of Civilization* (Boston: Little Brown & Company, 1990), 61–64; Dave Foreman, *Confessions of an Eco-Warrior* (New York: Crown, 1991), 6, 13–14.

7 Silvaggio, "The Forest Defense Movement," 66–68.

8 Foreman, *Confessions of an Eco-Warrior*, 25–26; Manes, *Green Rage*, 76.

9 Arne Naess, "A Defence of the Deep Ecology Movement," *Environmental Ethics* 6, no. 3 (1984): 265.

10 Dave Foreman, "Earth First!," *Earth First! Journal (EFJ)* 2, no. 3 (1982): 4–5; Scarce, *Eco-Warriors: Understanding the Radical Environmental Movement*, 58–59.

11 Bron Taylor, "Earth First! And Global Narratives of Popular Ecological Resistance," in *Ecological Resistance Movements*, ed. Bron Taylor (Albany, NY: SUNY Press, 1995): 27–28.

12 Susan Zakin, *Coyotes and Town Dogs: Earth First! and the Environmental Movement* (New York: Viking, 1993), 116–46.

13 Dave Foreman, "Welcome to EF," *EFJ* 5, no. 5 (1985): 16.

14 "Editorial," *Earth First!* 1, no. 1 (1980): 1; Scarce, *Eco-Warriors*, 61.

15 Foreman, "Earth First!," 4–5; Lee, *Earth First!: Environmental Apocalypse* 78–80.

16 Silvaggio, "The Forest Defense Movement" 57–70; Timothy Ingalsbee, "Earth First! Activism: Ecological Postmodern Praxis in Radical Environmentalist Identities," *Sociological Perspectives* 39 no.2 (1996): 264–65.
17 Zakin, *Coyotes and Town Dogs*, 191.
18 Quoted in ibid., 194.
19 Lee, *Earth First!* 120–21.
20 Douglas Martin. Obituary: James Phillips, 70, Environmentalist Who Was Called the Fox, Available [Online]: www.nytimes.com/2001/10/22/us/james-phillips-70-environmentalist-who-was-called-the-fox.html?_r=0 (Accessed 16 December 2015); Travis Wagner, "Reframing Ecotage as Ecoterrorism: News and the Discourse of Fear," *Environmental Communication* 2, no. 1 (2008): 26.
21 Sam Love and David Obst, *Ecotage* (New York: Pyramid, 1972), 1.
22 Ibid., 13–14.
23 Ibid., 174.
24 Peter A Coates and DJS Morris, "Support Your Right to Arm Bears (and Peccadillos)": The Higher Ground and Further Shores of American Environmentalism," *Journal of American Studies* 23, no. 3 (1989): 441–42; Edward Abbey, *The Monkey Wrench Gang* (Philadelphia: Lippincott, 1975).
25 Dave Foreman, "Editorial," *Earth First!* 1, no. 5 (1981): 1.
26 "Where Were You When We Cracked Glen Canyon Dam?," *Earth First!* 1, no. 4 (1981): 1–2; "22 Things to Do as an Earth Firster!," *Earth First!* 2, no. 6 (1982): 4.
27 Wang and Piazza, "The Use of Disruptive Tactics in Protest as a Trade-Off", 1675.
28 In 1990 Assistant Attorney General Bernie Hubley admitted to a court that the history of EF! demonstrated it had been more involved in civil disobedience than violence or sabotage. Associated Press, "Some Earth Firsters Shed Spikes," *Spokane Chronicle*, 11 July 1990, A10; Manes, *Green Rage*, 7–13.
29 Epstein, *Political Protest and Cultural Revolution*, 21–156; Melvyn Dubofsky, *We Shall Be All: A History of the Industrial Workers of the World* (Urbana, IL: University of Illinois Press, 2000), 89–97.
30 Manes, *Green Rage*, 78–83.
31 "Salt Creek Arrests," *EFJ* 3, no. 3 (1983): 1, 5.
32 Bart Koehler, "The Battle of Salt Creek," *EFJ* 3, no. 2 (1982): 1; Bart Koehler, "Bisti Mass Trespass," *EFJ* 3, no. 2 (1982): 1, 11; Karen Brown, "Bisti Circus," *EFJ* 3, no. 3 (1983): 5.
33 Dave Foreman, "Editorial: The Lesson of Salt Creek," *EFJ* 3, no. 3 (1983): 3.
34 Dave Foreman, "Around the Campfire," *EFJ* 3, no. 2 (1982): 2; "EF Local Groups and Contacts," *EFJ* 3, no. 3 (1983): 8.
35 Tim Palmer, *Stanislaus: The Struggle for a River* (Berkeley: University of Chicago Press, 1982), 160–85, 245–48; Jeff Jardine, "Water War of Yore Still Resonates with New Melones Protester," *Modesto Bee*, 8 July 2015, 15.
36 Seed Interview.
37 John Seed, "Australia Reports In!," *EFJ* 8, no. 2 (1982): 9.
38 Foreman, "Around the Campfire," 2; Mike Roselle, *Tree Spiker: From Earth First! To Lowbagging, My Struggles in Radical Environmental Action* (New York: St. Martins Press, 2009), 67, 73; Zakin, *Coyotes and Town Dogs*, 249–50.
39 "700 Arrested in Australia," *EFJ* 3, no. 3 (1983): 1.
40 Examples of such material include: Ric Bailey, "Bald Mountain in Retrospect," *EFJ* 4, no. 1 (1983): 6; Bill Devall, "The Edge: The Ecology Movement in Australia," *EFJ* 4, no. 5 (1984): 12–13; Brian Health, "What Did You Expect to Accomplish Anyway?," *EFJ* 5, no. 1 (1984): 5.

41 "Damn Hetchy Dam!," *EFJ* 4, no. 5 (1984): 19; Mike Roselle, "Middle Santium Heats Up," *EFJ* 4, no. 5 (1984): 1; Mikal Jakubal, Interviewed 2 May 2016.

42 Hattie Clark, "If a Tree Falls, John Seed Hears It," *Christian Science Monitor*, 13 August 1987, Available [Online]: http://search.proquest.com.ezp.lib.unimelb.edu.au/docview/1034959490?accountid=12372 (Accessed 22 May 2015); "Australian John Seed to Join EF Roadshow," *EFJ* 4, no. 2 (1983): 13.

43 Roselle, *Tree Spiker*, 73–75; John Seed, "Introduction," *World Rainforest Report*, no. 2 (1984): 1.

44 Silvaggio, "The Forest Defense Movement," 113–14.

45 Karen Coulter, Interviewed 9 May 2016; Jakubal Interview.

46 Ambrosia Krinsky, "Mike Roselle, Earth First Founder 2013 Interview," www.youtube.com/watch?v=qfBQc6HO0Ys, 2013.

47 Karen Wood, Interviewed 27 June 2016; Coulter Interview.

48 Beverley Brown, *In Timber Country* (Philadelphia: Temple University Press, 1995), 5–7.

49 Ibid, 23–35; Jakubal Interview.

50 "Kalmiopsis Blockade Begins," *EFJ* May(1983): 1, 6–7.

51 Durbin, *Tree Huggers*, 57–61.

52 Manes, *Green Rage*, 84–87.

53 Bailey, "Bald Mountain in Retrospect," 6; Zakin, *Coyotes and Town Dogs*, 232–41.

54 Freddie in EF slang is a pejorative applied to US Forest Service officers and other government employees involved in the exploitation of biodiverse areas. "Reforming the Freddies," *EFJ* 3, no. 5 (1983): 16; "Blockade Updates," *EFJ* 3 no.3 (1983): 1, 5.

55 Zakin, *Coyotes and Town Dogs*, 245–46.

56 "No G-O Road," *EFJ* 3 no.5 (1983): 15.

57 Paul Fattig. Future of Bald Mountain Remains Uncertain, Available [Online]: www.mailtribune.com/article/20030427/BIZ/304279999 (Accessed 5 November 2015).

58 "Resistance Gets Hairy at Bald Mountain," *EFJ* 21 no. 1 (2000): 11.

59 "Bald Mountain Road Stopped!!," *EFJ* 3 no.7 (1983): 1; Bailey, "Bald Mountain in Retrospect," 6–7.

60 "Wilderness War in Oregon," *EFJ* 3, no. 5 (1983): 1,4.

61 Dave Foreman, "Around the Campfire," *Earth First!* 3, no. 6 (1983): 2.

62 Coulter Interview.

63 Bailey, "Bald Mountain in Retrospect," 6.

64 Coulter Interview; Barbara Epstein, *Political Protest and Cultural Revolution*, 129–34,42–44.

65 Judith Stiehm, "Nonviolence Is Two," *Sociological Inquiry* 38, no. 1 (1968): 23–29; Thomas Weber, "Nonviolence Is Who? Gene Sharp and Gandhi," *Peace & Change* 28, no. 2 (2003): 250–65.

66 Jakubal Interview; Coulter Interview; Orin Langelle, Interviewed 13 May 2016.

67 "Kalmiopsis Blockade Begins," 1; Bailey, "Bald Mountain in Retrospect," 6–7.

68 Jakubal Interview.

69 Coulter Interview.

70 Karen Pickett et al., "Blockade Personal Accounts," *EFJ* 3 no. 5 (1983): 6–7; UPI. Protesters Remain in Jail, Available [Online]: www.upi.com/Archives/1983/07/01/Protesters-remain-in-jail/6156425880000/ (Accessed 4 November 2015).

71 Bailey, "Bald Mountain in Retrospect," 6.

72 T.A., "Blockade #6," *EFJ* 3 no.6 (1983): 9.

73 "Forest Service Says It Won't Push Bald Mountain Road," *Register-Guard*, 30 January 1985, 10B; Durbin, *Tree Huggers*, 57.

74 Manes, *Green Rage*, 88; Howie Wolke, "The Grizzly Den," *EFJ* 3, no. 7 (1983): 12.

75 "Round River Rendezvous," *EFJ* 3, no. 6 (1983): 1.

76 Associated Press. Environmentalists Protest Fiddler Timber Sale, Available [Online]: http://tdn.com/news/state-and-regional/environmentalists-protest-fiddler-timber-sale/article_990c41cf-3b37-5f13-87de-b733f5db5c91.html (Accessed 7 November 2015).

77 Bailey, "Bald Mountain in Retrospect," 6.

78 Tuatha De Danan, "Is EF! Selling Out?," *EFJ* 4 No.1 (1983): 4; Kris Maenz, "The Life and Times of Our Beloved Journal," *EFJ* 21 no. 1 (2000): 76–77.

79 Jakubal Interview.

80 Dave Foreman, "EF and Nonviolence: A Discussion," *EFJ* 3, no. 7 (1983): 11.

81 Coulter Interview.

82 Peter Swanson, "EF: Violence or Non-Violence?," *EFJ* 3, no. 7 (1983): 12.

83 Silvaggio, "The Forest Defense Movement," 121–25.

84 Mike Roselle, "Tree Huggers Save Redwoods," *EFJ* 4, no. 1 (1983): 1, 4.

85 Mike Roselle, "Sinkyone Struggle Continues," *EFJ* 4, no. 1 (1983): 10.

86 Keith Easthouse. EPIC Changes, Available [Online]: www.northcoastjournal.com/humboldt/epic-changes/Content?oid=2132384 (Accessed 5 November 2015).

87 Marcy Willow, "Last Stand for the Last Stand," *EFJ* 4, no. 5 (1984): 7.

88 Dan Wyant, "Protesters Vow to Continue Fight," *Register-Guard*, 8 June 1984, 1A; George Draffan, "Cathedral Forest Action Group Fights for Oregon Old Growth," *EFJ* 4, no. 5 (1984): 5.

89 Lee, *Earth First!: Environmental Apocalypse*, 83.

90 Dan Wyant, "15 Protesters Arrested on Logging Road," *Register-Guard*, 5 June 1984, 1C; Mike Roselle, "Earth First! Takes Regional Forester's Office," *EFJ* November(1984): 1.

91 There appears to have been no repercussions regarding these police actions. Matt Veenker, "Blockaders Roughed up in Middle Santium," *EFJ* 4 no. 7 (1984): 1; Brian Foulkes. State V. Hund, Available [Online]: www.aclu-or.org/content/state-v-hund (Accessed 18 November 2015).

92 Roselle, "Middle Santium Heats Up," 1, 5.

93 Foulkes, "State V. Hund"; "Editorial: Two Wrongs Don't Make a 'Riot'," *Register-Guard*, 5 August 1984, 16A.

94 Shabecoff, "The Acre by Acre Effort to Save the Environment," *New York Times*, 30 December 1984, A2; Cecelia Ostrow, "Letter from Oregon Jail," *EFJ* 5, no. 2 (1984): 7.

95 Mike Roselle, "Oregon Trials," *EFJ* 5, no. 2 (1984): 6.

96 "10 Protesters Praised, Convicted," *The Bulletin*, 28 October 1984, A9.

97 Ibid.

98 "Protesters Told to Pay Damages," *Register-Guard*, 4 September 1985, 5A.

99 Sharon Beder, "SLAPPS: Strategic Lawsuits against Public Participation: Coming to a Controversy near You," *Current Affairs Bulletin* 72, no. 3 (1995): 22.

100 Silvaggio, "The Forest Defense Movement"," 136–40; Wood Interview.

101 Scarce, *Eco-Warriors*, 76.

102 "'Fairly Extensive' Tree Spiking Found," *Register-Guard*, 12 May 1985, 5A.

103 "Eugene Woman Pleads Innocent to Trespassing on Native Forest," *Register-Guard*, 2 May 1985, 14C; Marcy Willow, "Foresticide in Middle Santium," *EFJ* 5, no. 5 (1985): 10.

104 Jakubal Interview.

105 Jakubal Interview.

106 Mike Jakubal, The Postcard and the Portaledge. Available [Online]: http://civilizeddisobedience.com/2012/03/21/the-postcard-and-the-portaledge/. (Accessed 31 December 2015).

107 UPI, "Logging Protester Arrested," *Washington Post*, 22 May 1985, A13; Ron Huber, "Tree Climbing Hero," *EFJ* 5, no. 6 (1985): 1.

108 Jakubal Interview.
109 Coulter Interview.
110 Aries, "Go Climb a Tree!," *EFJ* 5, no. 6 (1985): 7.
111 Coulter Interview.
112 "6 Stage Treetop Sit-in, Protest Forest Harvest," *Houston Chronicle*, 3 July 1985, 34; Australopithicus, "Nemesis News Net," *EFJ* 5, no. 7 (1985): 12.
113 Martin Walker and David Rowan, "Shock Troops in the Eco-War," *Guardian*, 17 August 1990, 21.
114 Coulter Interview.
115 Jakubal Interview.
116 Dan Hinson, "Staying Put," *Orlando Sentinel*, 13 July 1985, A16; UPI, "Oregon Logging Sends Protester up a Tree," *Chicago Tribune*, 13 July 1985, 4.
117 Dave Freeman. Linn County Deputy Sheriff's Report, Available [Online]: www.penbay. org/ef/ronhuber_sheriffrp85.html (Accessed 17 November 2015); Ron Huber, "Giant Crane Attacks Tree Sitter," *EFJ* 5, no. 7 (1985): 4–6.
118 "Interspecies Action" (Unknown, 1985), 1.
119 Ron Huber. Yggdrasil Survives, Available [Online]: www.penbay.org/ef/yggdrasil_ rebirth85.html (Accessed 17 November 2015).
120 Coulter Interview.
121 "Logging Protest Gives Alec a 'Birds-Eye' View of the World," *Mercury*, 28 February 1986, 1.
122 Alec Marr, "The Picton Blockades of 1986 and 1987," in *For the Forests: A History of the Tasmanian Forest Campaigns*, 174–76.
123 George Draffan and Mitch Freedman, "War in the Wenatchee," *EFJ* 6, no. 8 (1986): 8.
124 AP, "Protest Continues as Old Firs Cut," *Spokesman-Review*, 1 April 1986, A10; Mike O'Rizay, "Freddies Murder Millenium Grove," *EFJ* 6, no. 6 (1986): 13.
125 Karen Pickett, "Day of Outrage Shakes Forest Service Nationwide!," *EFJ* 8, no. 6 (1988): 1, 19.
126 Ibid.
127 Jakubal Interview.
128 Jakubal Interview.
129 Ibid; DAM Collective, *Earth First! Direct Action Manual* (Eugene, OR: Cascadia Summer, 1997), 50–51; "Peace Protests Roll Call," *New Statesmen* 109 (1985): 6.
130 P. Kahn (pseudonym). "Last Stand on Boggy Creek," *EFJ* 6, no. 3 (1986): 5.
131 "Texas EF Fights Freddie Godzilla," *EFJ* 7, no. 1 (1986): 1.
132 "Togetherness," *Newsweek*, 4 June 1990, 4; Tim Vercellotti, "'Lock and Block' Tactic Is Spreading Amongst Groups Protesting Abortions," *Pittsburgh Press*, 25 June 1989, A12; Patricia Baird-Windle and Eleanor J Bader, *Targets of Hatred: Anti-Abortion Terrorism* (New York: Palgrave Macmillan, 2001), 94–146.
133 Coulter Interview; Jakubal Interview; Wood Interview; Langelle Interview.

4

"WE SHUT 'EM DOWN"

Environmental blockading in the United States extends and entrenches, 1987–1990

From 1987 to 1990 internal and external factors promoted the action template across the United States and built capacity to the point where activists could hold sustained blockades and mass campaigns. Increasing rates of diffusion and use further refined many of the tactics and approaches that had been developed in previous years.

This chapter situates these developments in the changing political economy of the timber industry as well as increasing contradictions between government policies regarding conservation and resource extraction. It provides a history of the emergence of the "Timber Wars" in PNW and California and examines the causes and outcomes of deepening divisions between environmentalists and logging workers.

The changing context of blockading

From 1987 onwards the scale of ODA increased and its context shifted as a confluence of factors intensified campaigning and made concerns regarding endangered species and their habitats popular national issues. The main force driving EB remained the overcutting of forests. Since the 1940s the FS had followed a policy of "sustained yield", that is that any trees removed would be matched by regrowth. From the 1960s it was also directed to facilitate "multiple use" of national forests through combining recreation, conservation, and other outcomes with logging. In practice felling remained the priority, partially due to the internal culture of the organisation, but also because the FS's funding was mainly drawn from timber sales.[1]

The FS had first been warned in the 1930s that old-growth forests in PNW were being cut faster than trees were regrowing. By 1976 a report concluded that a 20-year supply gap would emerge by the 1990s and later studies showed that the agency had long overestimated levels of remaining old-growth forest. Despite internal and external warnings that logging was threatening the ability of the FS to meet its overall responsibilities, by 1987 record levels of timber were being cut in Washington and Oregon at 5.6 billion board feet a year, the majority of which was old-growth.[2]

The timber industry included several competing interests and stakeholders, from small independent millers through to large corporations, with differing degrees of focus on national and export markets as well as concern for the welfare of workers and local communities. Technological and structural shifts led to an overall increase in demand for timber alongside reductions in wages and job security. Much of this stemmed from fluctuating prices and the entry of new investors, as companies such as Crown Zellerbach, Crown Pacific, and Maxxam leveraged massive debts in the process of carrying out or foiling takeovers. The servicing of these involved a steep increase in logging on private land and increased lobbying for access to public land as reserves ran out. In turn 30,000 jobs were lost in Oregon and Washington between 1979 and 1989 as mill productivity increased by 50 per cent, stockpiling rose, and exports of unmilled timber reached 40 per cent of the overall cut.[3]

Conflicting pressures from Congress also played a role. Modest amounts of forest were protected via Wilderness Bills passed in the mid-1980s, but at the same time politicians from PNW, under pressure from industry sponsors and concerned workforces, facilitated increased exploitation of what remained. Targets mandating the FS sell up to one billion board feet more than the organisation itself had recommended were set for each year between 1987 and 1991.[4]

Political closure also followed the Wilderness Bills as key power brokers ruled out further legislative protection. Combined with a continuing lack of interest, and occasional hostility, from national conservation organisations the number of activists seeking to use unconventional means increased. Alongside a rise in ODA this also encouraged the formation of what Douglas Bevington describes as "Grassroots Biodiversity Groups". These monitored timber sales and FS activities, scientifically documented the health of ecosystems and species, and aggressively employed bureaucratic appeals and litigation. Current and former EF! members established and worked within these small, formal, regionally based organisations as well as took part in attempts to reform national bodies such as the Sierra Club.[5]

Among the pressures upon federal and state bodies was legislation requiring they protect endangered species. Laws under the Endangered Species Act, National Environment Policy Act, and National Forest Management Act required environmental assessments be carried out and made the survival of species of equal if not greater importance than economic considerations. Despite this, underfunding, political compulsion, loopholes allowing the Federal Wildlife Service to carry out studies incrementally, and a lack of public pressure, meant that many species were ignored. Disquiet within the bureaucracy became increasingly public from the late 1980s fuelling demands for reform.[6]

As there were few protections mandated for old growth ecosystems as a whole, PNW environmentalists strategically chose to focus on laws that provided protection for specific forms of wildlife within ecosystems while bureaucratically appealing timber sales in order to enforce rules, stretch FS and opponents' resources, and overload the system. These tactics had previously been sparingly used, but from 1987 onwards this changed. The PNW's Spotted Owl became a key focus as the species was unique to old-growth forests and required swathes of it to remain intact in order to survive.[7]

1987: Innovations spread and a backlash begins

Much of the ODA carried out at the point of destruction from 1987 to 1990 would form part of what became known as the PNW and California's "Timber Wars", as environmentalists responded to overcutting with a suite of strategies and tactics. Over this period a key point of contention became the redwood forests of Northern California, particularly those on private land, which was acquired through the takeover of the Pacific-Lumber company by Texan "corporate raider" Maxxam in 1986. The legality and nature of Maxxam's acquisition, its redeployment of workers' pension funds, and its subsequent policy of switching from selective logging to clear-cutting in order to increase the rate of cutting by 100 per cent or more, created much controversy amidst various court cases and government inquiries.[8]

Already quiet on the issue of old growth forests generally, national environmental organisations were even more reluctant to act over the issue of logging on private land. Explaining their inaction Sally Kabisch, Northern California field representative for the Sierra Club, stated in 1987, "The legal reality [regarding private land] is that, except for the Forest Practices Act, the owner has the basic constitutional right to do what he wants with it."[9]

EFers and their allies had less respect for property rights. Seeking to supplement the work being done by the Environmental Protection and Information Centre (EPIC) on the legal front with blockading, local journalist Greg King and musician Daryl Cherney founded the Redwood Action Team in 1986 to carry out protests targeting Pacific-Lumber. According to researcher Steve Ongerth, "EPIC members and volunteers had attended the initial Earth First! meetings, for strategic purposes, although they often worked alongside of and in concert with Earth First!, they kept their legal game plan independent of the latter."[10]

A series of protests in 1986 and a California EF! rendezvous brought a core of activists together who began surveilling likely logging sites. The following year King and activist Larry Rivers discovered 3,000 acres of redwood old-growth forest near Mendocino, which they dubbed the "Headwaters Grove" as it formed a point at which various streams met. Comparable in size to stands within national parks this pristine Spotted-Owl habitat became the focal point and symbol of a campaign targeting clearcutting in general.[11]

With only a small number of activists, and litigation from EPIC stalling the logging of what was dubbed the "All Species Grove" within Headwaters, Northern Californian EFers initially focused on actions designed to raise public awareness. During 1987 this included a one-day tree-sit on Pacific-Lumber land on 17 May, which was brought to an end through the use of security guards and a spur climber. Coordinated protests against Maxxam followed the next day at a number of locations along the West Coast. Timed to coincide with the end of student examinations at Northern California's Humboldt University, these actions, the largest of which involved women trespassing to shut down the transportation of timber from a logging deck, resulted in TV and newspaper coverage in several states.[12]

By late August, media interest in the issue was diminishing so activists embarked on a second tree-sit. Two platforms were assembled by a crew of 15–20 who carried 500 pounds of climbing gear, food, and clothing eight miles into the forest. As part of the Californian EFers focus on publicity, Greg King and a woman using the pseudonym "Jane Cope" occupied trees at a point where the clear-cut could be contrasted with surrounding old-growth forest by photographers.[13]

Due to forest fires in other parts of the state only minimal coverage ensued, but the pair remained in place for six days communicating with journalists and supporters via Citizens Band radio and a radio telephone. Logging was moved to the other side of the grove and Pacific-Lumber brought in security to protect equipment and harass the activists around the clock.

In order to directly transmit their messages via media to the public, sitters regularly dropped banners as part of their actions. An attempt to prevent this was foiled after the pair replaced one removed by a Pacific-Lumber climber reading "Free The Redwoods" with another stating "2000 Years Old/Respect Your Elders". The use of spotlights, and the constant noise from generators, proved more successful as they forced the pair to leave after a week in the trees. Underlining the dangers associated with climbing at such heights King, in a rush to withdraw before he could be arrested, became caught up in ropes and his waist was dangerously compressed by his harness.[14]

Despite such a perilous conclusion, King and "Cope" gained valuable experience and engaged in another tree-sit a month later. This time a five-day stint attracted greater media coverage. This was not least because, to the pair's chagrin, Cherney, who had previously worked in the advertising industry, pitched them to journalists as a modern-day "Tarzan and Jane".[15]

EFNAG contributed support to these actions by helping to fund the purchase of supplies and equipment via the recently founded Earth First Direct Action Fund (EFDAF). In June 1987, an advertisement appeared in the *EFJ* calling on network members to join EFNAG and contribute to a fund which would finance activist travel and support campaigns and actions, particularly in areas that lacked financing. This development was partially practical, as an existing tax-deductible EF! Foundation could only legally subsidise education and research. It also reflected growing internal divisions over the direction the network should take and the degree to which the *EFJ* and its resources should be used to promote and support blockading. Administered primarily by San Francisco-based EF! activist Karen Pickett and spruiked heavily by Mike Roselle, the EFDAF disbursed $30,000, generally in quantities of $100 to $200, in its first year and increasing amounts in those that followed.[16]

EFNAG's other main contribution was in training and assisting the Californians to undertake tree-sits. EFNAG members had taken part in anti-uranium actions and protests in Colorado, Arizona, and Utah during early 1987, but Jakubal, Roselle, and other members spent much of the year in Oregon.[17] Following an announcement by the FS of 24 new timber sales, activity focused on the Siskiyou National Forest, near where EF! had carried out blockading at Bald Mountain four years earlier.[18] Non-violence training, meetings, and a gathering were held in the build up to action commencing in April and consensus reached on the strategy of employing a series of intermittent actions. The reasoning behind this was outlined by David Barron in the *EFJ*:

> Since our overall strategy is to generate publicity and keep North Kalmiopsis in the news, we don't want to get everybody arrested at once and then fade out of the media. A series of direct actions with a few people arrested each time could provide prolonged pressure on the Forest Service and on-going education of the public.[19]

Although media coverage was a key concern, activists also employed tactics with a view to remaining in place for as long as possible in order to disrupt work and frustrate their opponents. During five actions carried out at logging sites between April and July they employed a suite of means including tree-sitting, employing kryptonite locks to secure legs and necks to access gates, and "soft blockading", using their bodies alone to occupy space and stand in front of vehicles and workers. Two innovations emerged during the summer. The first involved activists from a community who had been involved in the 1983 Bald Mountain protests. To fully cover an entrance, they chained two of their number to a gate and then buried them up to their necks in gravel.[20]

The second came during an action at the Sapphire Timber Sale on 23 July in which Roselle boasted, "We've not only made a statement and generated publicity, we shut 'em down."[21] Loggers were forced to work around the immediate site after a group of five men occupied tree platforms, with one remaining in place for 18 days. Cutting was also halted for 12 hours after five women and a man locked themselves to the high-lead yarding unit. Most of the group were connected with handcuffs and kryptonite locks to cables and other points in the lower section of the machine, while "direct action innovator 'Rhody-Dendron' sat harnessed at the summit of the 92-foot tower".[22] This was the first time that activists had used lock-ons to prevent workers from using such equipment to remove logs from a work site.[23]

Following an incident during a June action, in which a logging worker dangerously cut into a tree bearing an activist on a platform, it was reported in the *EFJ* that logging companies in the area had adopted a policy of instructing workers to avoid contact with activists. Nevertheless, EFers claimed that workers threatened women with rape and assault during the 23 July occupation.[24]

Hostility toward activists from timber workers and their associated communities steadily increased across the PNW and California during the late 1980s due to a range of factors. In a 1995 study covering the period, researcher and oral historian Beverley Brown argues that economic and social precarity increased in Oregon due to job losses in the timber industry and increased in-migration by wealthy residents from Washington and California. As a result, rural working-class rural Oregonians faced "a quadruple burden: loss of income and benefits, the threat of being priced or taxed out of their own homes, diminished access to forests and rivers, and an increasing dissolution of neighbourhood and community coherence".[25]

Although radical environmentalists had little to do with these changes their demand to end logging in the Kalmiopsis and other old growth forests caused them to be grouped with other "outsiders" as a threat. Employers harnessed this anger in the pursuit of their own interests and led a concerted media and grassroots campaign across the PNW and California.[26]

Well aware of the toll that structural changes were taking, with 13,000 jobs lost in Oregon alone between 1976 and 1986, timber workers often held ambivalent if not antagonistic views towards employers.[27] Although concerted efforts by EFers to build alliances were yet to come, there was some level of sympathy for arguments opposing clear-felling on the basis that it threatened both the environment and jobs. Cultural differences and a lack of contact between the two parties nevertheless proved an ongoing impediment to better relationships. In a media interview, logger and son of Pacific Lumber's former owner Woody Murphy stated that he opposed increased harvesting:

> But Earth First!'s a radical group, and a lot of that I just can't associate myself with. [They look like] a bunch of college kids with ponytails. One guy looked like he got dragged out of a sewer pipe. You've got to look like the people you're trying to convince.[28]

The co-founder of the Headwaters environmental organisation, Art Downing, also argued that EF!'s call for large-scale reductions in logging was a key contributor to tensions. The activist, who had moved to rural Oregon as part of the back-to-the-land movement of the 1970s, stated:

> I was really concerned and disapproved in a basic way of the positions I saw EF! taking. Not so much the positions, but the kind of polarization they can catalyse … We've got people who are just scared to death of being out of work. And you don't do yourself or your cause or the woods any good by alienating those people … Until we get together a real cogent strategy that addresses the bottom-line issues of fear and ignorance that underlie all the issues, we're not going to make any real changes.[29]

The perception of environmentalism as an active and malicious threat was further honed by the issue of tree-spiking. Supporters of the tactic claimed it had positive effects in protecting forest through the imposition of costs as well as by drawing attention to ecological destruction and creating a radical flank. It clearly also had negative effects for EF! These came to the fore in May 1987 after Mendocino County, California, mill worker George Alexander sustained major injuries, including having his jaw broken and jugular vein cut, after a spike embedded in a tree tore through a bandsaw he was operating. The story attracted widespread media interest and pro-logging interests were quick to link EF! to the incident. Tree spiking would be banned under various state and federal laws in the two years following.[30]

Activists such as Dave Foreman and Betty Ball continued to argue that the tactic was non-violent in that it did not deliberately seek to injure others. They also cast doubt on EF! involvement in the incident as it did not involve old-growth forest or a warning to workers.[31] Some PNW EFers, many of whom still supported ecotage against machinery, began to oppose tree-spiking, whether in the vicinity of a blockade or not, on the basis that even if physical harm was not intended it remained possible. This was particularly so, they argued, as the tactic would only work if employers prioritised the health and safety of workers by refusing to mill spiked timber, something they saw as unlikely in a period of intensifying exploitation. These activists, who would become more vocal in the coming years, also felt any association with the practice would undermine EF! claims to non-violence, inevitably damaging their public reputation and placing campaigners in danger.[32]

Despite coming under pressure from the many in the network who supported spiking, activists operating near where the incident occurred issued a statement reading:

> North Coast California Earth First! strongly condemns the timber industry's recent heavy-handed tactics designed to bring woodworkers wrath upon environmentalists. Our efforts have not and will not involve tree spiking, destruction of private property, or devices that threaten harm to any life form, including humans.[33]

For his part Alexander demonstrated that timber workers' views were often more nuanced than the mainstream media and many environmentalists assumed when he told the *San Francisco Examiner*, "I don't agree with spiking trees but I don't like clearcutting either." He rejected offers to become an anti-environmental spokesperson and was highly critical of his employer Louisiana-Pacific for not responding to repeated safety complaints he had made in the lead up to his injuries.[34]

Alexander's complaints, demands by union officials, and the failure of employers to install metal detectors and upgrade equipment did not generate as much anger and distress as spiking itself. EF! increasingly became the overt focus of logging communities' disquiet. A sense that issues around continuing access to forests could be won where those regarding automation could not, and the fact that unlike employers environmentalists could be confronted with relative impunity, possibly made activists a convenient target. Everyday harassment of activists and the use of covert "dirty tricks" operations, particularly leaflets and statements falsely linking activists to tree-spiking, became common in timber regions. Public opposition during 1987 peaked with Oregon's "Silver Fire Roundup" in which 1,226 trucks and up to 10,000 pro-industry demonstrators mobilised on 27 August to demand increased "salvage" logging following the burning of close to 100,000 acres of forest by wildfires.[35]

In the context of this backlash the risks involved in ODA were heightened during 1987. In California Greg King was fired upon while addressing journalists on Maxxam-owned land and there were claims of rough handling and assaults from FS and police officers in Oregon where various EFers served jail sentences of 10–15 days after refusing to pay fines or restitution costs. The region involved had a long history of racial violence and Ku Klux Klan activity. Four women arrested during the 23 July action complained of being placed in a cell with prisoners who assaulted them after abusing an African-American EFer.[36]

In addition to the protests already canvassed, EF! carried out a week of ODA in Washington's Cascade forests in which the combination of a sympathetic sheriff and the local FS's desire to avoid publicity meant no arrests took place despite roading and logging operations being halted. In San Bruno, California EFers extended ODA to preventing suburban development for the first time by locking onto a gate to protect endangered butterfly habitat while two separate protests against uranium mining, involving lock-ons and the blockading of trucks, were carried out in the Grand Canyon, Arizona.[37] While protest activity across the country led to increasing coverage of EF!, many of the resulting stories framed the network in terms of its connection to tree-spiking and allegations of "ecoterrorism".[38]

Barricading increases, 1988

From 1988 terminology regarding the value of "Old Growth" and "Ancient Forests" began to enter public discourse and garner increased media attention.[39] In line with this EF! continued to grow, as evidenced by new chapters and around 400 people attending that year's Round River Rendezvous. An "Earth First! Speakers Bureau" offering 38 performers and orators was also founded.[40] Although active from the network's beginning, women increasingly came to the fore, particularly in California where that state's rendezvous included a number of female speakers and performers as well as a women's caucus.[41]

Amidst continuing tensions over the direction of the network and concerns that the editorial group running the *EFJ* was failing to give NVDA enough prominence, 33 took part in an EF! activists' conference in Colorado. Some resolutions, such as setting up an internet-based bulletin board, did not come to fruition, but information sharing increased. A push for greater coordination of ODA with legal strategies and closer cooperation with other activists did not manifest across the entire network, but this would increase in coming years, particularly as EFers began to set up their own Grassroots Biodiversity Groups.[42]

The increase in EF! activity overall was reflected in the level of environmental blockading and the spread of its existing body of tactics. Lock-ons involving chain and kryptonite locks began to be more regularly used during office occupations and were also employed during forest blockades in Oregon and Washington. A blockade early in the year against logging on private land in the Skagit Delta, near La Conner in Washington, built on this through the first combination of barricading with lock-ons. This campaign brought together concerned locals, some of whom were members of nearby communes, with EFers from other parts of Washington. Blockading, litigation concerning Native American burial sites, and the loss of road access, due to the use of environmental laws related to the discovery of an eagle's nest, held up logging for a month. During a two-week vigil an old car was towed onto the site and piled up with tires and board. When police arrived an activist slowed down the vehicle's removal by chaining himself to its axle. In an ironic twist one of those detained in the local jail after taking part in a road occupation was the architect who had designed it.[43]

Mike Jakubal was one of the roving EFers present during this campaign and in late October he and fellow EFer Jeff Miller helped to fully introduce barricading into the American tactical repertoire. The opportunity arose during three days of blockading at Cahto Peak in Mendocino County, California, an area with a heavy concentration of back-to-the-landers. Logging on a parcel of land under BLM control had been opposed for a decade based on threats to Native American sacred sites, old-growth Spotted Owl habitat and water quality. Another action, involving up to 150 people attempting to stall logging ahead of a court case, was occurring simultaneously in the same region at Goshawk Grove. As a result, only 20–30 protesters were available. Favourable terrain however lent itself to the use a new set of tactics.[44] As Jakubal recalls:

> The only way in was via an old and very narrow logging road. They had just reopened it so there were piles of brush, boulders and stumps and debris around. On one side was a super steep drop off and a cut bank on the other so the whole road was just one long tactical choke point.
>
> I had been reading all these stories of revolution and insurrection from around the world including May '68 and the Spanish anarchists and the first thing they all did was build barricades in the street. It's a statement, it's a presence and it provides cover from whichever side of the road [opponents] come from. The need for EF! to be building barricades was totally in my head. It was an action looking for a place and what do you know, here was the spot.[45]

Earlier debates regarding minor levels of sabotage resurfaced here. According to Jakubal:

> Consensus regarding what we should do totally bogged down as there were a couple of militant pacifists present who argued that barricading would be violent and a form of property destruction, property destruction and "violence" being forbidden by the nonviolence code. The meeting ended up going on until the early hours of the morning, by which time I'd gone to

bed. There were only three people who were willing to be arrested and it was eventually agreed that it would be okay for there to be one log on the road if they sat down in front of it.

I hadn't been part of the formal "consensus" decision so decided to go off first thing in the morning with a friend as an affinity group of two. Our logic, admittedly somewhat intellectually dishonest, was that since we didn't consense, we weren't bound by others' decisions. We started piling up as much stuff as we could once we got far enough up the road that we were out of sight of the other protesters. The loggers came and left and then a local marijuana grower demanded we move the protest and barricades off private land up to where the BLM land was. By this time something had shifted and the majority of the group decided to join in. Suddenly we had this big happy direct action family of hill people, Cahto tribespeople and EFers pushing boulders and logs into the road that the two of us could never have managed.[46]

According to a report by Daryl Cherney the following day "a convoy of Mendocino County sheriffs, federal marshals, CAMP goose steppers (Campaign Against Marijuana Planting), BLM officials, Louisiana-Pacific security (L-P has adjacent holdings), and California Highway Patrol men rolled up the Jack of Hearts Road."[47] Despite this show of force, and one arrest, logging was stalled by a combination of tactics. As Jakubal recalls:

Among our 20 people we only had a few voluntary arrestees, but the cops didn't know that. We stood in their way and kept telling them that they'd have to wait until Daryl Cherney and Judi Bari had finished singing songs and then we began holding consensus meetings about everything. We kept ignoring the cops until they interjected themselves. For once this tedious decision making process came in really handy (laughter).

When they eventually forced us to go we had an [elderly] woman with us so we walked as slowly as possible with 4WD police vehicles creeping along behind us. We had a big banner that spanned the width of the road so the cops couldn't see what was ahead. Each time we reached a barricade, we'd lift the banner over it, then stop a short way up the hill behind it while the deputies dismantled it by hand. Then we'd take more time to get moving again. This slow motion cat and mouse game went on for hours.[48]

Although media coverage had been minimal, logging was successfully disrupted at Cahto Peak and stopped entirely at two other sites. Following two days of blockading, which included protesters being assaulted and sprayed with mace, EPIC gained a TRO on 27 October at Goshawk Grove. News that Maxxam, likely hoping to exploit the fact that activists were busy elsewhere, had begun logging at All Species Grove also led to a physical intercession by two activists which forced the company to honour an earlier court ordered halt to work.[49]

Buoyed by these successes and the efficacy of their new tactic, EFers at Cahto Peak barricaded through the night, adding deep ditches, and once more employed the tactic of slowly walking between each pile. In the process of dealing with this authorities were informed that operations at Cahto Peak were also to be suspended as pressure from the Cahto Tribe and Democratic Senator Alan Cranston had forced the BLM to carry out further investigations into historically and spiritually significant sites.[50] These victories convinced Californian EFers that public campaigns combining ODA with more conventional forms of campaigning promised likely success.[51]

Barricading had occasionally featured at previous EF! blockades, but thereafter became a consistent and central part of the US repertoire, spawning a series of adaptations of its own. Objections to the tactic on the grounds that it breached the parameters of nonviolence soon faded as it did not trigger the kind of backlash tree-spiking had.[52]

Northern Californian activists continued to use soft blockading and tree-sits against Maxxam's Headwaters operations in All Species Grove in four separate actions during 1988 and also introduced the tactic, likely drawn from the peace movement, of attempting to perform citizens arrests on loggers they considered were performing illegal acts. They promoted blockading's publicity value to others in the network with Greg King claiming that

> Major stories in the *New York Times,* the *Los Angeles Times,* the *Philadelphia Inquirer, Newsweek,* nearly all San Francisco Bay Area publications, countless television news programs, and hundreds of other radio and print accounts, would likely never have been written if not for the Earth First! campaign.[53]

With smaller numbers of activists available than in California, tree-sitting was the primary tactic deployed in Oregon's Kalmiopsis against "salvage" logging. The issue of whether logging crews should be able to remove timber from forests recently burnt by wildfires was a highly divisive one in the state. Having encountered ODA during consecutive summers the FS and other authorities put in a major effort to counter it.[54]

Despite roadblocks, patrols, surveillance by helicopters and the deployment of paramilitary Forest Service officers, ostensibly armed and trained to counter marijuana growing, activists carried out three waves of tree-sitting actions. The use of hired climbers to cut down platforms and provisions limited the ability of activists to remain in place. For the first time in two years the FS was able to actively remove a tree-sitter by sending climbers up while an officer trained a rifle on the activist and ordered him not to resist.[55]

Following on from protests in San Bruno the previous year, environmentalists in Florida and Texas used ODA to counter threats from urban development. In the former, city authorities called a halt to the clearing of a 45-acre tract adjoining the Paynes Prairie nature reserve after a week-long occupation of the area by up to 100 protesters culminated with them blocking equipment with their bodies and climbing and chaining themselves to trees.[56]

In Texas, a highly active EF! chapter in Austin would combine strategies involving ODA, litigation, lobbying, and intervention in city elections over the coming years. Texan activists had previously pioneered the tactic of using kryptonite locks to attach body parts to machinery. This was effectively neutralised in 1988 during an April action in which three activists locked onto equipment to disrupt the clearance of rare bird habitat on the city's fringes. Illustrating the diversity of responses across the country, rather than drill out the lock mechanism or remove a section of machinery, police riskily used hydraulic rescue tools known as "the jaws of life" to pry apart the devices while the activists' necks remained locked within them.[57]

As noted earlier, alongside other influences a key determinant of tactical development can be the nature of the terrain involved. This was clearly illustrated when Austin EFers seeking to interrupt the blasting and cementing up of caves for suburban development near Barton Creek adapted the common tactic of occupying space to create manufactured vulnerability. Although the habitats and animals involved, five species of insects and spiders, were not as iconic as ancient forests or Spotted Owls, they were critically endangered and thus potentially subject to safeguards under federal law. Following four years of lobbying, and with developers poised to act before the Federal Wildlife Service completed the process of giving the animals formal protection, EFers entered three caves from 29 August. Having taken precautions to ensure minimal impact to the fragile ecosystem, one occupier was forced to leave after police threatened to spray tear gas into the cave. By rotating stints and using radios to communicate with supporters the other caves were held by at least four people at a time for 12 days. Exhaustion on the part of the small group brought an end to the occupation, but the publicity generated led the Federal Wildlife Service to speed up its investigation dramatically with the result that an emergency listing was granted.[58]

ODA accelerates: 1989

The pace of diffusion increased during 1989 with at least 27 ODA actions carried out at the point of destruction against mining, logging, and urban development operations in 13 states. EF! organised a national week of tree-sitting to publicise the scope of old-growth forest destruction. This introduced the tactic to Montana, New Mexico, Massachusetts, and Colorado. Beginning on 13 August, this event, dubbed "Save America's Forests", was auspiced by the EFDAF and also involved actions in Washington, California, and Oregon.[59] Activists in the latter two states carried out further blockading during the year. Similar actions occurred in Illinois, North Carolina, New Mexico, Hawaii, and Los Angeles for the first time. In Arizona work at Mount Graham was disrupted by a 12-day vigil and two separate lock-ons to road grading equipment, marking the first incidents in what would be a long campaign by environmentalists and Native Americans to prevent the clearing of forest for an Astrophysical facility.[60]

In Colorado a rare combination of a blockade and monkeywrenching occurred. There, a tree-sit and a lock-on to an entrance gate, as well as extensive tree spiking and $10,000 worth of damage to a logging vehicle, were credited with bringing logging at Bowen Gulch at a halt.[61]

In keeping with previous years, almost all of these blockades were intermittent and lasted only a day or two. Sometimes actions were deliberately planned this way, but in other cases their short duration was due to work being rapidly completed and/or all available activists being arrested.

Despite their inability to commonly close down operations for extended periods, activists continued to blockade for a variety of reasons. These included the common strategic aims of drawing attention to an issue, costing opponents time and money, and providing a radical flank for moderates. As the use of "paper wrenching", as EFers dubbed bureaucratic appeals, and litigation grew, activists also hoped to stall work ahead of potentially favourable legal and bureaucratic outcomes. Politicians had long been targets for the pressure protest could generate, but activists believed the controversy and difficulties associated with ODA could also sway judicial decisions. Such goals often remained uncoordinated and vaguely defined.[62]

In keeping with what Jasper describes as "tactical tastes", the form and goals of ODA also reflected the "artful, strategic component" of campaigning and the pleasures of action since "protesters do what they are good at, what they can do creatively or freshly".[63] In relation to tactical persistence, scholar Geoff Larson further notes that "once affixed to an identity, a tactic takes on totemic significance that can override concerns for efficiency and efficacy".[64] Ideas concerning the latter can also be based on "incomplete and potentially misleading information ... shaped by organizers' abilities and resources to gather and interpret information, their previous experiences, historical contexts, prevailing cultural models, and relations with other social movement groups."[65]

Although some short environmental blockades were primarily unconventional demonstrations of opposition with little ongoing impact, many successfully disrupted work, served wider causes, and initiated activity that would lead to more extended protest in the future. An example of this was in Southern Illinois where ODA was deployed against old-growth logging in the Trail of Tears State Forest. On Thursday, 28 September 1989, a group of 20 activists blocked two entry points with a car body and a kryptonite lock-on to a gate as well as by standing in the road and entwining their arms. This followed on from a rally the previous weekend and a tree-sit that had been held earlier in the year.[66] Work was stopped for half a day and the protesters immediately raised the profile of the issue, as recalled by Native Forest Network co-founder Orin Langelle:

> The common perception throughout Illinois was that there was no logging going on in state forests. Our group of EFers discovered that there was a tremendous amount and we wanted to bring attention to it as well as stop it... This kind of thing was totally new to this part of the country and the media

was shocked to see police and protesters in the forest. It was the lead story on TV news in Illinois and Missouri and provoked a great deal of outrage.[67]

Media coverage fed into a set of political opportunities that favoured local and state-wide support for conservation. Logging in state forests was not a major contributor to the economy or local tax base and most of the work was being done by workers from neighbouring Missouri. The previous year the FS, whose timber sales in Illinois were loss-making ventures, had been pressured into virtually eliminating clear-cutting for five years on the federal lands it controlled in the Shawnee National forest. By 1989 bids to extend these gains and obtain a moratorium on all logging in Illinois had largely stalled. Buoyed by their success in drawing attention to state logging, EFers would greatly extend their efforts the following year by launching an 80-day blockade of logging in the Shawnee.[68]

Illustrating the way in which emotional factors, historical associations, and ideological predilections feed into tactical and strategic decisions Langelle emphatically states that ODA was not carried out with the sole intention of generating publicity:

> All of us who came together in the Illinois chapter believed in the philosophy of direct action and that direct action was the way to accomplish things. Many of us in the chapter were anarchists, but there were also back-to-the-landers and other environmentalists. There were other people doing lobbying and that sort of thing, but from our point of view logging was going on then and there and we had no choice but to step in between the loggers and the trees. It wasn't a media stunt, it was deep and heartfelt. This wasn't just beautiful forest but was the area through which the Cherokee people had been forced to march into mass incarceration [during 1838–9].[69]

At times blockading was of short duration due to immediate success. An example of this came when developers and city authorities agreed to renegotiate the clearing of remnant forest in Los Angeles' Caballero Canyon in early June. This followed a series of resident protests that climaxed with an occupation of the site involving EFers and others chaining themselves to bulldozers.[70]

As with other EF! chapters, the Los Angeles group had previously carried out office occupations and demonstrations covering issues that ranged from cattle ranching to nuclear weapons and the hunting of wolves. Commentary from TV news reporters about these activities illustrate common tensions in regard to mainstream media coverage of protests and ODA in general. Asked about the criteria he applied to choosing stories involving EF! and other radicals Channel 8 TV news director Jim Holtzman argued that he was at pains to avoid taking part in "publicity stunts" stating, "When groups like [Earth First!] say they'll be at a certain place at a certain time you got to ask yourself 'Is there a story only because we're going to be there, or would there be a legitimate story to report even if we're not there?'"[71]

As during other ODA-based protests, activists wishing to remain newsworthy needed to meet demands for novelty and conflict while simultaneously attempting to steer content and frame their story as "genuine" rather than "inauthentic". If they failed to balance these demands they not only risked having their key messages swamped by controversies related to protest means rather than environmental threats, but also not being covered at all.

Examples of activists consciously designing actions as short "hit and run" events based on their assessment of available resources can be seen in two actions held in the Jemez Mountains of New Mexico during 1989. The first in April targeted a pumice mine in the Santa Fe forest and involved three protesters, supported by a group of 16 others, locking themselves to a bulldozer, backhoe, and log skidder before being taken to jail. While the use of kryptonite locks maximised the disruptive potential of the small number of local protesters present, the next action in June, carried out after EF's annual RRR was held nearby, involved over 100 people. This time protesters were able to close down cable logging operations through the use of road occupations, barricading and lock-ons at multiple points. Despite the numbers involved, the fact that few could remain in the area ensured that this too was a one-day action. Following its success local activists hoped "that a protracted campout be considered in the future". Lacking another event with the ability of EF!'s national gathering to bring in activists, this was not realised.[72]

The year's longest forest blockade was carried out in Oregon's North Santiam River watershed in the Willamette National Forest. The location of the alternative Breitenbush Community near the logging site had enabled activists to carry out a week of blockading three years previously. With local support to draw on and the protest coming over the 1989 Easter break, up to 50 people carried out daily blockading until logging was completed six days later. While the protest slowed rather than ended logging, those involved claimed success on the basis that it received coverage on national and international TV networks as well as in national publications such as *Life* and the *New Yorker*. Although this had been an unusually sustained protest, Australian examples were still clearly on some EFer's minds with one participant writing "The goal is to work toward truly large-scale, Franklin River-type blockades which, considering the ever-increasing interest in the ancient forests, may not be too far away."[73]

During the blockade activists extended tactics devised in recent years with a group of six using interconnected kryptonite locks to chain themselves by the neck around a tree. Trenches and barricades made of debris were repeatedly used to obstruct access and were augmented by a protester buried up to his neck in rocks. A bonfire was lit on a bridge, which further disrupted traffic to the logging site as the crossing had to be subsequently tested for structural soundness. Despite an incident involving monkeywrenching, relationships were cordial enough for 20 loggers to join protesters around the bonfire while waiting for the road to be cleared.[74]

Three months later, an action involving roving activists in Oregon's Kalmiopsis included further innovation when, during a two-day action, three activists combined a road occupation with barricading by cementing their feet into a road. Frustrated police arrested all in the area, whether trespassing or not, and threatened to shoot off a kryptonite lock.[75]

Political developments, 1989–1990

The successful use of litigation during 1989 by local and regional environmental groups brought contradictions to a head within the FS and other agencies regarding their mandate to preserve habitat while facilitating profitable timber operations. Beginning in 1987 the Sierra Club Legal Defence Fund, an organisation which had split from the nationally dominant Sierra Club in the 1970s, began representing local conservationists in a series of lawsuits concerning the management of owl habitat. On the basis of provisions protecting endangered species US District Court Judge William Dwyer eventually issued an injunction in 1989 against the FS's sale of much of the old-growth federal timber in Oregon and Washington not already protected by a prior suit against BLM.[76]

With their financial supporters in the timber industry facing a major shutdown, Senators Mark Hatfield and Brock Adams pushed through a new federal rider. This was the ninth time they had used such means to exempt timber sales and federal bodies from statutory requirements. The amendment, passed in October 1989, referred to the importance of old-growth forest and required the FS to produce a plan for owl management, for which it drew praise from the mainstream organisation that had helped insert such provisions, the Sierra Club. Its main impact however was to nullify recent protections by exempting 1.1 billion board feet from the Dwyer injunction and a further 8 billion from other lawsuits for a year. In doing so it allowed most of the 1990 timber sales to go ahead. However, the controversy associated with the rider, and the political capital expended in getting it passed, meant that similar means would be unavailable again for years.[77]

Such legislative moves, continuing intensification of logging on private lands and frustration at the piecemeal approach to protecting forests, convinced Californian environmentalists to take a different tack and seek reform of the state's laws. In September 1989 a coalition of organisations launched the Forests and Wildlife Protection Initiative and Bond Act of 1990, which would become widely known as "Forests Forever". This sought electors' approval to raise $742 million to buy the Headwaters forest from Pacific-Lumber, ban most clear-cutting, protect all old-growth forest and slow timber harvesting to a rate at which trees could be sustainably replaced.[78]

Successfully gaining the 600,000 signatures required for it to be put to a referendum during 1990's November state election, the proposal triggered a major industry backlash. Creating a counter-initiative titled the Global Warming and Clear Cutting Reduction, Wildlife Protection and Reafforestation Act the timber lobby attempted not only to confuse voters, but also realise their own long-term goals of almost total deregulation. In the midst of the Federal Wildlife Service finally designating the Spotted Owl a threatened species, the industry's proposals were supported by a major upsurge in "Yellow Ribbon" rallies and campaigning, which claimed that Forests Forever would cost over 176,000 jobs. EF! had not sponsored Forests Forever, but as the radical edge of the conservation movement its members became the focus of opposition. Harassment, death threats, and attacks on EFers became common in 1989 and 1990 with, as will be discussed later, near fatal consequences.[79]

FBI operations, environmentalist–worker alliances and anti-environmentalist violence

Growing public support for action on ancient forests, and the escalating debate regarding related issues, affected EF! internally in a number of ways. For some, frustration with a lack of overall strategy, negatives associated with the network's image, and a sense that change via conventional means was increasingly possible led them to either join more moderate organisations or apply EF's confrontational approach to conventional means via Grassroots Biodiversity Groups. For others it consolidated their view that environmental blockading could play a vital role, coordinated or not, within campaigns employing a suite of tactics. Confidence among those in Northern California grew to the point where they announced plans for an event named 'Redwood Summer' in February 1990. Drawing inspiration from the iconic civil rights "Mississippi Summer" voting drive of the 1960s, EF!'s first mass campaign of rolling protests called on activists around the country to join in two months of educational activities, demonstrations, and NVDA from July onwards.[80]

The call for Redwood Summer came at a time at which the network was not only being pressured by industry opponents, but also the Federal Bureau of Investigation (FBI). On 31 May 1989, paramilitary agents arrested three Arizonan monkeywrenchers for attempting to cut down power poles supplying electricity to an irrigation project. Five activists would eventually be indicted and four jailed following what had been a lengthy $2 million operation. This had involved infiltration and monitoring via recorded conversations, wiretaps, and aerial surveillance.[81]

The operation was ostensibly designed to bring a halt to a campaign of sabotage but was also used to target EF! more widely through attempts by undercover agent Mike Fain to draw EF! founder Dave Foreman into the group's activities. Arrested in a raid on his home on 1 July Foreman's case was eventually downgraded and separated from the others as he had not directly taken part in any of their actions.[82]

During the trial, which concluded in 1991, a conversation inadvertently recorded between Fain and another agent was released. During this the infiltrator, referring to Foreman, admitted that "This isn't the guy we need to pop in terms of actual perpetrator", but was vital in terms of him being "the guy we need to pop to send a message, and that's all we're doing … If we don't nail this guy … we're not sendin' any message …"[83]

The message most EFers, and those engaging in sabotage outside the network, appear to have received was not to abandon their activities, but to tighten security. History concerning FBI operations designed to infiltrate and disrupt a variety of radical organisations was drawn upon in activist discussions and articles. Despite warnings that increased distrust and paranoia could weaken the movement, and thereby aid its opponents, these responses became widespread.[84]

Raids on EFers in Montana and the subsequent serving of grand jury subpoenas to seven activists in connection with alleged knowledge regarding tree-spiking followed in October 1989 as did overt FBI monitoring and questioning of activists in a variety of states. These impacted heavily on those involved and along with media coverage of the Arizona arrests reinforced public perceptions of EF's connection to ecotage.[85]

By April 1990, several prominent EFers based in Northern California became concerned enough about the negative implications of tree-spiking to publicly release a statement renouncing its use. Most of these had previously disassociated themselves from the practice. However, in the build up to Redwood Summer they felt compelled to publicise their position as widely as possible. These activists' critique of tree-spiking held that it was ineffective in imposing costs as "Through the coalitions we have been building with lumber workers, we have learned that the timber corporations care no more for the lives of their employees than they do for the life of the forest."[86] Arguing that spiking was divisive harmful to environmentalists in terms of inviting aggression from timber communities they released a public statement declaring:

> In response to the concerns of loggers and millworkers, Northern California Earth First! organizers are renouncing the tactic of tree spiking in our area …These companies would think nothing of sending a spiked tree through a mill, and relish the anti-Earth First! publicity that an injury would cause.[87]

As the statement made clear these EFers' position regarding normative protester behaviour was largely a result of their attempts to build a new repertoire of action within the network – that of supporting direct action at the point of production. This reflected their identification with their fellow residents and workers in Northern Californian communities as well as a willingness to experiment with a suite of strategies to change the economic direction of the region. Local union activists and members of EF!, primarily led by prominent activist Judi Bari, formed a Northern Californian chapter of the Industrial Workers of the World (IWW) in 1989. The radical, syndicalist union had enjoyed widespread support amongst timber workers 70 years earlier but had not existed in the area for decades. This history and the organisation's radical politics led activists to believe it could be a vehicle for genuinely addressing issues commonly faced by environmentalists and timber workers. Throughout 1989 and 1990 radical environmentalists in the IWW worked with and supported a small number of dissident timber company employees. Support was given to campaigns and legal challenges aimed at improving working conditions and addressing toxic hazards, workplace deaths, and injuries, as well as pressuring existing union leaderships to take more effective action.[88]

The potential to link up with workers existed on the basis of aforementioned timber community hostility towards employers during increasing precarity and outsourcing. There were occasional strikes and other protest actions during 1989 and 1990 regarding company policies, but unions for the most part were either weak, pro-employer or non-existent. Most importantly for environmentalists, some timber workers continued to express an understanding that overcutting was threatening their long-term interests.[89]

These attempts to forge alliances largely failed to overcome mutual distrust and cultural differences. The potential impact on jobs that decreases in logging threatened and the immediate lack of viable economic alternatives beyond low-paying and insecure tourism work mitigated against cooperation. Existing divisions were exploited by employers and the message that environmentalists were malicious outsiders intent on destroying their livelihood continued to resonate with many in timber communities.[90]

Among environmentalists there were also major divisions. Many within EF! refused to countenance any suggestion that working with loggers could be ethical or tactically advantageous. Sea Shepherd founder and long-time tree-spiker Paul Watson was not alone in equally condemning all involved in the industry, pronouncing "the logger is a rot, a disease and an aberration against nature".[91] While some in the network made statements in favour of economic reform to create alternative jobs, few prioritised or invested in outreach beyond this.[92]

While repudiating tree-spiking the Northern Californians' continued support for other forms of ODA also exacerbated divisions with workers. Blockading carried out in the region before, during, and after Redwood Summer involved obstruction and as such created polarisation and antagonism. As Sierra Braggs writes in her Master's thesis regarding the evolution of EF! ideology and repertoires, "During encounters facilitated by direct action, the worker, who may have seen him/herself in opposition to Maxxam Inc. as an exploited labourer, then becomes the logger, who sees his/her work and lifestyle threatened by [EF!]."[93]

Despite this dynamic, Californian activists did not resile from their support for environmental blockading. They continued to employ NVDA in the years that followed in the belief that it would generate media coverage, slow destruction, train activists, build campaigns, and ultimately protect forests while continuing to maintain that tree-spiking could not. Roselle argues that this was because while they sought alliances with timber workers Bari and her supporters were "unwilling to compromise even one redwood to gain their support".[94]

Messages regarding nonviolence and common interests were also muddied and undermined by statements, satirical songs, media interviews, leaflets, and images released by those involved before and after the official renunciation. These included a 1988 photo of Judi Bari posing with an Uzi machinegun, which spoofed an infamous photo of Patty Hearst, and the lyrics to the 1987 Daryl Cherney song "They Sure Don't Make Hippies Like They Used To", which lauded tree-spiking. Further to this the renunciation of tree-spiking did not extend to other forms of ecotage so long as they took place away from the site of NVDA.[95]

Movement humour and distinctions between different EFers and their changing opinions concerning appropriate behaviour were generally not understood by those outside the network. Satire was taken seriously and ODA and minor property damage widely equated with major sabotage and tree-spiking. Any statements, regardless of the source, which potentially or tacitly endorsed monkeywrenching were exploited by EF!'s opponents. Where these did not already exist, they were created and a series of fake press releases and leaflets calling for violence against timber workers were distributed in California.[96] Efforts to forge alliances between workers and environmentalists over the issue of log exports in Olympia, Washington, were similarly undermined by the distribution of forged EF! minutes regarding sabotage.[97]

Northern Californian EFers had believed that a renunciation of tree spiking would help reduce the violence they faced, but this was unfortunately not the case. In May 1990, Judi Bari and Daryl Cherney were the victims of a car bombing in Oakland, California, which came close to killing them. Bari, whose pelvis was fractured and tailbone crushed, was hospitalised with critical injuries and would suffer pain and difficulty walking for the rest of her life. Carried out by persons unknown, the attack removed three key EF! organisers, as Greg King pulled out of campaigning in fear of his life. Ironically it bolstered EF!'s reputation as "ecoterrorists" as Bari and Cherney were subsequently arrested on the basis of accusations by the FBI and Oakland police that their injuries had been incurred in the transportation of a bomb. The charges were dropped during July 1990 and the FBI forced to pay out $4.4 million for misconduct twelve years later.[98]

The bombing scuppered EF!-IWW attempts to build alliances with timber workers, but they would continue in the future via organisations such as The Alliance for Sustainable Jobs and the Environment.[99] Although the bombing spread alarm throughout EF!, Redwood Summer went ahead in its wake as others stepped into key organising roles and the campaign received new support from other organisations and movements. Solidarity rallies were held around the country and Illinois activists carried out an action in which they visited the FBI to turn in their weapons. These included typewriters, telephones, and a tyre.[100]

Environmental blockading and the Earth First! split

The decision of Californian EFers to launch the network's first mass campaign and the debates discussed above fed into and reflected other divisions, which ultimately led in 1990 to the departure of key figures from the network. Although there were a variety of positions, many of which were fuzzily defined, the primary differences concerned appropriate tactical repertoires and the type of issues the movement should focus upon. EF!'s rapid growth had largely drawn on those either belonging to or oriented towards feminist, left-wing, and anarchist movements. Such activists had been involved in the network from its foundation, but their influence and the expansion of the network made those with a narrower and more socially conservative agenda uncomfortable. The 15,000 subscribers to the *EFJ*, the level of campaign activity, and the amount of media interest devoted to the split all signalled that the network had moved past being what Foreman had envisaged as a "kamikaze or sacrifice group that would introduce new tactics and then disappear".[101]

The *EFJ* editorial staff, Foreman, and other members of what became dubbed the "old guard" held that the network should primarily be concerned with issues explicitly related to biodiversity and wilderness. As part of preventing humans from impacting on such ecosystems they generally advocated that activists continue to carry out monkeywrenching, including tree-spiking. Due to its radicalism and lack of resources some held that EF! would best play a catalyst role in drawing attention to an issue and creating a radical flank before mainstream groups came in to complete the process of change. There was also a fear expressed that an outward looking focus and too rapid a growth could lead to compromise and a weakening of commitment to biocentrism as well as require that a "bureaucracy" form within the network.[102]

Those among what became variously known as the "humanist", "social justice", and "leftist" grouping argued that economic and social relations rather than humans per se were the main cause of environmental destruction. As the fate of the environment was inextricably tied to whatever direction society took, addressing such determinants would require EF! to connect with a range of issues. Some of these activists also rejected what they characterised as an individualist "macho" and "saviour" mentality, rooted in Edward Abbey's novel *The Monkeywrench Gang*, and asserted that it would take more than audacious and daring acts by a hardy few to protect biodiverse areas. Instead, they argued that EF! needed to ally itself with other movements, such as those addressing gender inequality, Indigenous sovereignty and class exploitation, and build its own campaigning capacity rather than cede that ground to narrowly focused and reformist mainstream organisations.[103]

Given that these positions were largely predicated on differing viewpoints concerning the role of humans in ecological destruction, it is unsurprising that a series of misanthropic articles and opinion pieces published in the *EFJ* and elsewhere became a lightning rod for debate. For the "old guard" overpopulation was a key issue and this was reflected in works written by Foreman, Christopher Manes, and Edward Abbey in 1986 and 1987 that argued for restrictions on immigration into the United States and satirically celebrated famine and the spread of AIDS. Articles and statements that followed in a similar vein carried an apocalyptic tone with EF! activity intended to slow down society's destruction of wilderness before its inevitable crash, rather to enact changes to avoid such a cataclysm.[104]

Social-justice-oriented activists found these positions offensive and also argued they were strategically flawed as they misdirected attention from corporations and elites. Those in EF! who, in Roselle's words, "no longer wanted to be defined by the antics of its founders", were concerned at the effect such articles could have on attempts to work with activists from other movements. Debates over the use of patriotic symbols and incidents in which EFers clashed over racism and whether to fly or burn the US flag at a RRR further exacerbated divisions. Following much debate in the *EFJ* and various meetings, control of the journal moved over to those more sympathetic to social justice concerns and several "old guard" members left the network.[105]

Previous analyses have primarily focused on individual personalities and the ideological component of these developments. This is exemplified in works produced by Martha Lee and Susan Zakin. Both tend to ignore what most of the network's membership was doing in terms of campaigns, blockades, and protests and how this played into disputes. Where work has focused on the interplay of ideology and practice it has overemphasised the role of Northern Californian activists, overlooking how their activity built on and consolidated that previously carried out elsewhere as well as the close connections between activists from different states and the movement of activists between them.[106]

In addition to the factors discussed above, it is important to recognise that dynamics also existed in which the practical demands, use and outcomes of differing repertoires affected and promoted broader ideas and currents within EF! This can be clearly seen in the turn against tree-spiking by those who had previously supported, if not employed it. Prominent EFers such as Roselle, Bari, and others primarily attributed their changing position to everyday experiences of forest activism. Out of this they claimed to have come to the conclusion that NVDA, worker outreach, and working in concert with those employing aggressive litigation and public campaigning were more efficacious strategies than tree-spiking, regardless of the role it may have played in bringing attention to the destruction of forests. Drawing on practical examples they argued that the needs of various strategies, in terms of recruiting, public outreach and avoiding violence, were being undermined by spiking.[107]

By 1990 a desire to use different tactics and reach beyond the existing cohort of EFers and their immediate circles, as well as to avoid negatives associated with tree-spiking and controversial statements in the *EFJ*, had already led some activists to leave the network or operate under another name. This further stimulated activists within EF! to address what they saw as the source of such losses.[108]

The use of blockading and other repertoires associated with public campaigning had also played a major role in recruiting those with a social justice bent to the network, tipping the balance in their favour. Unlike clandestine activity – which involves individual or small cell activity and requires tight security – the dynamics involved in blockading and campaigning encourage activists to publicly gather resources and continually recruit new members in order to overcome the use of arrests and other counter-tactics. Since environmentalism was primarily associated with progressive milieus, and most EFers employing NVDA in the United States already belonged to them, it is hardly surprising that recruitment among this sector was most fruitful. Those wishing to build large-scale public campaigns believed that newcomers would be attracted by the opportunity to radically and publicly participate in successful environmental activism in ways that neither the more controversial and clandestine practice of monkeywrenching nor the reformism of mainstream environmentalism allowed.[109] As such, the use of blockading and the influx of social-justice-oriented activists can be seen as mutually reinforcing trends.

Arguments have been made by some former EFers, as well as scholars and writers such as Lee and Zakin, that the original or "true" EF!, effectively ended in 1990.

From this point of view, a "social justice" oriented network wholly supplanted that which had been founded in 1982.[110] This interpretation is simplistic and overly dichotomous, reduces the network's early identity to that of the minority who exited, and ignores continuities in membership and ideas. As was argued by activists at the time, the debates and changes canvassed above marked an evolution in direction rather than a decisive break with the past. The humour, songs, and confrontational attitude of EF!'s founders remained extant. Some members, including those active in blockades and opposed to the "old guard", did not renounce monkeywrenching and the *EFJ* continued to run stories concerning it, albeit to a lesser degree than before.[111]

Debates concerning appropriate tactics, controversial material in the *EFJ* and related issues did not end with the split and the *EFJ* remained overwhelmingly focused on environmental issues, particularly those involving wildlife and biodiverse areas. Despite an increased orientation towards social justice and NVDA, EF! continued to include activists with a wide gamut of cultural orientations and spiritual, political, and tactical positions.[112] Amid predictions that the "new" network would soon wither or dissolve into a wider leftist milieu, EF! launched its biggest ODA-based campaigns during 1990 and continued on as a distinctive force within American environmentalism.

1990: The spread of environmental ODA continues

In addition to the major campaigns of Redwood Summer and Illinois's Shawnee blockade (covered below), the geographical spread of environmental blockading continued. NVDA actions including lock-ons and tree-sitting took place in South Carolina for the first time, successfully bringing road building in the Sumpter National Forest to a halt. In Hawaii, initial ODA in 1989 grew to include successful protests of up to 1,000 people against the clearing of the Wao Kele o Puna rainforest for a geothermal powerplant.[113] In New York state, two activists used a boat to delay the poisoning of a pond and Montanan activists held their second tree-sit.[114]

In Austin, urban development and associated road building remained a focus. Texan authorities appear to have not been sharing information with each other as during a protest on 5 January officers failed to follow the previously successful use of hydraulic equipment. Instead, they improvised an equally risky countertactic by wrapping protesters' heads in spark-proof, bullet proof vests before using a grinder to remove kryptonite locks. By the time this solution was found those fastened to equipment had gained publicity and held up work long enough for litigation to secure an injunction against clearing on the city's outskirts.[115] The following month activists claimed that preparations for blockading had provided the final pressure required to persuade city authorities to prevent the removal of rare bird habitat.[116]

In northern Oregon two separate days of blockading involving lock-ons and road occupations targeted the expansion of logging under the Hatfield Rider with a May action bolstering negotiations by non-EFers that led to the curtailing of logging near Forest Park. Northern California saw various demonstrations and actions carried out prior to Redwood Summer, including a lock-on action to prevent the aerial spraying of forest with pesticides and the blockading of trucks carrying redwood timber. Following Redwood Summer blockading against clearing for road construction was also carried out.[117]

A campaign in Washington specifically responded to the effects of Senators Mark Hatfield and Brock Adams' legislative overriding of environmental protections. The dispute concerned logging in the 9,495-acre Cedar River Watershed, which provided 1.2 million Seattle residents with water. City authorities had historically sold timber from their share of the area, but in 1989 environmental and health concerns had led them to reduce supply. Although mainstream environmentalists had capitulated, the Mayor's office challenged the expansion of FS harvesting to meet newly mandated federal targets.[118]

The dispute came to focus on the logging of 73 acres of Spotted Owl habitat. In late May, three days before negotiations were to begin, the FS surprised the City by closing the area and allowing the Emunclaw logging company to enter. Under the name of the Cedar River Action Group (CRAG) radical environmentalists immediately responded with a series of rolling actions, which involved site and office occupations, truck blockades, and tree-sits.[119] Following further protest, and citing past tree-spiking incidents, Emunclaw deployed armed guards, dogs, and surveillance equipment including radar, remote-control cameras, and ground sensors.[120] CRAG distanced itself from tree-spiking and continued with occasional trespass actions. At the end of August, the company ceded its right to log the remaining 25 acres in the tract in return for access to second-growth forest elsewhere and the FS agreed to leave 3,500 acres of key habitat untouched for a minimum of three years.[121]

Although activists in other areas had already used blockading to complement other strategies, CRAG employed unusually close integration for the period. As Karen Coulter recalls:

> Although I can see in retrospect how they popularised concepts such as "old growth", many of the campaigns I was involved in that used blockading lost the battle they were fighting. The Cedar River campaign was quite different in that it was highly strategic. We chose that area because we knew that a lot of people were concerned about their drinking water. Although EFers started the campaign we used the name CRAG. We drew in new people and once we realised the group was heavily infiltrated we split our activities so that one affinity group was doing nothing but legal stuff like petitioning the city to take action, holding pickets outside council meetings, writing letters to the editor, etc.
>
> Meanwhile a separate affinity group planned for actions. When the time came the targets were specifically chosen to highlight different aspects that would help with what the first group were doing. We did a tree-sit on the

edge of the forest to show how beautiful it was, a lock-down at the Forest Service to show who was responsible, and then actions where we scattered through the forest during felling to slow things down and show what was happening with the logging. We had a reporter come in with us at night and he got arrested for being in a closure area so that showed what was going on with that. All of this kept the issue on prime-time TV and in the newspapers. By the time they had security set up so that we could no longer get into the forest we'd pretty much won.[122]

As part of the second in a four-year campaign to protect endangered species habitat, which had been exempted from federal regulations, activists returned to Mount Graham in Arizona where they camped from late September until mid-November.[123] Along with the Shawnee and Redwood Summer campaigns discussed below this was one of the first times that environmental blockaders had maintained a long-term protest camp in the United States.

Three days of sustained blockading in October, involving between 15 and 35 activists at a time, demonstrated how existing tactics were being combined to maximise disruption. On 2 October roads were blocked with timber and three-foot-deep trenches, culverts jammed to encourage flooding, and survey stakes removed. Road spikes made from reinforced steel and nails were used against vehicles for the first time in a US environmental blockade. In addition, backhoe tyres were let down, activists chained to trees and people ran through worksites. The locking of a gate necessitated its complete removal. Altogether work was delayed for eight hours before seven arrestees successfully secured their release and overcame high bail conditions by engaging in a hunger strike. Further work was disrupted as protesters, in the style of Cahto Peak, erected new barricades at night and walked slowly in front of vehicles during the day. Clearing and other preparatory work was eventually finished, but a week-long tree-sit and intermittent barricading and soft blockading continued to interrupt it for weeks.[124]

The completion of work despite litigation, lobbying and, as one activist put it, the use of EF!'s "entire bag of tricks" brought differing responses from those involved. According to an article in *EFJ* written by Erik Ryland, some pointed to positives such as "the energy of the activists, the monumental odds, the likelihood next year's actions will be even bigger". In contrast the author argued that the time taken erecting extensive barricades had produced a minimal level of disruption and called on EFers to "recognise that there are already mechanisms in place which will see to it that our moral indignation and righteous civil disobedience will be impotent". Despite encouraging his fellow activists to consider options beyond the "strategies of 60's civil rights and anti-war protests", none were detailed.[125]

The destruction of forests and other habitats to which they have become emotionally attached is distressing, if not traumatic, for activists. As in other movements and places, they react to the failure of militant tactics in various ways. In some cases, loss during this period led to a withdrawal from activism altogether. Among those who remained, some responded uncritically by, as Ryberg argued, satisfying themselves with hopes of future success. Alternatively, disruption was defined as a victory in itself with ODA essentially acting as an unconventional form of symbolic protest.[126]

Most, however, strove for tactical and strategic improvement. Some moved into mainstream groups or Grassroots Biodiversity Groups while others, particularly in the 1990s with groups such as the Earth Liberation Front, once more embraced monkeywrenching as well as insurrectionist politics. Many sought to maintain a long-term perspective by accepting losses and limits while using individual campaigns and blockading strategically to serve wider campaign goals. As Karen Wood recalls:

> There were instances where people said we had to stop the logging, period. That we had to do whatever it took to save a particular timber sale. The problem was, yeah, you could stop them, but for how long? How many people could you keep bringing in?
>
> Later on, in the 1990s people mounted blockades that halted logging for extended periods. Many of us felt in the 1980s that since we couldn't stop every timber sale physically, we had to stop them politically. Civil disobedience was useful because it pressured public officials and stalled things and made these practices public. All of that spurred a mounting sense of outrage. The logging of old growth went from being something that most people didn't even know about to being one of the most discussed issues in the Pacific Northwest. It got there because we were constantly out there making trouble, blocking work, and getting in the way.[127]

The Shawnee blockade

A more immediately successful environmental blockade incorporating a long-term protest camp occurred earlier in the year within Southern Illinois' FS controlled Shawnee forest. Members of the Big River and Southern Illinois EF! chapters had originally made a commitment to travel to Redwood Summer. After being approached by a member of a mainstream group about an upcoming timber sale they decided to build on their previous year's efforts and focus on logging in their own bioregion.[128] Conditions for protest were considered favourable as the Shawnee was the largest forest in Southern Illinois and a popular leisure place. Public opposition to logging was already substantial and pressure on the FS mounting as evidenced by Illinois Congressmen Glen Poshard and Richard Yates proposing, and then after ODA was underway, introducing measures to restrict logging on federal lands.[129]

Following demonstrations, an office occupation, and the holding of a regional rendezvous EFers reclaimed the proposed Fairview Timber Sale as a "hiking trail" on 20 June by setting up a protest camp. In the coming months blockaders worked closely with locally based Grassroots Biodiversity Group the Regional Association of Concerned Environmentalists (RACE). Despite sharing members, the two maintained a public separation of roles with RACE focused on securing a legal injunction while EF! continued their vigil and prepared for possible ODA.[130]

This was the first US environmental blockade to maintain a sustained protest camp directly in the path of an access road. The profile of the issue and the support of local environmentalists, as well as the determination and militancy of core occupiers, led blockaders to conclude that on this occasion they could safely maintain a long-term presence. Similar infrastructure to that seen at anti-nuclear protests and in Australian campaigns was put in place including a kitchen, toilets, and information tent. Informal consensus processes were employed for decision making with a flow of people fulfilling different roles and organisations such as Seeds of Peace and Food Not Bombs assisting with food at major events. Despite points of confrontation with authorities, a relaxed and at times festive atmosphere was generally maintained. Numbers at the site were at times as low as 15 with hundreds mobilised for specific protests.[131]

Core occupiers were largely drawn from Illinois and Missouri's anarchist and EF! circles. Some did not subscribe to the ideological precepts of non-violence or believe that it was tactically wise for activists to avoid violence in all situations. In this case however consensus was maintained around the need to adhere to the standard codes associated with EF! blockades in order to avoid negative publicity, maintain local support, and minimise confrontations. Activists were impressed by the outreach efforts of Bari and others but did not try to forge alliances beyond informing timber workers that they did not consider them adversaries. When a sign went up in another part of the forest claiming trees had been spiked occupiers distanced themselves from the act. Unlike their Northern Californian peers, they maintained a position of neither "condoning nor condemning" such tactics.[132]

With a temporary court stay on work imposed in early July, local officers inexperienced in dealing with blockading, and the story receiving heavy coverage in regional media, the FS waited until 3 August to institute an official closure of the sale area. A demonstration of 150 at the site was immediately held in response. A photo taken of two activists buried up to their necks in a logging road alongside a dummy of Ronald Reagan circulated in media across the country, demonstrating the ability of humour and innovative protest to generate publicity.[133]

FS officers with a more antagonistic attitude than local staff towards protesters were eventually brought in from Texas along with armed paramilitary CAMP agents from California. Large numbers of personnel patrolled the woods at times, but their presence had minimal impact.[134] As Langelle recalls:

> We were ready for anything and took it as a joke. We knew we were getting our message across and this was confirmed by the fact that they were so afraid of the effect this rag-tag bunch of anarcho-hippies were having. We refused to be intimidated and met their psychological warfare with our own. We messed with them by having people run around the forest at night yelling to send them running in circles. Since local law enforcement resented all this

Federal interference we actually got more sympathy from them and received some good intel.[135]

Protests were held at local ranger stations and the FS waited until mid-August before carrying out road works. These were held up by lock-ons and a road occupation. Following its demolition, the protest camp was quickly replaced and a car body placed in the road as a barricade.[136]

Logging did not commence until 4 September and was further disrupted by activists surrounding vehicles and workers, resulting in arrests. The varied and uncoordinated responses of authorities to lock-ons continued with FS officers responding to one by covering a man's head with a reflective fire shelter blanket. They then doused him with water while using a blowtorch to cut through the kryptonite lock around his neck. During this improvised process the activist sang "God Bless America" for the benefit of assembled journalists and camera people.[137]

The following day RACE secured an injunction from the 7th US Court of Appeals pending a judicial review of their lawsuit challenging the original FS sale. This brought an immediate end to logging and another court finding stalled further confrontation until the following summer.[138]

Regional newspaper *The Southern Illinoisan* followed up previous criticism of EF! by hailing RACE's litigation as proof that "a good attorney and a commitment to work within the system will get you a lot further than stupid stunts".[139] EF! members and supporters argued that the forest would likely have been gone by the time of the court decision had it not been for the 80-day blockade.[140] For his part FS spokesperson Tom Hagerty told reporters that "I think Earth First!, by virtue of its rather direct tactics, stimulated public discussion among people who hadn't spoken very loudly before about our timber practices."[141]

The ongoing impact of the campaign was evident by late September when the FS, at a meeting attended by 200 people, announced that it would be cancelling more than 80 per cent of its planned harvest.[142] Having previously refused to campaign on the issue the Sierra Club, members of whom had individually attended the blockade, subsequently successfully appealed further sales.[143]

Redwood Summer

The most significant event of the year concerning forest defence was Redwood Summer. The campaign's main innovation was to integrate blockading into one of the most sustained bursts of coordinated protest activity since the civil rights and anti-war movements of the 1960s and early 1970s. In the run up to the event campus speaking tours were organised and 450 information packets sent to colleges and newspapers. Co-sponsorship and funding came from a variety of organisations including the IWW, Earth Action Network, EFDAF, Mendocino Environmental Center, San Francisco Environmental Action Center, and Arcata Action Center. A sizeable minority of those involved travelled across the country to join the thousands of PNW and Californian residents who took part in more than 50 events and actions from June to September.[144]

Protest activities involved regular demonstrations, leafletting and public meetings in a variety of towns and cities, interspersed with one-day pickets, site occupations, and ODA carried out at lumber mills, corporate offices, and other points in the chain of production. Large events in towns and mills involving up to 1,500 people were held intermittently throughout the summer. Work in forests was disrupted on more than a dozen occasions, sometimes for four days at a time.[145]

This pattern was planned from the beginning for a variety of reasons. The consensus among the event's organisers was that the protection of old-growth forest would require a wide-ranging political solution rather than the continuation of largely uncoordinated rear-guard defence campaigns. EF! was not a sponsor of the "Forests Forever" referendum and actively resisted industry attempts to describe it as such.[146] This was partially because negativity towards EF! might cost votes, but also because EFers did not want to be bound to any one solution and believed they could best contribute by keeping the public's focus on the overall issue while operating as a radical flank. It was also consistent with the movement's long held rejection of parliamentary solutions and insider political deal making. Nevertheless, the fact that EF! were carrying out near daily protests meant that many casual onlookers conflated its work with Forests Forever, a connection timber interests worked hard to maintain.[147]

The campaign's desire to highlight the range of areas threatened by logging meant that its protest camp was moved every few weeks and activities rotated throughout the region. Due to the number of new participants being recruited, the high degree of hostility from opponents, and a desire to reach beyond existing EF! members, activities were deliberately designed to allow for a range of commitment by including varying degrees of length, risk, and arrestability. In providing such variety, organisers hoped to maintain media interest and the flow of activists. With Maxxam and others accelerating logging in fear of coming restrictions, EFers saw the blockading not only as a means of drawing attention, but also as a means of slowing clear-felling itself.[148]

Unlike most EF! campaigns the general time and location of actions was largely directed by a small core of organisers with the specifics decided upon by those carrying them out. In keeping with this, a timetable of events was organised ahead of the campaign's launch.[149]

With Bari and Cherney incapacitated, and King withdrawing, new organisers were drafted in. Ongoing coordination largely fell to a core group of 20 to 30. The majority of these were women, making Redwood Summer the first EF! campaign in which they had publicly played such a predominant role. Bari later argued that this marked "the feminization of Earth First!", a development which she claimed "profoundly affected the movement".[150] Veteran non-violence activists Seeds of Peace took on responsibility for feeding the protest camps and, following the bombing, Greenpeace assisted with preparations and training.[151]

Other than serving as accommodation and a base of operations, the protest camps ran regular NVDA training sessions. Reflecting the more centrally organised aspects of this campaign, and its focus on maintaining safety and avoiding negative publicity, involvement in these was deemed mandatory. As previously canvassed, EFers had long applied a standard code of non-violence to environmental block-ading, but in this case it was particularly emphasised with King declaring at the campaign's launch, "Any participant not in full agreement with non-violence as *the* principal concern during the actions will not take part in Redwood Summer."[152]

The campaign's official launch came on 20 June with a rally of 750 and the occupation of the road leading into Louisiana-Pacific's mill in the town of Samoa. This resulted in national media coverage and 44 arrests. Preliminary blockading in the forest had already taken place with 15 disrupting logging at the Tail Frog Grove on 6 June and a two-day tree-sit involving four occupiers set up near Fortuna on 19 June.[153]

The next wave of forest-based action began on 18 July after activists responded to a call for assistance from residents neighbouring a 10-acre stand, dubbed Osprey Grove, owned by Louisiana-Pacific near Mendocino County. Over the following days 39 were arrested and a group of six were beaten by loggers during an occu-pation of the site. A court order on 23 July halted work on the basis of threats to Spotted Owl habitat. It was lifted eight days later, but with Redwood Summer still occurring Louisiana-Pacific declared its own moratorium. Company employees returned in September when, despite further blockading, the trees were clear-cut.[154]

The day following the court injunction, and for two days afterwards, up to 2,000 people rallied in Fort Bragg; 17 activists were arrested for carrying out road occupations and locking onto gates on two different logging roads. Reports of fur-ther violence came after four activists alleged their heads were forcibly shaved by Humboldt County police officers and Redwood Summer attorney Mark Harris complained of an attempt to force his vehicle off the road.[155]

Such harassment did not deter protest. On 30 July the largest environmental blockade of the campaign involved hundreds of activists targeting logging on public lands. Entering the Sequoia National Forest, they stopped all cutting at eight different sites. On the same day three tree-sitters were arrested in the Maxxam-owned Murrelet Grove near Eureka and the following day a logging deck was locked onto.[156]

The most sustained blockading of the campaign took place a few weeks later when activists returned to Murrelet Grove. Around 70 barricaded roads with debris, slash piles and rocks for days while others locked on to machinery and gates. The creation of manufactured vulnerability, most starkly illustrated by a man placing his hand under an active chainsaw blade and others circling their arms around trees being felled, led to work stoppages. Confrontations included shots being fired by an angry driver and a protester hit by a truck. After 37 activists were arrested the remainder moved on to barricade another clear-cut a few miles away.[157]

A blockade using bicycles in the Sherwood area of Mendocino County and forest occupations at Elk and Gualala rounded out the campaign's blockading component in late August. The Elk protest involved local residents targeting an L-P operation based on threats to the town water supply. Four days of ODA, included the holding of a champagne breakfast in the middle of a logging road, ensued before an injunction on work was gained. Following further public protests Redwood Summer officially ended on 3 September with a march of 600 through the town of Fortuna to a mill owned by Pacific-Lumber.[158]

Organisers claimed Redwood Summer a success on the basis that despite regular violence being levelled against protesters, sustained campaigning had gone ahead, gaining unprecedented coverage for the issue of old-growth logging. The involvement and training of thousands of participants was seen as an achievement in itself, one which organisers believed could have a positive effect on the level of future activity.[159] While a small number of areas were saved through a combination of blockading and litigation, organisers argued that the campaign had primarily affected the rate of felling by mobilising the public and slowing logging approval processes through sustained pressure on decision makers.[160]

With so many political, legal and other influences impacting at the time it is difficult to gauge the specific effect any one had. Certainly the campaign attracted a huge amount of publicity and the amount of protest activity and blockading in Northern California vastly increased from that carried out in previous years. The timber industry and its supporters claimed that the campaign was a failure and argued that the attention it had gained played in their favour. Critics from within EF! and among mainstream environmentalists claimed that their long-standing criticism of blockading had been vindicated as the media regularly focused on incidents of violence and the countercultural identity of many involved.[161]

Come November, Forests Forever was defeated 48 per cent to 52 per cent and the timber industry's initiative 29 per cent to 71 per cent. Partisans continued to argue over the effect of Redwood Summer on the results, but the campaign strengthened Northern California's EFers commitment to grassroots campaigning. They launched "Corporate Fall", a series of national events targeting company offices and Wall Street, during late 1990. In years to come they and allies such as EPIC would continue to oppose clear-felling on private land via a combination of blockading, litigation and public protest.[162]

The embedding of the environmental blockading repertoire

From 1983 to 1990 EF!'s experimental culture, and the desire of significant numbers of its activists to defeat opponents in individual battles and create novel media stories, led to innovation and the embedding of a repertoire of environmental blockading tactics in the United States. These factors also saw the introduction of strategies such as mass campaigning and the pursuit of worker alliances and played a major role in the changing composition and political direction of the network. Such developments facilitated and fed off capacity building which allowed innovations to diffuse across the United States alongside a steady increase in the number and duration of blockades. Although actions were often regionally focused and rarely envisaged as part of a well-structured strategy, they played a role in popularising concepts and issues related to the preservation of biodiverse habitats.

Many of these processes echoed those that had taken place in Australia from 1979 to 1984, in that militants focused on creating obstruction and employing a loose version of non-violence primarily drove tactical innovation. Underlining the way in which movements, their opponents, and other stakeholders interact, factors such as opposition from major mainstream organisations and higher levels of hostility in logging areas played a role in US activists taking seven years to launch sustained blockades. Over time ODA came to be employed in a larger number of locations and situations than Australia. In part due to differing political circumstances, but also aided by the existence of a network organising a regular series of activities and publications, the United States did not suffer the major drop off in environmental blockading that Australia did during the mid-1980s.

Notes

1 William Dietrich, *The Final Forest: The Battle for the Last Great Trees of the Pacific Northwest* (New York: Simon & Schuster, 1992), 169–88, 232.
2 Ibid, 74,124; Mark Bonnett and Kurt Zimmerman, "Politics and Preservation: The Endangered Species Act and the Northern Spotted Owl," *Ecology Law Quarterly* 18 (1991): 115–18; Durbin, *Tree Huggers*, 76.
3 Clayton W. Dumont, "The Demise of Community and Ecology in the Pacific Northwest: Historical Roots of the Ancient Forest Conflict," *Sociological Perspectives* 39, no. 2 (1996): 280–82.
4 Dietrich, *The Final Forest*, 176.
5 Bevington, *The Rebirth of Environmentalism*, 17, 34–40.
6 Ibid, 78–83,102–5; Dietrich, *The Final Forest*, 161–65.
7 Durbin, *Tree Huggers*, 87–91.
8 David Harris, *The Last Stand* (New York: Times Books, 1996), 200–10.
9 Ilana DeBare, "A Tale of Two Owners: Old Redwoods, Traditions Felled in Race for Profits," *Los Angeles Times*, 20 April 1987, 1.
10 Steve Ongerth. Redwood Uprising: Chapter 4, Maxxam's on the Horizon, Available [Online]: http://ecology.iww.org/texts/SteveOngerth/RedwoodUprising/4 (Accessed 20 March 2016).
11 Greg King, "Old Growth Redwood," *EFJ* 7, no. 2 (1987): 9.
12 Harris, *The Last Stand*, 186–87; Socratrees, "Tactical Thoughts on the Maxxam Protests," *EFJ* 7, no. 6 (1987): 5.
13 Steve Ongerth. Redwood Uprising, Chapter 7: Way up High in the Redwood Giants, Available [Online]: http://ecology.iww.org/texts/SteveOngerth/RedwoodUprising/7 (Accessed 20 March 2016).
14 Greg King, "Redwood Tree Climbers," *EFJ* 7, no. 8 (1987): 1, 6.
15 Bill Israel, "'Tarzan and Jane' Try to Delay Loggers," *San Francisco Chronicle*, 3 September 1987, 25.
16 Mike Roselle, "Guest Editorial: Nomadic Action Group," *EFJ* 7, no. 8 (1987): 3; Mike Roselle and Karen Pickett, "Direct Action Fund: The Year in Review," *EFJ* 9, no. 4 (1989): 26.
17 Roger Featherstone, "Grand Canyon Uranium Battle," *EFJ* 7, no. 4 (1987): 1.

18 Chant Thomas, "Return to Bald Mountain," *EFJ* 7, no. 4 (1987): 1.
19 David Barron, "CD Begins Anew in Kalmiopsis," *EFJ* 7, no. 5 (1987): 5.
20 "18 Arrested in Three Actions in North Kalmiopsis," *EFJ* 7, no. 6 (1987): 6.
21 Jericho Clearwater, "Kalmiopsis Shutdown," *EFJ* 7, no. 7 (1987): 1.
22 Ibid., 9; Valeri Wade, "Kalimopsis Kangaroo Court," *EFJ* 7, no. 8 (1987).
23 A small sample of the news stories covering the campaign can be seen in footage compiled by EF camera-person Andy Caffrey. Andy Caffrey. 1980s Earth First!: The Antidote for Despair, Available [Online]: www.youtube.com/watch?v=DMxX7twsYNo (Accessed 6 April 2015).
24 "Spiked Trees Found in Oregon – Cutting Perils a Protester," *San Francisco Chronicle*, 23 June 1987, 8.
25 Brown, *In Timber Country*, 24–29.
26 Richard Widick, *Trouble in the Forest: California's Redwood Timber Wars* (Minneapolis: University of Minnesota Press, 2009), 227–48.
27 DeBare, "A Tale of Two Owners: Old Redwoods, Traditions Felled in Race for Profits," *LA Times*, 20 April 1987, 1; Brown, *In Timber Country*, 35, 245.
28 Ibid.
29 Ibid, 261–62.
30 Larry Stammer, "Environment Radicals Target of Probe into Lumber Mill Accident," *Los Angeles Times*, 15 May 1987, 3; Amy Gamerman, "New Drug Bill Spells out Penalties for Sabotaging Logging Operations," *Colorado Springs Gazette*, 7 December 1988, B7.
31 James Coates, "Terrorists for Nature Proclaim Earth First!," *Chicago Tribune*, 2 August 1987, 21.
32 Judi Bari, *Timber Wars* (Monroe: Common Courage Press, 1994), 264–82.
33 North Coast EF!, "Press Release," July 1987, 1.
34 Brazil, "Tree Spikers Draw Sawmill Blood," 24 June 1987, D12.
35 "Roundup Message Clear," *Eugene Register-Guard*, 18 September 1987, 22A; Brown, *In Timber Country*, 27–29.
36 AP, "Seven Logging Protesters Arrested," *Register-Guard*, 5 May 1987, 3C; Wade, "Kalimopsis Kangaroo Court," 9.
37 Daniel Kirkpatrick, "Washington Old Growth Campaign," *Earth First!* 7, no. 8 (1987): 1,4; John Patterson and Jean Ravine, "EF Shuts Down Uranium Mine," *EFJ* 7, no. 7 (1987): 1; Bill Workman, "San Bruno Mountain Condos: Chained Protesters Cut Free," *San Francisco Chronicle*, 25 August 1987, 5.
38 A search of 30 US newspapers using the Proquest News & Newspapers database found over 200 separate stories ran on Earth First! in 1987 with the majority of stories running in California and Washington. This was more than double the previous year. The newspapers sampled did not include smaller titles in areas where ODA was taking place. Proquest News & Newspapers 1987 Search Results. Available [Online]: http://search.proquest.com.ezp.lib.unimelb.edu.au/news/results/C81B065313A841A7PQ/1?accountid=12372 (Accessed 14 March 2016).
39 This is a claim made by various EFers which is reflected in search results using Proquest's newspaper archives. Mitch Freedman, "Old Growth Strategy Revisited," *Earth First!* 9, no. 2 (1988): 7; Proquest News & Newspapers Archive Search. Available [Online]: http://search.proquest.com.ezp.lib.unimelb.edu.au/news/results/EDD0BD2A7D444126PQ/1?accountid=12372 (Accessed 16 May 2016).
40 Dave Foreman, "The Question of Growth in Earth First!," *EFJ* 8, no. 6 (1988): 32; "Earth First! Bulletins," *EFJ* 8, no. 6 (1988): 17.
41 Judi Bari, "California Rendezvous," *EFJ* 9, no. 1 (1988): 4.

42 Sister Extraterrestial, "Earth First! Activists Conference," *EFJ* 8, no. 4 (1988): 9; Coulter Interview.

43 Tony Flynn, "Fishtown Protesters Concede," *Skagit Argus*, 9 February 1988, 3; Bruce Budworm, "Battle for Fish Town Woods," *EFJ* 8, no. 4 (1988): 6.

44 Daryl Cherney, "Triple Victory in Three Day Revolution," *EFJ* 9, no. 2 (1988): 1,6.

45 Jakubal Interview.

46 Ibid.

47 Ibid.

48 Ibid.

49 "12 Arrests Made During Two Logging Protests," *Ukiah Daily Journal*, 28 October 1988, 3; Steve Ongerth. Redwood Uprising, Chapter 11: I Knew Nothin' Till I Met Judi, Available [Online]: http://ecology.iww.org/texts/SteveOngerth/RedwoodUprising/11 [Date Accessed 20 May 2106].

50 "Indian Tribe's Protest Halts Logging Operation," *San Francisco Chronicle*, 29 October 1988, A15.

51 Greg King, "New Battles in Maxxam Campaign," *EFJ* 8, no. 6 (1988): 5; "12 Arrests Made During Two Logging Protests," 28 October 1988, 3; Cherney, "Triple Victory in Three Day Revolution," 1, 6.

52 Jakubal Interview.

53 King, "New Battles in Maxxam Campaign," 3; "3 Nest in Humboldt Redwoods to Protest Building of Road," *San Francisco Chronicle*, 21 May 1988, A2.

54 Caffrey, 1980s Earth First!: The Antidote for Despair, 1,5; Greg King, "Freddies Set Their Sights High: Kalmiopsis Tree-Sitters Targeted," *EFJ* 8, no. 8 (1988), 1,5.

55 Ibid.

56 Elaine Ellis, "In Defense of Mother Earth," *Sun Sentinel*, 6 August 1989, 1E; Myra Noss, "Florida EF! Saves Paynes Prairie," *EFJ* 9, no. 2 (1988): 1,4.

57 Christi Stevens, "Daybreak Dozer Occupation," *EFJ* 8, no. 8 (1988): 7; "Developers Attack Vireo," *EFJ* 8, no. 8 (1988): 6–7.

58 Christi Stevens and Barbara Dugelby, "Cavebugs Saved from Oblivion!," *EFJ* 9, no. 1 (1988): 1,4.

59 Mark Harden, "'Tree Sitters' Protest Logging Earth First! Environmentalists Camp out in Branches near Granby," *Denver Post*, 15 August 1989, 1; Judi Bari, "Californians Start a New Fad: Tree-Sitting Becomes a Pastime," *EFJ* 9, no. 8 (1989): 4.

60 Iain McIntyre, Environmental Blockading Timeline, 1974–1997.

61 'Canyon Wolf', "Victory!!! Earth First! Saves Colorado Old Growth," *Earth First! Journal* 10, no. 2 (1989): 8.

62 Coulter Interview; Jakubal Interview; Wood Interview.

63 Jasper, *The Art of Moral Protest*, 237.

64 Larson, "Social Movements and Tactical Choice," 871.

65 Ibid, 871–72.

66 Orin Langelle, "EF!ers Face Jail for Defending Illinois Hardwoods," *EFJ* 10, no. 1 (1989): 7,8; Norm Heikens, "Earth First! Vows Intensified Anti-Clearcutting Maneuvers," *Southern Illinoisian*, 28 August 1989, 3.

67 Langelle Interview.

68 Paul De La Garza, "Hopes for Clear-Cutting Ban Rooted in Shawnee Dispute," *Chicago Tribune*, 22 October 1989, 16; Christopher Batio, "Shawnee Battle Reaches State Capital," *Southern Illinoisan*, 27 February 1990, 1.

69 Langelle Interview.

70 "Foes of Road Win Delay after Chaining Selves to Earthmovers," *Los Angeles Times*, 6 June 1989, A3; Peter Bralver and Dan Strachan, "Los Angeles EF! Wins Fight for LA's Last Wilderness," *EFJ* 9, no. 6 (1989): 1.

71 Gorman, "Earth First! Tactics in Fight to Save Planet Anger Some, Tickle Others," 14 August 1988, 9.

72 Big Bark, "3 New Mexicans Arrested at Copar Strip Mine," *EFJ* 9, no. 5 (1989): 20; "Post Rendezvous Action Shuts Downtimber Sale," *EFJ* 9, no. 7 (1989): 1,19.

73 Paul Roland, "Breitenbush Blockade Draws National Attention to Ancient Forests," *EFJ* 9, no. 5 (1989): 6.

74 Mike Burge, "13 Arrested for Blocking Entry to Timber Sale," *Register-Guard*, 27 March 1989, 1C.

75 Lone Wolf Circles, Karen Wood, and Moss, "Escalation! The Kalmiopsis 24," *EFJ* 9, no. 7 (1989): 6.

76 Brendon Swedlow, "Scientists, Judges, and Spotted Owls: Policymakers in the Pacific Northwest," *Duke Environmental Law and Policy Forum* 13, no. 1 (2002): 194–228.

77 Kathleen M Vanderziel, "Hatfield Riders & Environmental Preservation: What Process Is Due," *Boston College Environmental Affairs Law Review* 19, no. 2 (1991): 434–50; Michael C Blumm, "Ancient Forests and the Supreme Court: Issuing a Blank Check for Appropriation Riders," *Urban Law Annual; Journal of Urban and Contemporary Law* 43(1993): 40–43.

78 Alston Chase, *In a Dark Wood: The Fight over Forests and the Myths of Nature* (New Brunswick and London: Transaction, 2009), 298–300.

79 Dale Turner, "So You Got a Death Threat..." *EFJ* 10, no. 6 (1990): 6.

80 Daryl Cherney, "Freedom Riders Needed to Save the Forest," *EFJ* 10, no. 5 (1990): 1, 6.

81 Ilse Asplund, "Evan Mecham Eco Tea-Sippers International Conspiracy," *EFJ* 31, no. 2 (2011): 18–22.

82 Zakin, *Coyotes and Town Dogs*, 316–41, 420–43.

83 Patt Morrison, "Terrorists or Saviors?," *LA Times*, 16 June 1991, A24.

84 John Davis, "Ramblings," *EFJ* 10, no. 6 (1990): 2.

85 Dale Turner, "Montana Earth Firsters Get Federal Subpoenas," *EFJ* 10, no. 1 (1989): 1; Coulter Interview.

86 "Tree Spiking Renounced Behind Redwood Curtain," *EFJ* 10, no. 5 (1990): 12.

87 Ibid.

88 Ongerth, "Redwood Uprising, Chapter 11 ".

89 Brown, *In Timber Country*, 22–25; Steve Ongerth. Redwood Uprising Chapter 3: He Could Clearcut Forests Like No Other, Available [Online]: http://ecology.iww.org/ texts/SteveOngerth/RedwoodUprising/3 (Accessed 30 March 2016).

90 Brown, *In Timber Country*, 35, 245; Bari, *Timber Wars*, 13.

91 Paul Watson, "In Defense of Tree Spiking," *EFJ* 10, no. 8 (1990): 8.

92 Bari, *Timber Wars*, 13; Langelle Interview.

93 Braggs, "Earth First!," 115.

94 Roselle, *Tree Spiker*, 132.

95 "Gun-Toting Photo a Joke, Friends Claim," *Ukiah Daily Journal*, 12 June 1990, 1; Daryl Cherney Music. Available [Online]: http://asis.com/users/dced/hippieslyrics.htm (Accessed 14 June 2016).

96 Steve Ongerth. Redwood Uprising, Chapter 33: The Ghosts of Mississippi Will Be Watchin', Available [Online]: http://ecology.iww.org/texts/SteveOngerth/Redwood Uprising/33 (Accessed 20 May 2016).

97 Mitch Freedman, "Old Growth vs. Old Mindsets", *EFJ* 9 no.5, 1989.

98 Mike Geniella, "Bari Juror Explains Verdicts, Marathon Deliberations," *Press Democrat*, 14 June 2002, A1; Roselle, *Tree Spiker*, 126.

99 Braggs, "Earth First!," 117–19.

100 Langelle Interview.

101 "Earth First! Founder Turns to Persuasion," *Arizona Republic*, 1 June 1991, D3; Mike Geniella, "Leadership Dispute Splits Earth First!," *Press Democrat*, 12 August 1990, A1.

102 Foreman, "The Question of Growth in Earth First!," 32; Nancy Zierenberg, "Time to Move On," *Earth First!* 10, no. 8 (1990): 2–3; Barry Noreen, "Earth First! Members Divided by Differences in Philosophy," *Colorado Spring Gazette*, 14 October 1990, B1.

103 Mike Roselle, "Roadkill," *EFJ* 10, no. 3 (1990): 27–28; Judi Bari, "Expand Earth First!," *EFJ* 10, no. 8 (1990): 5–6; Langelle Interview.

104 Edward Abbey, "Letter " *EFJ* 8, no. 2 (1987): 3; Christopher Manes [Pseudonym: Miss Ann Thropy], "Overpopulation and Industrialism," *EFJ* 7, no. 4 (1987): 29; Christopher Manes [Pseudonym: Miss Ann Thropy], "Population and AIDS," *EFJ* 7, no. 5 (1987): 14–16.

105 Roselle, *Tree Spiker*, 135; John Davis, "The Successors of EFJ," *EFJ* 11, no. 2 (1990): 2; Coulter Interview; Nancy Morton and Dave Foreman, "Good Luck, Darlin'. It's Been Great.," *EFJ* 10, no. 8 (1990): 5; Loose Hip Circles, "Riotous Rendezvous Remembered," *EFJ* 9, no. 7 (1989): 19.

106 Braggs, "Earth First!," 66–67, 81–83; Zakin, *Coyotes and Town Dogs*, 303–409; Lee, *Earth First!*, 96–127.

107 Alexander Cockburn, "Beat the Devil," *The Nation*, December 1990, 670–72; Bari, *Timber Wars*, 219–25, 71–83.

108 Ibid.; Bevington, *The Rebirth of Environmentalism*, 21–40.

109 Coulter Interview; Wood Interview.

110 Lee, *Earth First!: Environmental Apocalypse*, 128, 40; Zakin, *Coyotes and Town Dogs*, 442.

111 This summary is based on EFJ and related materials from 1990 to 1997. Examples of the argument for continuity and evolution can be found in: Karen Pickett, "Breaking up or Breaking Apart?," *EFJ* 11, no. 1 (1990): 2; "Earth First! Founder Quits Group over Rhetoric," *Arizona Republic*, 15 August 1990, 20.

112 Major debates concerning the *EFJ* and tactics which reflected continuing diversity in the network occurred over the publication of, likely satirical, articles regarding the killing of cows and hunters. A Nony Moose, "Shooting Cows: A Novel Idea," *EFJ* 11, no. 8 (1991): 10; Robert Marten, "A Hunting We Will Go," *EFJ* 12, no. 1 (1991): 26–27.

113 Annie Szveteca, "Activists· Defend Hawaii's Last Rainforest," *EFJ* 10, no. 7 (1990).

114 Andy Molloy, "Activists Assaulted at Tract Pond," *EFJ* 11, no. 2 (1990): 1; Karen Pickett, "Direct Action Fund 1990 Report," *EFJ* 12, no. 4 (1991): 34.

115 'Savannah Underdog', "Texas Earth First!, Locks up Outer Loop Construction," *EFJ* 10, no. 4 (1990): 12–13.

116 John Harris, "City Halts Removal of Trees," *Austin American Statesman*, 9 February 1990, B2.

117 Scott Wyland, "Demonstrators Delay Clear-Cutting near Forest Park," *Oregonian*, 4 May 1990, C4; Nelson Pickett, "Earth First! Padlock Fails to Halt Clear-Cut Start," *Oregonian*, 1 August 1990, B2; "5 Who Chained Themselves to Logging Trucks Are Arrested," *San Francisco Chronicle*, 14 February 1990, A20"Ecotrans – Collective & Individual Action," *EFJ* 11, no. 3 (1991): 6.

118 Robert Nelson, "Watershed Timber Sparking City Debate," *Seattle Times*, 1 March 1990, H1; William Dietrich, "New Law Forces Old-Growth Sale," *Seattle Times*, 21 November 1989, C3.

119 "Old-Growth Protest," *Seattle Times*, 30 May 1990, E1; "Old-Growth Backers Chain Themselves Up," *Spokane Chronicle*, 31 May 1990, B4; AP, "Old Growth Logging Foes Sit in Trees," *Spokane Chronicle*, 7 June 1990, C8.

120 Louis Corsaletti, "Security Gantlet Thrown up to Discourage Logging Foes," *Seattle Times*, 6 August 1990, 3.

121 "Saving Sugar Bear," *Seattle Times*, 31 August 1990, 2.

122 Coulter Interview.

123 Erik Ryberg, "Lessons from Mt. Graham," *EFJ* 10, no. 3 (1990): 8.

124 "Protesters Try to Block Route but Loggers Reach Mount Graham Site," *Mohave Daily Miner*, 3 October 1990, 2; Jim Leonard, "Trees Cut on Mt. Graham," *EFJ* 11, no. 1 (1990): 12.

125 Ryberg, "Lessons from Mt. Graham," 8–9.

126 Ibid.

127 Wood Interview.

128 Langelle Interview.

129 Paula Davenport, "Protesters Hope to Save Shawnee Trees," *St. Louis Post-Dispatch*, 2 July 1990, 113; "Lawmakers Seeks to Save Wilds of Shawnee Forest," *St. Louis Post-Dispatch*, 25 June 1990, 24.

130 Wes Smith, "Tangled Woods " *Chicago Tribune*, 17 May 1990, 1; Orin Langelle and John Wallace, "Showdown on the Shawnee," *EFJ* 10, no. 7 (1990): 1

131 Langelle Interview; Linda Sickler, "Activists Fight the Sale of Timber," *Southern Illinoisan*, 16 July 1990.

132 Phil Brinkman, "Spiking Incident Draws Investigation and Denials," *Southern Illinoisan*, 10 August 1990, 1–2; Langelle Interview.

133 Phil Brinkman, "Temporary Clamp Placed on Fairview Timber Sale," *Southern Illinoisan*, 4 July 1990, 1; John Curley, "Protesters Vow to Bar Logging," *St. Louis Post-Dispatch*, 6 August 1990, 1,6.

134 "Shawnee," *Southern Illinoisan*, 5 September 1990, 8.

135 Langelle Interview.

136 Orin Langelle, "Shawnee Saga Continues," *EFJ* 10, no. 8 (1990): 8; Myers Linnet, "Two Arrested in Bid to Block Logging in Illinois," *Chicago Tribune*, 17 August 1990, 20.

137 Phil Brinkman, "Shawnee Showdown," *Southern Illinoisan*, 5 September 1990, A1, A8.

138 Myers Linnet, "In Illinois Forest, Court Halts Cutting," *Chicago Tribune*, 6 September 1990, 17; "Bar against Logging Set to Expire Today," *St. Louis Post-Dispatch*, 19 December 1990, 7.

139 "Editorial: Working within the System Gets RACE Results," *Southern Illinoisan*, 6 September 1990, 8.

140 Paula Davenport, "Logging Opponents Dig in at Shawnee," *St. Louis Post-Dispatch*, 19 August 1990, 4; David Colombo, "Letter: Earth First! Had Part in Victory," *Southern Illinoisan*, 24 September 1990, 4.

141 Paula Davenport, "Timber Is Spared in Shawnee," *St. Louis Post-Dispatch*, 30 September 1990, 38.

142 Paula Davenport, "Forest Service Cancels Sale of Shawnee Harvest," *St. Louis Post-Dispatch*, 29 September 1990, 6.

143 Safir Ahmed, "Ban on All Tree Cutting in Shawnee Sought," *St. Louis Post-Dispatch*, 3 October 1990, 14; Paula Davenport, "Rival Groups Join to Fight Shawnee Logging," *St. Louis Post-Dispatch*, 25 April 1990, 101.

144 "Redwood Summer Chronology," *EFJ* 11, no. 1 (1990): 7; John Woestendiek, "Anti-Logging Protest Still Spreading Its Roots," *Philadelphia Inquirer*, 12 August 1990, A2; Pickett, "Direct Action Fund 1990 Report," 134.

145 "Redwood Summer and Beyond," *Polemicist* 2, no. 2 (1990): 13.

146 Chris Calder, "Redwood Struggle Expected to Continue," *Ukiah Daily Journal*, 7 September 1990, 10.

147 Bari, *Timber Wars*, 37–38; "Earth First! Link Cut to Ballot Issue," *Ukiah Daily Journal*, 24 July 1990, 1.

148 Bari, *Timber Wars*, 37–38.

149 "Econet Post: Redwood Summer Calendar," 24 July 1990, 1; Coulter Interview.

150 Bari, *Timber Wars*, 73, 225.

151 Karen Pickett and Woody Joe, "Redwood Summer Goes On!," *EFJ* 10, no. 6 (1990): 1; "Redwood Summer Chronology," 7,10.

152 Emphasis in original article. Cherney, "Freedom Riders Needed to Save the Forest," 1; David Foster, "Earth First! Takes Non-Violent Tack in California," *Phoenix Gazette*, 9 July 1990, B4.

153 "Redwood Summer Chronology," 7; "44 Arrested at L-P Mill," *EFJ* 10, no. 7 (1990): 1.

154 "Redwood Summer Activists Assaulted by Loggers," *EFJ* 10, no. 7 (1990): 7; Zack Stentz, "Osprey Grove Falls," *EFJ* 11, no. 1 (1990): 9–10.

155 "17 Protesters Arrested at Logging Roads," *San Francisco Chronicle*, 25 July 1990, A4; "Redwood Summer Activists Harassed by Police," *EFJ* 10, no. 7 (1990): 7; "Jeers Greet 1500 Logging Protesters at 'Redwood Summer'," *The Pantagraph*, 23 July 1990, 4.

156 AP, "Redwood Summer' Rolls Along," *Orange County Register*, 31 July 1990, A3; "Earth First! Protesters in Sequoia," *Santa Cruz Sentinel*, 31 July 1990, 26; Stein, "Redwood Summer," 2 September 1990, 3.

157 Michael Robinson, "The Battle for Murrelet Grove," *EFJ* 10, no. 8 (1990): 15.

158 "Redwood Summer Chronology," 10; "Timber Group, Activists Clash on North Coast," *Ukiah Daily Journal*, 4 September 1990, 1; "Fourth Day of Elk Logging Blockade," *Ukiah Daily Journal*, 22 August 1990, 1.

159 Ken Fireman, "Green Measures Compete in Calif. Voters Will Face 4 Options," *Newsday*, 3 September 1990, 7.

160 "Redwood Summer and Beyond," 33.

161 Calder, "Redwood Struggle Expected to Continue," 1,10.

162 Harris, *The Last Stand*, 340–41; Zakin, *Coyotes and Town Dogs*, 380–81.

5

"YOU'RE WELCOME TO VISIT OUR PARK, BUT LEAVE YOUR SAWS IN THE BOAT"

Canadian First Nations and conservationist activism, 1983–1984

Founded in Vancouver in 1971, Greenpeace's use of ODA regarding nuclear and animal rights issues had a major global impact in the decades that followed.[1] Despite generating such impact, Canada's first major wave of environmental blockading aimed at forestry, mining, and other forms of development started later than in Australia and the United States.[2] This chapter analyses campaigns in Canada that began to establish the template during 1984 and 1985, while the next discusses how that template became firmly entrenched over the four years that followed. A number of key differences, both within Canadian campaigns and with trends in environmental blockading in Australia and the United States, will be explored. Due to the extent of Indigenous community involvement and leadership in Canadian campaigns, the chapters pay particular attention to confluence and conflict with non-Indigenous environmentalists regarding campaign goals, understandings of biodiverse places, tactical development, and other issues.

The political and economic context of forestry and resource extraction in Canada

Since colonisation began in earnest in the 1600s, Canada's settler economy has been heavily dependent on resource extraction. As a result, conflict with First Nations people regarding ownership and treatment of land has long been a feature of Canadian society. From the 1960s onwards, new movements drawn from the non-Indigenous population emerged to critique and contest the environmental effects of industrial society. Although opposition to mining, dam construction and other forms of exploitation generated grievances, as in the United States and Australia contestation over biodiverse areas was primarily related to the logging of old-growth forests.

Ten per cent of the world's forests reside within Canada, making up almost half of the country's land mass. Jurisdiction regarding land and natural resources has primarily been the authority of the ten provincial governments who control 87 per cent of forests with a remaining 5 per cent under federal control and 8 per cent privately owned.[3] Despite variation regarding policy-making and the pressures exerted on administrations, resource extraction has followed a similar pattern across the country. Exploitation of forests has generally involved the leasing of government-controlled land to private interests, usually on a 5- to 20-year basis with tenure easily renewed. During the 1980 and 1990s, federal and provincial departments covering wildlife, fisheries, pollution, and other overlapping issues had limited influence over mining and logging decisions and there was little scope for public input.[4]

Scientific, policy, and technical research and debate has been principally devoted to commercial timber extraction in a process that led governments, industry, and unions to ignore evidence regarding over-exploitation. Values involving recreation and hunting have generally been of a much lower priority than in countries such as the United States, something that has generally been attributed to Canada's low population in relation to its size.[5] By the 1980s the maintenance of biodiverse ecosystems, outside of what Canadians have generally dubbed "special places" in National Parks and reserves, rarely trumped extraction. Unlike Australia and the United States, there was also a greater focus on converting forests into managed plantations.[6]

Despite acting as landlords, provincial governments' reliance on primary resource extraction and resource exports, combined with a reluctance to invest in timber management, meant they largely relinquished responsibilities concerning harvesting, management, and processing. They have subsidised the private sector directly and indirectly via infrastructure projects and the imposition of royalty rates that at times were less than the cost of collection. Industry and government figures and bodies based in cities have historically acted in a corporatist fashion.[7]

First Nations and environmentalist opposition has overlapped with communities and businesses negatively affected by resource extraction. The removal of viewscapes and forest posed obvious threats to tourism while the contamination of rivers was damaging spawning grounds and fisheries. Increasing pressure from these stakeholders led government departments to adopt policies that recognised uses beyond extraction in the 1970s. These rarely enabled critics to mount effective appeals or challenge the authority of ministries and managers committed to maintaining timber supplies.[8]

A combination of factors beyond long-term political closure led to old-growth forests becoming a major issue, particularly in the provinces of British Columbia (BC) and Ontario. Following a boom in the late 1970s, the forest industry suffered a major downturn, resulting in tens of thousands of job losses. Although caused in part by over-cutting, the main response was to facilitate renewed expansion and exploitation via the lowering of costs through increased productivity and deregulation. Combined with budget cuts, this led to several public incidents involving mud slides and erosion, which increased public concerns at a time when the industry's ability to economically justify its practices was diminished. Although the sector recovered by 1987 governments and industry remained unable to find solutions that would allow them to maintain historical levels of profits while sustaining a necessary resource base and public support.[9]

First Nations activism regarding biodiverse places

Unlike in Australia and the United States, activism in the 1980s concerning biodiverse places was dominated by First Nations activists. In part this stemmed from demographic differences, as while less than 10 per cent of Canada's population resides in forest regions, close to half of those that do are Aboriginal. It also resulted from a broad upswing in Aboriginal protest and campaigning in response to ongoing deadlocks regarding land rights and sovereignty.[10]

Colonial control in Canada was generally achieved through negotiation rather than military conquest with more than 300 agreements signed with First Nations. Many communities had terms forced upon them, believed they had entered into different agreements to those written down, or had their rights restricted by later laws regarding acquisition and ownership of land.[11]

Despite the diminution of access, in comparison to Australia and the United States, a much larger number of Indigenous communities continued to use land for hunting, trapping, and other purposes, thereby increasing the possibility of conflict with resource extraction industries. The existence of treaties provided some communities with, in the terms of settler society, avenues of action and legal and moral validation. Their common breaching by mining and forestry interests added further layers of contention and resentment.[12]

Many First Nations in BC and the northern regions of Canada had not entered into treaties by the time the provinces joined the confederated Canadian state. A continuing refusal by governments to negotiate them added to grievances regarding land use. At the same time the fact that ownership of land and water had not been formally ceded opened legal avenues for opposing government and industry decisions. Regardless of treaty status, community support for activism was increasingly mobilised in the 1980s through the merging of specific land use and environmental issues with broader ones around sovereignty and discrimination.[13]

Indigenous people have faced common issues but categories imposed since colonisation came to differently structure relationships between various groups as well as with the state. A major compartmentalisation flows from whether people are recognised by federal authorities as members of First Nations or are accorded other status, such as Metis people of mixed Aboriginal and European descent, or no Aboriginal status at all.[14]

The complexity of Aboriginal experiences and activism has been further extended by differences in jurisdictional authority. Provinces control much of the land and resources subject to disputation, but the federal government is largely responsible for constitutional reform regarding such matters. Many First Nations people effectively remained wards of the federal state into the 1980s and 1990s, with most living on small areas of reserve land designated for use and occupancy to which they could claim no ownership, while the non-status and Metis people living beside them were governed by separate bodies. As a result, there has been much inertia and "buck passing" between authorities.[15]

Aboriginal people in Canada have had a long history of covert and overt resistance to colonisation. A major increase in activism during the 1960s led to litigation and collective action during the 1970s and 1980s, which established court findings recognising in principle that they possessed rights regarding land and resources. The interpretation and meaning of those rights remain largely unresolved, particularly in terms of "comprehensive claims" regarding the assertion of rights and titles to unceded land. The resolution of "specific claims" related to existing treaties has also been incremental.[16]

Opposition to development and resource extraction has never been universal among First Nations. Where it has occurred, it has generally incorporated interconnected concerns regarding the ecological, social, and economic impacts on fishing, hunting, and cultural practices. Land ownership, the distribution of wealth, and the possibility of resources being exhausted and land destroyed before native title claims are resolved have also generated grievances. As a result, demands for sovereignty have been entwined with those concerning the protection of important places, Aboriginal involvement in management, and the application of traditional environmental knowledge.[17]

In some cases, campaigns and blockades that emerged from the issues canvased above were carried out in coalition with non-Aboriginal environmentalists. Different understandings and conceptions of nature, culture, and society, and the connections between them, became manifest during these and at times generated tensions. Non-native environmentalists, and many of the wider publics they draw support from, have primarily valued places under threat in terms of their ecological and aesthetic values.[18]

Aboriginal peoples' demands have generally stressed the resolution of land claims alongside protection, while some non-Aboriginal activists have been more content with solutions, such as the creation of National Parks, which potentially undermine Indigenous management and control and proscribe activities such as hunting and fishing.[19] As Teme-Augama Anishnabai activist Mary Laronde recalled in relation to the anti-clear-felling blockades her community carried out in Ontario during 1988 and 1989: "There is no such thing as wilderness, it's home. Our chief Gary Potts used to say the environmentalists want to make our homeland a zoo and the loggers and miners want to make it a desert."[20]

As will be seen in the campaigns that follow, some local environmentalists did respect First Nations' concerns, but the focus of the media, publics, and many activists outside of the region tended to privilege understandings of "wilderness" as a stable, self-regulating place in which culture or people are absent, except in idealised representations of "traditional" Aboriginal harmony. As a result, the potential for preservationist concerns to marginalise Aboriginal understandings and goals became a recurring issue.[21] Such differences have regularly emerged during environmental campaigns in other settler nations. However, with the dominance of First Nations in campaigns in Canada, they were present to a greater degree than in Australia or the United States during the 1980s.

Historical precedents for blockading among First Nations

In Canada the main precedents for engaging in ODA at the point of extraction came from First Nations' experiences of activism. Initial colonisation had been met by force in parts of the country. Blockades of roads, rivers, and other transport routes with bodies and barricades were later used in the nineteenth and early twentieth century to challenge settlement, protest government actions, and demand access to traditional lands.[22]

As part of an upswing of activism during the 1960s and 1970s blockades came to be employed once more. Many of these were initiated or carried out by First Nations militants, often acting unilaterally and not always with the broad support of their communities. During the 1980s the strategy came to be more commonly promoted by elected and hereditary band leaders and carried out by communities following discussions within bands and larger bodies.[23] A 2004 event analysis study by sociologist Rima Wilkes found that amongst the 22 per cent of Aboriginal bands who took part in non-routine and non-institutionalised collective action from 1981 to 2000, blockading was the most popular tactic, used in close to 50 per cent of 266 events. They were employed more than twice as often as demonstrations, and far more commonly than boycotts and fish and log-ins, with actions primarily based on violence or property destruction making up less than 2 per cent. Annually there were few events from 1981 to 1985 with a steady rise thereafter until a major spike in 1990.[24]

Generally, ODA was aimed at disrupting non-Aboriginal mobility in order to bring attention to grievances, directly contest work in or movement through traditional territory, and/or disrupt the flow of resources and commodities out of a territory.[25] The creation of transportation routes historically advanced the interests of colonisation by facilitating dispossession and an economy based upon it as well as by challenging First Nation' concepts of time and space. However, they also assisted Aboriginal struggles by connecting communities and providing a target and means of resistance. In terms of creating disruption, many blockades have made a virtue of the geographic dispersion and social isolation of Aboriginal communities as this has made it difficult for authorities and opponents to predict and counter ODA or mobilise resources against it. The reliance of some provinces on the movement of primary resources along a small number of remote routes and the placement of roads and railways near or through reserves has created further vulnerabilities and allowed activists to draw on nearby bases of support.[26]

The popularity of blockading appears to have been drawn from both its instrumental effectiveness as well as its deep resonance with First Nations political culture in terms of clearly delineating and expressing sovereignty. Although a minority of blockaders have sought revolutionary, secessionist, and confrontational ends, during the period this chapter will cover, the aim was more typically to restart or influence negotiations with authorities.[27]

As with those carried out elsewhere, blockades involved means and ends that have been interpreted in a variety of ways by wider audiences as well as those involved. Issues of property ownership and control of space are implicit in environmental ODA, but the goals and meaning of blockades for Aboriginal people have meant they are more commonly and explicitly linked to fundamental challenges regarding the existing political and economic order.[28] Similarly, while direct action often has a pre-figurative dimension, in that it enacts elements of a desired future in the present, in the case of Indigenous action it goes beyond this to form a reshaping and extension of past powers and conditions into the present.[29]

Conservation movements in Canada and precedents for blockades among non-Indigenous communities

While the employment of ODA regarding land-use disputes was already well established among First Nations by the 1980s, this was not the case amongst non-Indigenous conservationists. Its use by labour, anti-nuclear, and other movements however did have a long history. As in the United States and Australia, Canada's environmental and anti-nuclear movements were interlinked, with Greenpeace addressing both issues, as well as animal welfare, within one formal organisation. Greenpeace, which remains one of the most high-profile exponents of ODA, was founded as part of a wave of new environmental activism, which grew out of the New Left and counter-cultural movements of the 1960s. Vancouver, the largest city in BC, was the site of much activism as it also incorporated a large number of American draft-resisters whom Canada would not allow to be extradited to the United States.[30]

Following some instances of environmental ODA, including the occupation of a park slated for hotel development and four protesters successfully preventing road construction from destroying a Vancouver beach, BC activists formed the Don't Make A Wave Committee to oppose nuclear testing. In 1971 members sailed a ship renamed the *Greenpeace* to Amchitka Island in Alaska with the intention of disrupting a detonation that was to take place there. Although two separate attempts to enter the testing zone failed, the publicity helped delay the operation and led to the US government cancelling further testing in the area.[31]

Out of this success the Greenpeace Foundation was formed and carried out further high-profile actions aimed at internationally disrupting nuclear testing, whaling, sealing, and other activities. By the 1980s it had been restructured along formal and hierarchal lines into Greenpeace International and developed a formula of combining spectacular ODA and lobbying run by small, experienced teams supported by extensive fundraising carried out by other employees. In most countries this led to a generally passive membership with minimal involvement in ODA. In contrast opportunities for involvement in blockades and other protests in Canada appear to have been higher, with grassroots members taking part in anti-nuclear campaigns involving the disruption of cruise missile testing and associated blockades of military bases during 1983 and 1984.[32]

As with Greenpeace the rest of the environmental movement steadily grew from the 1960s onwards. An initial focus on pollution and energy issues rapidly diversified, as new groups began to engage with those regarding forestry, pesticides, mining, and wildlife. New laws concerning air and water pollution were passed in the 1960s and early 1970s and Environment Canada formed in 1971 as the primary federal ministry responsible for parks, endangered species, regulating pollution, and later conducting environmental assessments.[33] During the 1970s there were occasional instances of ODA or the threat of it being employed, such as part of a successful campaign to ban uranium mining in BC, but these did not lead to the application of blockading to other environmental issues or embed the repertoire in the way the actions of the 1980s would come to do.[34]

Although larger formal organisations focused on multiple issues played a support role, most environmental blockades carried out during the 1980s were initiated by local, grassroots groups. A similar pattern of protest emerged to that in Australia and the United States, with campaigners primarily challenging logging and resource extraction on a reactive basis as threats arose. In part this pattern was a result of provincial governments deliberately avoiding consolidated decision making in order to splinter their opponents' efforts. In Canada these became known as "valley-to-valley" conflicts due to the accessibility and concentration of resources in those areas. The overall period of contestation regarding forestry in the 1980s and 1990s was eventually dubbed the "War in the Woods".[35]

Canada had historically experienced a low level of activism regarding preservation and a "frontier mentality" remained extant in resource rich areas such as BC. Policy regimes in the United States and Australia during the 1970s and 1980s were also heavily dominated by understandings favouring extractive industries, but wider repertoires of opposition existed than in Canada. Here, opportunities for recourse to local and federal authorities and courts regarding environmental issues and regulations were minimal or untested. Relevant powers were generally concentrated in the hands of provincial ministers and officials with great discretionary powers. Activists could rarely call on federal governments to intervene, as they did with mixed success in Australia, or utilise local and national laws regarding conservation and the protection of endangered species, as they would in the United States. The location of the majority of campaigns in provinces supportive of industry and the existence of "first past the post" elections and unicameral parliaments further meant that campaigners had limited or no opportunity to deploy electoral preferences and blocs or means such as referenda to pressure relevant decision makers. From 1984, these factors combined with expansion in clear-felling to fuel the formation of new organisations as well as the greater use of blockading.[36]

Meares Island

On the morning of 20 November 1984 the *Kennedy Queen* began to cross BC's Clayoquot Sound, a 262,000 ha area located on the west coast of Vancouver Island. Featuring pristine waterways, islands, creeks, and much of the region's remaining old-growth temperate rainforest, those aboard were to start work on a log dump that would enable a new round of clear-felling to be carried out. Made up of officials, engineers and a work crew employed by forestry corporation MacMillan Bloedel, the boat's passengers were bound for Meares Island's Tisaquis/Heelboom Bay.[37]

As the vessel entered a channel near the bay the message "Coffee's On" went out across radio receivers and a small flotilla of boats captained by environmentalists and members of the Tla-o-qui-aht and Ahousaht bands, part of the Nuuchah-nulth people who make up 45 per cent of Vancouver Island's population, poured in. Negotiations between the Royal Canadian Mounted Police (RCMP)[38] and protesters began while helicopters and boats carrying media representatives captured images contrasting the stranded logging vessel with the island's beauty.[39]

With the way eventually cleared by consent of the blockaders, MacMillan Bloedel representatives reached land by 1 p.m. Here they were met by a crowd of 60 protesters including Chief Councillor Moses Martin. The Tla-o-qui-aht leader read from Land Act provisions and the original 1905 timber license covering the area, underlining provisions requiring the holder to "respect all Indian grounds, plots, gardens, Crown and other reserves". Having declared the island a "Tribal Park" some months earlier, Martin informed the party that his people considered it just such a garden, telling MacMillan Bloedel officials: "You're welcome to visit our park, but leave your saws in the boat." Unwilling to comply, and clearly in no position to begin work, company representatives left.[40]

These events constituted classic political theatre and extended a strategy and tactics previously employed by First Nations in land and fishing disputes to logging for the first time.[41] After more than a decade of campaigning, the blockade represented a tactic of last resort with Friends of Clayoquot Sound (FOCS) co-founder Michael Mullin telling reporters,

> We did sit on a planning team for naught. We did send a petition with 17,000 signatures to Parliament for naught. We did request a moratorium on logging for naught. After you've been through each of these steps there's nothing left but to stand in front of a tree.[42]

For the Nuuchah-nulth both the conditional welcome and the blockade formed a powerful public assertion of sovereignty.[43]

MacMillan Bloedel's arrival provided the company with a legal basis to initiate civil practices previously established during land disputes and First Nation's blockades.[44] The use of civil rather than criminal law, which dominated in Australian and US responses to ODA, reflected the intention of authorities to maintain an appearance of neutrality. As one media source, in reference to a separate blockade, argued in 1985, "As a criminal force, the RCMP tries to avoid involvement in civil disputes until clear evidence of criminal code infractions emerges."[45]

Following this logic, the usual response of authorities during the 1980s and early 1990s was to only employ criminal law when dealing with the closure of major transport routes covered by specific legislation such as the Highway Act. The obstruction of ungazetted roads and remote sites was generally treated as a "civil" issue and met with attempts to mediate between parties rather than carry out arrests. Where this failed the obstructed party could then apply for a court injunction against the blockaders. This generally only applied to named individuals and organisations, as judges were reluctant to grant blanket injunctions. The process that followed allowed blockaders to dispute the case in court and potentially file a counter-injunction to halt work. A decision could be made within a few days, but sometimes took months. If the complainants received an injunction they could return to the site where police were obliged to announce that anyone not complying with an order to disperse would be arrested. If the area was not vacated, then police could clear any obstruction and carry out arrests for contempt of court.[46]

Having filmed their opponents and taken their names, MacMillan Bloedel withdrew from the area on 20 November and applied for just such an injunction in the BC Supreme Court to prevent ten named individuals from interfering with work. Nuu-chah-nulth representatives filed a counter-injunction requesting a moratorium be put on all work until an ongoing claim for Aboriginal title was resolved. On 3 December MacMillan Bloedel were granted their injunction, but at the same time the finding restricted them to minor survey work until the counter-injunction was heard a fortnight later.[47]

That the first in a series of environmental blockades bringing First Nations and environmentalists together occurred over land in BC subject to a comprehensive treaty claim was unsurprising. The factors described in earlier sections as driving grievances were particularly extant in the province. Forestry provided 17 per cent of provincial income and 14 per cent of employment during the 1980s. More than 5 million hectares of forest were clear-cut and 120,000 kilometres of logging roads constructed between 1969 and 1996. Almost all the timber was transported out of the area from which it originated for processing and little of the fees paid to the provincial government flowed back to local, let alone Aboriginal, communities.[48]

Only a small number of treaties had been struck in the province during the early days of colonisation and, as in Australia, the provincial government had long cleaved to the concept of "Terra Nullius". Holding that native title had not existed prior to colonisation, the government argued that BC was exempt from British laws regarding the need to make treaties.[49] In the early 1980s the Nuu-Chah-Nulth Tribal Council presented to the federal government a claim for the west coast of Vancouver Island, adjacent islands and surrounding waters. Opposed to the continuing removal of wealth from the region and the effect of clear-felling on traditional lands, as well as hunting, trapping, and other cultural activities, they requested Meares Island and other areas remain unlogged until the issue of title was settled.[50]

Concerns regarding logging were also shared by some non-Aboriginal locals, particularly those based in the coastal town of Tofino. Many of these were relative newcomers to the region. As in other countries, trends in the 1960s and 1970s counterculture had led to an influx of people with environmental values and activist experience into rural areas that had previously favoured conventional forms of agriculture and resource extraction.[51]

While tourism was on the rise, and would become central to Tofino's roughly 1,000 residents, its value was yet to acknowledged, let alone measured. Although some were materially reliant on logging, most First Nations and non-Aboriginal people living in the area were generally marginalised, if not directly threatened, culturally, economically, and politically by existing forestry practices and their long-term effect on ecosystems.[52]

During the 1980s, the conservative British Columbia Social Credit Party administrations' primary response to the emerging conservation movement was to ignore it. Where this failed, they contained their opponents by involving them in industry dominated decision-making bodies in a pattern that environmentalists came to describe as "talk and log".[53]

Threats to old-growth forests had prompted the formation of FOCS in 1979. An informally organised group, it operated with volunteer staff, minimal administration, and few resources. As elsewhere this organisational form, member backgrounds in countercultural and political milieux, and the close connection of activists to the lands under threat predisposed the group towards tactical experimentation, agility and resilience.[54] As in the United States and Australia, emotional connections to the land were a primary motivation for action. As FOCS co-founder Susanne Hare Lawson recalls:

> The harm to the Earth was being felt by many if not all beings in the area … The land looked and felt like a war zone with armies of hard hats, trucks, blasting, and chainsaws going on … People felt it and some were collectively moved to do something including my husband and myself, even our children urged us.[55]

Concerted opposition from FOCS and the Nuu-chah-nulth to a new logging plan concerning Meares Island had stymied cutting since 1980, but long-running negotiations were eventually overridden by a direct appeal from MacMillan Bloedel to the BC cabinet.[56] This, and the moral shock associated with the indifference and collusion of corporate and government interests, emboldened rather than dispirited residents to the point where they added litigation and ODA to their existing strategies.[57] FOCS member and local General Practitioner Ron Aspinall recalls that "We had gone into the process with some faith, but it soon became clear that MacMillan Bloedel were out for blood." Despite Meares Island being small and its timber of negligible value, "We eventually realised they didn't want anything to be given away because they didn't want a wedge that could open the door [to further claims]."[58]

The higher density of pro-conservation forces in logging regions, and overall calmer social and political environment, meant that involvement in Canadian activism generally posed fewer costs for activists that in the United States. While incidents of violence, countermovement organisation, and the levying of heavy penalties for ODA occurred, they were initially at a much lower level than in the United States. Divisions nevertheless opened between communities. Government and business offices, such as those belonging to MacMillan Bloedel and the Bank of Commerce, were also relocated from Tofino, primarily to the nearby pro-logging town of Ucelet.[59]

The substantial strategic divisions between formal "insiders" and informal "outsiders" in US environmentalism were also largely absent in Canada during the 1980s. The long-term involvement of local groups with close personal ties in conventional campaigning allowed activists in areas such as Clayoquot Sound to build expertise and experience in research, monitoring, and negotiations as well as deepen roots in their communities while gathering support from others across the province.[60] These factors would enable a series of campaigns across BC to mobilise enough committed supporters to launch larger and more open ended blockades than US activists had initially been able to.

Unlike in Australia, where the first use of ODA at Terania Creek had been largely spontaneous, activists at Clayoquot Sound engaged in months of preparation. Personal relationships, shared interests, and the lengthy period of campaigning allowed Aboriginal and non-Aboriginal residents to build close alliances. These were unusual for the time with Clayoquot band council member Joe Martin stating in 1985 that, "It's the first time the whites and natives have gotten together on anything that's worthwhile."[61]

Crucial to coalition building was the primacy accorded to Aboriginal concerns. Potential disagreements related to the typical environmentalist demand to protect endangered areas via national park status were circumvented through the innovative designation of the area as a "Tribal Park". There are differing claims as to who first employed the concept, but it had been deployed by BC's Haida nation in 1981 when they declared 150,000 ha of their traditional lands under protection. Named "Duu Guusd", that tribal park's creation involved a feast held with forestry workers, during which they were informed they could no longer log the area. This was ignored and four years later, as detailed below, blockading commenced.[62]

While activists in Australia and the United States had previously renamed areas and unilaterally declared environmental parks, Duu Guusd and Meares Island appear to be the first time such a concept was integrated with Indigenous sovereignty. At Tofino's Meares Island Festival on 21 April 1984, 600 logging opponents witnessed a declaration that included total preservation and protections for Nuu-chah-nulth use and ownership. Further Tribal Parks would be dedicated in Canadian campaigns and the model extended internationally with bodies such as the Indigenous Peoples' and Community Conserved Territories and Areas consortium later visiting Meares Island.[63]

Clear warnings that a blockade would ensue should logging commence were issued to MacMillan Bloedel. A cabin was built at Tisaquis and a month-long occupation maintained. As with non-Aboriginal blockades, the protest camps created at this and later events provided a location for cultural and political activities. In the case of First Nation blockades these not only generated new meanings but tapped into existing geographical and historical understandings and provided opportunities for the practice of traditional activities such as canoe carving.[64]

The Nuu-chah-nulth leadership made it publicly clear from the beginning that any blockading would be "peaceful". A strict code of normative protester behaviour based on MNS/ONV or other established principles was not imported. Instead, a loose set of guidelines were used. Band Council Chairman George Watts told an 21 October rally of 1,200 outside parliament in Vancouver that "We are not going to start fighting. We are not going to fight another human being."[65] This reflected ethical, image, and safety concerns and was also aimed at ensuring cordial and respectful relations with local workers and residents.[66]

Differences regarding sabotage publicly emerged after the media ran stories in September 1984 concerning tree-spiking. At least one warning letter sent to MacMillan Bloedel on 16 September 1984 — containing a spike, a receipt for a box of nails, and a vow that protesters would never let the company log the island — was signed by local activist Carl Hinke. The activist viewed spiking as an inexpensive form of "insurance" that would permanently render trees worthless to MacMillan Bloedel due to the potential cost of damage to their sawmills.[67]

Articles that appeared in the *Earth First! Journal* and elsewhere subsequently claimed that up to 23,000 trees were treated and the situation was later cited by US EFers as an example of efficacious deployment. Spikes were placed above where cutting would occur and signs posted warning of their presence. After the company marked trees they believed were spiked, to indicate they should not be cut, activists copied the symbols onto trees which had not.[68]

Hinke maintains that he was initially unaware of the tactic's previous use and first learnt of it from an Ahousaht elder who spoke of logging workers using it during industrial action earlier in the century.[69] For his part Sea Shepherd founder Paul Watson later claimed to have been the first to have successfully used the tactic against logging in Canada, at Grouse Mountain near Vancouver in 1983.[70]

Opinions differ regarding the degree to which spiking stalled logging, thereby giving blockaders time to set up their encampment and build support for ODA.[71] Although there appears to have been some local support for spiking, Aboriginal groups and FOCS officially distanced themselves from the tactic.[72] Despite deepening divisions with logging workers, and criticism from the timber industry and courts, tree-spiking did not lead to the amount of controversy or odium experienced during US campaigns. This was possibly due to the fact that alongside environmentalists, companies and the BC government were also in a period of learning, so that counter-movements were not yet capable of fully exploiting and promoting the issue.[73]

Spiking was not yet illegal, so no police action was taken. Attempts by MacMillan Bloedel to use the incidents to convince courts to ban all protesters from the island failed. Tree spiking, bridge burning, and other forms of major sabotage occurred during later forest campaigns. However, radical Canadian environmentalists did not typically advocate them and unlike in the United States they did not become closely identified with them.[74]

Following MacMillan Bloedel's withdrawal logging opponents continued to camp on the island as the issue worked its way through various levels of the court system. Initial findings against the Nuu-chah-nulth focused on threats to the province's economic practices. However, on 27 March the BC Supreme Court ruled in favour of the Nuu-chah-nulth, finding that clear-felling would leave them unable to enjoy Aboriginal rights if they were to win control of the island, while delays in logging would not significantly harm MacMillan Bloedel's interests. This ground-breaking decision, essentially the first time courts had overridden a provincial decision on the basis of a native title dispute, imposed a moratorium until November, when it was anticipated that hearings regarding title would begin. It was ultimately extended by delays in that case. The Nuu-chah-nulth and BC finally entered negotiations in 1993, which resulted in an Interim Measures Agreement the following year allowing for Aboriginal veto regarding some land use decisions. Tensions with environmentalists rose at this time due to the possibility that some bands might support non-conventional logging. Some Nuu-chah-nulth leaders would later claim they were left to complete campaigning work while environmentalists from outside the area moved on to other crises. Although native title issues have not been resolved at the time of writing, Meares island remains unlogged and is unlikely to be in the near future.[75]

The decision of the Supreme Court did not bring the flood of injunction applications, or the industry shut down and unemployment that the BC Social Credit government and logging companies warned of. Such cases were expensive to mount and the court had been clear in stating that Meares Island was a "special place so far as the Indians are concerned ... it is no ordinary logging site."[76] Despite this the campaign had demonstrated, according to Ric Careless, "that mass public protests worked in contrast with the terrible results of endless Forest Service processes".[77] The next environmental blockade in BC, held in Haida G'Waii/Queen Charlotte Islands in late 1985, would in many ways replicate the Meares Island campaign in terms of its grievances, tactics, and alliances. It would also involve a much longer period of ODA.

Lake Wollaston

Before the Haida blockades came to command national interest a smaller, four-day blockade was mounted at Lake Wollaston in northern Saskatchewan from 14 to 17 June 1985. Its outcomes illustrate how the nature of local and regional politics, and differences in the character and depth of relationships between non-Indigenous and Indigenous activists, can produce very different results.

The blockade was initiated by the Collins Bay Action Group, an alliance formed in December 1984 between anti-nuclear activists and members of the Lac Le Heche band, Chipewayan people within the Dene nation. Both had long opposed plans by mining company Eldorado to create a new uranium mine and use a former mine running adjacent and under Lake Wollaston to store waste. As in BC, grievances related to political closure and the overriding of local sovereignty were interlinked with threats to a biodiverse place, as well as in this case human health.[78]

Although non-Indigenous anti-nuclear activists – such as Mike Goldstick, whom the Lac Le Heche band hired as an advisor – had a major influence, it had been agreed at the Collins Bay Action Group's inception that local Aboriginal people would retain control over all major decisions. The majority of non-Indigenous people involved in Canadian blockades during the 1980s were local residents, many of whom did not have extensive prior experience with blockading. In contrast, during this blockade non-Indigenous members were drawn from anti-nuclear movements outside of the area. These lacked the long-standing links that might have allowed them to counter accusations of fearmongering and exploiting sovereignty issues to further their own agenda.[79]

Aboriginal support for the blockade was far from unanimous. Although hundreds of local Aboriginal people would take part in ODA and the cultural activities and protest meetings that preceded it, others supported the mine. Uranium mining was also promoted by most of the province's northern Aboriginal leadership, including the Federation of Saskatchewan Indian Nations (FSIN), a subsidiary of which provided trucking and security services to the industry.[80]

During the four days of ODA, the sole entry point to the construction site was blocked with bodies, cooking fires, rocks, and a barricade. Protesters eventually withdrew after the RCMP threatened arrests and announced that promised meetings with the mine's owners would be cancelled. Barring one short confrontation, in which police tried to apprehend a protester for spray painting anti-nuclear slogans, interference up to that point had been minimal.[81]

This rapid shift in protest policing contrasted with the way most Indigenous led blockades were dealt with in BC and Ontario. The decision to use criminal, rather than formalised, and often drawn out, civil processes likely stemmed from authorities being emboldened by the lower level of local and provincial support for environmental issues and divisions among Aboriginal people concerning them.[82]

Internal community debates were intensified by ODA. Combined with interventions from FSIN these resulted in a rapid abandonment of opposition to the mine. Following a meeting with government and company officials three days after the blockade the local band leadership, and FSIN representative Sol Sanderson, framed grievances solely in terms of communication and economic issues and agreed to involvement in committees tasked with solving them. In a press conference Lac La Heche Chief Kkalthier stepped back from strident anti-mining statements he had previously made, stating: "I'm for employment. I'm not Chief for fighting with the government or protesters or something like that …"[83]

Although some lobbying and networking efforts continued over the following decades, the ending of the blockade marked a major downturn in local Aboriginal anti-nuclear activism.[84] Studies undertaken by O'Neil, Elias, and Yassi argue that despite waste spills and few economic benefits accruing from the project, antagonism towards activists rather than miners remained strong into the 1990s. In part this stemmed from beliefs that the company was punishing the entire community for the blockade as well as claims that publicity around health risks had caused a reduction in demand for locally caught fish.[85]

Despite the publication of a book by Mike Goldstick in 1987, which further exacerbated local enmities,[86] the blockade appears to have exerted negligible, if any, influence on the development of Canadian environmental ODA. This is attributable to its remote location and minimal national media coverage as well as possibly the failure to stop the mine. Its composition, outcomes, and approach nevertheless serve as an example of the way in which ODA can intensify internal conflicts and break up existing alliances. The blockade also demonstrates how associated publicity and confrontation can bring in previously largely uninvolved or marginalised actors, such as the FSIN and pro-mining residents, thereby closing rather than opening opportunities for environmental protection.

Haida G'Waii and other BC campaigns

Opposition to the clear-felling of rainforests in the Haida G'Waii archipelago, a group of islands 170 kilometres offshore from the city of Prince Rupert, BC, began in the early 1970s. Dubbed the "Galapagos of the north", the area includes several endemic animal, tree and bird species, and a complex series of old-growth rainforest ecosystems.[87]

Sparsely populated, the area is home to the Haida people. Having long agitated to formalise their land ownership they registered a land claim with the Canadian state in 1981. At the same time, they challenged its authority by registering the traditional boundaries of Haida Gwaii with the UN, rejecting national claims to the Law of the Sea Convention giving Canada a 200-mile maritime jurisdiction. Combined with these assertions of sovereignty were arguments for environmental preservation with Council Chief Miles Richardson stating in 1985: "For us to exist as a people, for us to exist as a culture, there has to be a place on our homelands, on our domain, our only place in the world, that is left unspoiled."[88]

In response to an application to log South Moresby's Burnaby Island, Haida carver Gary Edenshaw/Guujaw and American kayaker Thom Henley formed the Island Protection Committee and drew up a proposal for a 145,000-hectare park in 1974. In alliance with members of the area's small countercultural scene and the Haida Nation Executive, a recently formed campaigning body, Guujaw and Henley managed to convince the Skidegate Band Council to reject the logging plan. With the progressive New Democratic Party (NDP) in power a five-year moratorium was placed on logging. In response Rayonier switched operations to nearby Lyell Island and from 1976 to 1986 cut around two million cubic metres a year from it.[89]

A decade of activism – including attending public hearings, carrying out ecological surveys and litigation, and touring the country with slideshows – failed to significantly affect logging but garnered support to the point where the creation of a South Moresby National Park became a national issue. In response the federal government announced in 1984 that it would be prepared to jointly fund this with the province.[90]

Sovereignty issues, distrust of government decision makers, and apprehension over increased visitor numbers meant that Haida activists held concerns regarding the creation of a new national park. These were not fully overcome, but the existence of federal regulations enabling an alternative National Park Reserve, which would allow for Aboriginal hunting, trapping, and other rights, led them and their allies within the Island Protection Society (IPS) to not publicly oppose the plan.[91]

Progress was brought to a halt by resistance from the BC government as well as bureaucratic and political turmoil within the federal environmental portfolio.[92] This, combined with a decision to log Lyell Island's Windy Bay, one of the few areas whose logging had previously been stalled through involvement in bureaucratic processes, fuelled a new round of contention.[93]

Building on their 1981 declaration of the area as a Tribal Park, the Haida leadership designated it a Haida Heritage Site under the authority of the Haida constitution in 1985. In line with this Haida representatives refused to take part in a new BC consultative body, the Wilderness Advisory Committee. In September they announced plans to block any attempt to begin logging with Richardson subsequently stating "The Haida people have made a decision that there is to be no further logging in that area. We fully intend to uphold that."[94]

Highlighting the enforcement of sovereign rights Guujaw prefers not use the term "blockade", instead characterising the ODA that followed as a "stand" against what the Haida considered illegal destruction of their lands.[95] Although they welcomed environmentalist support, the community directed that any blockading would be a Haida only affair. This was to send what Richardson later described as "a crystal clear, unmistakeable message that this was a Haida issue – it was an environmental issue, but a Haida responsibility."[96]

Although most residents in the area opposed clear-felling to some degree, support for it, as during the Meares dispute, centred on a town economically reliant on the practice. Hostility towards the Haida and environmentalists across the province was spurred by vitriolic attacks in a local magazine, *The Redneck News*, which was initially funded and printed by logging contractor Frank Beban. Paralleling the rise of Wise Use movements in the United States, the publication's editor would come to head up the pro-logging Moresby Island Concerned Citizens Group. This organisation's formation, campaigning, and framing of the issues as "jobs vs environment" would be replicated during other Canadian disputes. While assaults on environmentalists and boycotts of their businesses followed, the costs were not significant enough to deter the use of ODA.[97]

In October Beban, who was keen to maximise logging while he still could, increased his workforce on Lyell Island and requested RCMP support. At the end of the month Chief Dempsey Collison led a small group of Haida onto the road where they stood and refused to move when trucks arrived. No one initially knew how their opponents might respond, but all parties were keen to avoid conflict, with Beban stating: "We're not going to cause violence. We won't log as long as they're stopping us." Speaking on behalf of the Haida, Richardson told the media, "Our people value their relationship with the other people on the island."[98]

Beban's workers withdrew and his employer Western Forest Products sought an injunction against the blockaders. Before this was granted, legal proceedings held up logging for ten days and allowed the Haida, who eschewed the use of lawyers and represented themselves, to make arguments concerning sovereignty and sustainability to the court and media. The company was granted an injunction on 11 November and workers entered their worksite three days later. Little logging occurred for four days, as intermittent road occupations held up further passage until blockaders were presented with writs. At this point they voluntarily withdrew.[99]

Following much dialogue, employing traditional Haida methods of informal consensus decision making, it was determined that three elders would be the first to be arrested. This flowed from the trio's wish to lead the next phase of action and show support for the younger generation's activism, as well as to maximise poor publicity for the government. On the morning of 15 November they addressed the media and blocked the way. When politely asked to leave, they politely declined. After the six RCMP officers present had issued writs and allowed Ethel Jones to complete her prayers, the group was arrested for "mischief", gingerly escorted to waiting vehicles, and transported for processing 50 miles away. As would become common, both in this and later Canadian campaigns, the blockaders were subsequently freed after agreeing not to return to the site. Fifteen minutes after their removal Beban's crew drove over a symbolic row of cedar branches to begin work.[100]

Media interest in the situation had been high and grew further after cameras captured the first ever arrests in BC for anti-logging ODA.[101] The conduct involved appears to have been received by wider publics in the way it was intended: as a dignified stand in defence of traditional lands. In an interview conducted years later Guujaw revealed:

> We looked at all the different options, including armed confrontation ... We knew that we would be beat ... we knew that probably there were a lot of people in government who would like nothing better than for us to have made an armed blockade – and just done away with us.[102]

The restraint exercised by all parties was also related to personal relationships. These were evidenced by the fact that some loggers and many RCMP officers were either local residents or Haida themselves. Inspector Harry Wallace was quoted as stating that 15 November was "the worst day he had ever had" while another officer Allan Wilson, the nephew of elder Ethel Jones – who would later become a Hereditary Chief and agitate for land rights – openly cried during the arrests.[103]

Blockading did not occur the next day, but resumed on 18 November when ten young Haida, wearing traditional paint, were arrested after they stood in front of Beban's convoy at several points along the logging road. As would recur throughout the remainder of the blockade they calmly declined police requests to move and allowed themselves to be escorted away. Concerned that his workers would become aggressive in response, Beban gave them a two-day rest. When they returned, they were met once more by Haida community members, 12 of whom were arrested.[104]

Almost daily occupations continued thereafter, resulting in a total of 72 arrests by early December. In response the BC Supreme Court issued warnings and subsequently served 17 Haida with contempt of court notices. NDP member of Parliament Svend Robinson, one of the few non-Haida allowed to join the blockade, had not been arrested, but was also cited for participating in ODA and encouraging others to do so. He would eventually issue an apology to the court and pay a fine. Of the others, seven had charges dropped and the remainder received suspended sentences.[105]

After the first few road occupations little logging was disrupted. Despite this, activists did not introduce tactics of escalation – such as building barricades, changing the location of blockading, and avoiding or resisting arrest – as would later be employed in Canadian protests. There was one incident of sabotage, in which the tyres of logging trucks were slashed and fuel taken, but this received only minor media attention. Haida spokespeople immediately apologised and no charges were laid.[106] According to Guujaw, the Haida did not believe that they could force an end to logging through obstruction or the imposition of costs alone, favouring instead a strategy of building public awareness and support via media coverage while simultaneously using obstruction to assert their authority.[107]

This proved successful, as while journalistic interest in events at the blockade itself faded, overall coverage persisted due to the contempt of court charges and other developments that flowed from ODA. As the arrests continued, the situation took its toll on police morale. Rather than blame the Haida for the situation, the operational commander RCMP Superintendent Bob Currie, stated, "It's not my position to say the Government should step in, but it would certainly be a welcome step."[108]

By early December the Haida's continuing resolve had built support to the point that polls were showing a majority of BC residents were in favour of the province beginning negotiations regarding land rights.[109] With $200,000 already spent on policing, BC sought and failed to receive an injunction barring anyone from travelling to Lyell Island without first agreeing not to obstruct logging operations.[110] Blockading was only suspended in December after Social Credit ministers met with Haida representatives. Negotiations quickly stalled, but with a summer break in logging approaching, no new ODA was undertaken.[111]

During the same period, the provincial government found itself under increasing pressure from other conservationist campaigns as well as two more First Nations blockades – one at Saanichton Bay and the other in Kitwanga. The former saw ODA take place on 26 November 1985. The Tsawout people had long opposed the issuing of licenses enabling the construction of a breakwater, car park, and 500 berth marina on the basis they would destroy a traditional fishery and transform the only sheltered bay in the area not already subjected to major development. The dispute also concerned sovereignty issues, as while the group was one of the few in the province to have entered into the 1852 James Douglas treaty, the provincial government had consistently ignored and challenged its validity. When test dredging commenced, protesters deployed small boats but failed to block a barge before two Tsawout activists climbed aboard it. Amidst a snowstorm they clung to cables for two hours before lawyers for both parties agreed that protesters would withdraw in return for the company cancelling work until the following week.[112] Dredging never resumed after the Tsawout gained a court injunction.[113]

The other BC blockade involved members of the Gitwangak band, part of the Gitxsan nation, blocking trains near Kitwanga over a dispute regarding compensation for land that had been expropriated for railways in 1914. ODA took place on 29 November and 7 and 8 December and involved a wooden barricade, cars, and soft blockading. A Supreme Court injunction was sought and received by Canadian National Rail before the parties achieved a resolution. Although not involving conservation issues this dispute was another demonstration of BC First Nations' growing willingness to use ODA as well as the outcomes it could bring.[114]

Despite facing further contempt of court charges Haida activists returned to Lyell Island to block loggers on 20 January 1986. Three days earlier a scandal had forced the Forestry Minister to resign after it emerged that he and the Energy Minister owned shares in the company logging Windy Bay. The scandal, which Guujaw argues only came to light because of the increased scrutiny the Haida stand had brought, prompted the government to reopen negotiations and announce a moratorium on further permits for Lyell Island until the WAC issued its report. During this time activists lifted their blockade so that talks could be held in "good faith".[115]

Beyond the scandal and widening public pressure, which had now come to include US politicians and environmental organisations, it is likely that lobbying from industry played a role. Although activists often attempt to directly impose coercive financial costs on individual opponents, the use of disruptive measures can also be used to compel change by creating broader economic uncertainty. Displeased at the attention this dispute was drawing to forestry as a whole, and correctly warning that continued blockades and land disputes would deter investment over time, the Business Council of BC urged the government to find a solution.[116]

Wedged between competing demands the Social Credit government continued to adopt ad hoc responses. For the next 14 months the fate of the South Moresby issue lay in the hands of federal and provincial politicians, as the two negotiated intermittently to set the boundaries and funding of a park.[117] When logging resumed Haida activists briefly interrupted transportation, but thereafter focused on other means of asserting sovereignty. These included renouncing Canadian citizenship, launching a train-based "caravan", which held protests across the country, and building a traditional longhouse at Windy Bay in defiance of government zoning regulations. Unlike campaigns elsewhere, which had faltered when blockading failed to bring results, the Haida's long-term view and refusal to rely on any one strategy allowed them to adopt different tactics as required.[118]

A deal was finally struck on 7 July 1987 to create a 147,000 ha park roughly matching the boundaries drawn up by Guujaw and Henley 13 years earlier.[119] Activists had been preparing again for ODA, but instead held a celebration involving around 1,000 guests. Despite preservation having been achieved for the lower third of the archipelago they expressed wariness regarding the outcome's impact on their land claim. Decades later this remained unresolved, but an agreement covering Gwaii Hanas was reached, which recognises both Haida and government jurisdiction and involves power sharing arrangements and government-to-government negotiations. This is in contrast with co-management deals struck elsewhere, in which Canadian law and designations remain paramount.[120]

Some within the Haida community questioned the investment of time and effort in this issue, especially given problems of poverty and unemployment. In response Guujaw argued in 2010: "We've gone from having no say over the resources, to now having half of the landscape under protection, all of it to be co-managed. We have knocked down the logging to one-third of what it once was."[121] With the park costing the federal government $106 million in funding and $37 million in payouts to loggers, some non–Aboriginal conservationists claimed at the time that the investment involved would deter politicians from enacting further protections.[122] Nineteen years later Western Canada Wilderness Committee co-founder Paul George argued that this had not been the case, asserting that it "set the precedent that a large wilderness area could be protected", emboldening activists to compromise less thereafter.[123]

During the campaign some sections of the media, and national and international environmentalists, focused on what geographer David Rositer describes as ecological aspects, "all but divorced from Haida culture, except where it might serve as an aesthetic backdrop, through reference to tradition and harmony in nature."[124] Despite this "wilderness" framing, examination of coverage in Canadian newspapers during the period suggests that the media at times prioritised "First nations" claims over "ecological" ones. This was particularly true during the sections of the campaign over which the Haida were able to exert the most control, that is during the blockade and before provincial and federal government negotiations began in earnest. Even at this point, newspapers regularly carried reports of Haida concerns regarding ownership and management. The decision to have a Haida-only blockade therefore appears to have been successful in ensuring they were the visual focus and primary spokespeople during ODA while further campaigning allowed them to continue to publicly assert that land use decisions ultimately rested with them.[125]

Conclusion

The BC blockades of 1984 and 1985 applied an existing template of First Nation ODA to anti-logging and anti-development campaigns. In doing so they played a major role in drawing national and international attention to specific land disputes as well as forestry and land rights in the province more generally. Each of the campaigns acted and developed tactics largely independently. Although Haida activists attended Nuu-chah-nulth strategy sessions a year before they launched their own blockade, it appears that each blockade provided inspiration rather than a specific model for one another.[126]

Unlike in Australia and the United States, these early Canadian campaigns were primarily led by First Nations communities and blended environmental concerns with those regarding sovereignty. In keeping with existing Indigenous community practices ODA tactics were mainly of the soft blockade and barricade variety. While intentions to blockade were telegraphed long in advance and relationships between police and protesters generally civil, exact details of tactics were rarely provided. During this time approaches towards normative protester behaviour in BC did not involve formal codes of conduct.

The use of civil law in Canada to deal with blockading was advantageous in that it allowed soft blockading and barricading to create longer delays than elsewhere. It also allowed ODA to provoke civil action that facilitated counter-injunctions. In some cases, courts delivered the immediate outcomes that activists sought. At others, the combination of blockading and litigation generated enough publicity and pressure to force the provincial government to make major concessions. Although blockades and campaigning sapped community resources, the preservation of areas, and the legal and political precedents that accompanied it, encouraged an expansion in blockading over the coming years.

Notes

1 Rik Scarce, *Eco-Warriors*, 46–57.
2 Jeremy Wilson, *Talk and Log: Wilderness Politics in BC, 1965–96* (Vancouver: University of British Columbia Press, 1998), 194–98; George Hoberg and Edward Morawski, "Policy Change through Sector Intersection: Forest and Aboriginal Policy in Clayoquot Sound," *Canadian Public Administration* 40, no. 3 (1997): 398–99; Margaret Horsfield and Ian Kennedy, *Tofino and Clayoquot Sound: A History* (Madeira Park: Harbour Publsihing, 2014), 501–4.
3 Claudia Notzke, *Aboriginal Peoples and Natural Resources in Canada* (North York: Captus Press, 1994), 81–83; Steven Bernstein and Benjamin Cashore, "The International-Domestic Nexus: The Effects of International Trade and Environmental Politics on the Canadian Forest Sector," in *Canadian Forest Policy: Adapting to Change*, ed. Michael Howlett (Toronto: University of Toronto Press, 2001), 65.
4 Michael Howlett and Jeremy Rayner, "The Business and Government Nexus: Principal Elements and Dynamics of the Canadian Forest Policy Regime," in *Canadian Forest Policy: Adapting to Change*, ed. Michael Howlett (Toronto: University of Toronto Press, 2001): 24–25.
5 Andrea Olive, *The Canadian Environment in Politcal Context* (Toronto: University of Toronto Press, 2016), 151–74.
6 Howlett and Rayner, "The Business and Government Nexus," 29–31.
7 Jeremy Wilson, *Talk and Log*, 67–72.

8 Ibid., 64–72; Howlett and Rayner, "The Business and Government Nexus," 45–46.

9 Ibid; 127–28, 43–45.

10 James Frideres, "Circle of Influence: Social Location of Aboriginals in Canadian Society," in *Aboriginal Peoples and Forest Lands in Canada*, 33–34.

11 BC Treaty Commission, What's the Deal with Treaties? (Vancouver: BC Treaty Commission, 2007). 1.

12 Notzke, *Aboriginal Peoples and Natural Resources in Canada*, 2–5.

13 Ian Gill, *All That We Say Is Ours* (Vancouver: Douglas & McIntyre, 2009), 25–42.

14 Howard Ramos, "What Causes Canadian Aboriginal Protest? Examining Resources, Opportunities and Identity, 1951–2000," *The Canadian Journal of Sociology* 31, no. 2 (2006): 226–27; Kiera Ladner, "Aysaka'paykinit: Contesting the Rope around the Nations' Neck," in *Group Politics and Social Movements in Canada*, ed. Miriam Smith (Toronto: University of Toronto Press, 2008), 248.

15 Olive, *The Canadian Environment in Political Context*, 201–18; Commission, *What's the Deal with Treaties?* 19.

16 D.B. Tindall, Ronald Trosper, and Pamela Perreault, "The Social Context of Aboriginal Peoples and Forest Land Issues," in *Aboriginal Peoples and Forest Lands in Canada*, 4–7.

17 Frank Cassidy and Norman Dale, *After Native Claims?: The Implications of Comprehensive Claims Settlements for Natural Resources in British Columbia* (Lantzville: Oolichan Books, 1988), 91–96.

18 Bruce Willems-Braun, "Colonial Vestiges: Representing Forest Landscapes on Canada's West Coast," in *Troubles in the Rainforest* ed. Trevor Barnes and Roger Hayter (Victoria: Western Geographical Press, 1997), 100–19.

19 Notzke, *Aboriginal Peoples and Natural Resources in Canada*, 100–1.

20 Mary Laronde, Interviewed 20 September 2017.

21 Willems-Braun, "Colonial Vestiges," 112–19.

22 Yale D. Belanger and P. Whitney Lackenbauer, "Introduction," in *Blockades or Breakthroughs?: Aboriginal Peoples Confront the Canadian State*, ed. Yale D Belanger and P Whitney Lackenbauer (Montreal: McGill-Queen's Press, 2014), 33.

23 Ibid., 17.

24 Rima Wilkes, "A Systematic Approach to Studying Indigenous Politics: Band-Level Mobilization in Canada, 1981–2000," *The Social Science Journal* 41, no. 3 (2004): 449–52; "Zig Zag", *BC Native Blockades and Direct Action*, (Unknown: Warrior Publications, 2006), 3.

25 Nicholas Blomley, "'Shut the Province Down': First Nations Blockades in British Columbia, 1984-1995," *BC Studies: The British Columbian Quarterly*, no. 111 (1996): 11–12.

26 Ibid., 18–21.

27 Belanger and Lackenbauer, "Introduction," 14.

28 Ibid., 11–14.

29 Blomley, "'Shut the Province Down'," 24–25, 28.

30 Jim Bohlen, *Making Waves: The Origins and Future of Greenpeace* (Montreal: Black Rose Books, 2001), 21–26.

31 Ibid.

32 Frank Zelko, "Making Greenpeace: The Development of Direct Action Environmentalism in British Columbia," *BC Studies*, no. 142/143 (2004): 227; Bohlen, *Making Waves*, 31–101.

33 Robert Paehlke, "The Canadian Environmental Movement," in *Group Politics and Social Movements in Canada*, ed. Miriam Smith (Toronto: University of Toronto Press, 2014): 286–90.

34 Kathleen Rodgers, *Welcome to Resisterville: American Dissidents in British Columbia* (Vancouver: UBC Press, 2014), 143–49.

35 Rod Bantjes, *Social Movements in a Global Context: Canadian Perspectives* (Toronto: Canadian Scholars' Press, 2007), 234; Joseph G. Moore, "Two Struggles into One? Labour and Environmental Movement Relations and the Challenge to Capitalist Forestry in British Columbia, 1900–2000" (PhD thesis, McMaster University, 2002), 217–40.

36 W.R.D. Sewell, P. Dearden, and J. Dumbrell, "Wilderness Decisionmaking and the Role of Environmental Interest Groups," 168; Howlett and Rayner, "The Business and Government Nexus," 46–47.

37 George Hoberg and Edward Morawski, "Policy Change through Sector Intersection" *Canadian Public Administration* 40, no. 3 (1997): 398–99.

38 In Canada the RCMP operate as the primary police force on both a federal and provincial level in areas that do not maintain an existing police force. Ontario and Quebec have provincial police forces, but the remainder of provinces and territories generally employ the RCMP in rural areas and most cities under the direction of provincial governments. RCMP Organizational Structure, Available [Online]: www.rcmp-grc.gc.ca/about-ausujet/organi-eng.htm (Accessed 7 February 2017).

39 Margaret Horsfield and Ian Kennedy, *Tofino and Clayoquot Sound*, 504.

40 Shayne Morrow, "Chief Councillor Remembers the First Days of a Long Battle on Meares," *Ha-Shilth-Sa* 25 April 2014, Available [Online]: www.hashilthsa.com/news/2014-04-25/chief-councillor-remembers-first-days-long-battle-meares (Accessed 22 August 2016); Horsfield and Kennedy, *Tofino and Clayoquot Sound*, 504.

41 Rima Wilkes, "The Protest Actions of Indigenous Peoples: A Canadian-U.S. Comparison of Social Movement Emergence", *American Behavioral Scientist* 40, no.4 (2006): 514–19; Terry Glavin, Interviewed 29 August 2017.

42 Christopher Wren, "Canadian Battle Rages over Lovely Timbered Isle," *New York Times*, 17 May 1985, A2.

43 Morrow, "Chief Councillor Remembers the First Days of a Long Battle on Meares."

44 Ron Aspinall, Interviewed 22 October 2017.

45 John Cruickshank, "3 Haidas Charged with Defying Order on Island Logging," *Globe and Mail*, 18 November 1985, A5.

46 Blomley, "Shut the Province Down," 14–15; "Zig Zag," *BC Native Blockades and Direct Action*, 2.

47 "Logging Firm Gets Injunction against Meares Protesters," *Globe and Mail*, 4 December 1984, 3.

48 Trevor Barnes and Roger Hayter, "Introduction," in *Troubles in the Rainforest: British Columbia's Forest Economy in Transition*, ed. Trevor Barnes and Roger Hayter (Victoria: Western Geographical Press, 1997): 5; Wilson, *Talk and Log*, 1.

49 Hoberg and Morawski, "Policy Change through Sector Intersection," 394–95.

50 Notzke, *Aboriginal Peoples and Natural Resources in Canada*, 98; Wilson, *Talk and Log*, 195.

51 Ibid, 5, 250; Rodgers, *Welcome to Resisterville*, 4–17.

52 Trevor J Barnes, Roger Hayter, and Elizabeth Hay, "Stormy Weather: Cyclones, Harold Innis, and Port Alberni, BC," *Environment and Planning* 33, no. 12 (2001): 2129–45.

53 Ric Careless, *To Save the Wild Earth* (Vancouver: The Mountaineers, 1997), 132.

54 Horsfield and Kennedy, *Tofino and Clayoquot Sound*, 501–3; Susanne Hare Lawson, Interviewed 15 October 2017.

55 Ibid.

56 Astrid Wallner, "The Aboriginal Peoples' Position in Land-Use Conflicts in British Columbia, Canada," *Bulletin* 62(1998): 62; Cassidy and Dale, *After Native Claims?*, 93; Aspinall Interview.

57 Horsfield and Kennedy, *Tofino and Clayoquot Sound*, 502.

58 Aspinall Interview.

59 Horsfield and Kennedy, *Tofino and Clayoquot Sound*, 506.

60 Valerie Langer, "It Happened Suddnely (over a Long Period of Time)," in *Witness to Wilderness: The Clayoquot Sound Anthology*, ed. Howard Breen-Needham et al. (Vancouver: Arsenal Press, 1994): 251–53.

61 Wren, "Canadian Battle Rages over Lovely Timbered Isle," A2.

62 Gill, *All That We Say Is Ours*, 102–03; Author correspondence with Guujaw, 2017.

63 First Tribal Park in BC/Indigenous Relations, Meares Island, Turns 30 Years Old and Is Expanded Available [Online]: http://fnbc.info/news/first-tribal-park-bcindigenous-relations-meares-island-turns-30-years-old-and-expanded (Accessed 22 August 2016).

64 Morrow, "Chief Councillor Remembers the First Days of a Long Battle on Meares"; Horsfield and Kennedy, *Tofino and Clayoquot Sound*, 502–3.

65 "Indians Take Stand on Logging in Park," *Globe and Mail*, 22 October 1984, 4.

66 Morrow, "Chief Councillor Remembers the First Days of a Long Battle on Meares".

67 Mike Roselle, "Meares Island: Canada's Old Growth Struggle," *Earth First! Journal* 5 no.3 (1985): 5; C.J. Hinke, Interviewed 26 October 2017.

68 "Logging Firm Gets Injunction against Meares Protesters,", 4; Roselle, "Meares Island," 5; Hinke Interview.

69 Ibid.

70 Paul Watson. Letter to Ha-Shilth-Sa Reporter, Available [Online]: http://forestcouncil.org/paul-watson-on-clayoquot-tree-spiking/ (Accessed 11 November 2016).

71 Ibid; Aspinall Interview.

72 Roselle, "Meares Island," 5.

73 Hinke Interview; Aspinall Interview.

74 "Judge Won't Ban Activists from Island Logging Site," *The Globe and Mail*, 18 December 1984, 5.

75 "Indians Win Appeal as B.C. Island Gets Reprieve from Loggers' Saws," *The Gazette*, 28 March 1984, B1; Notzke, *Aboriginal Peoples and Natural Resources in Canada*, 98–101.

76 "Indians Win Appeal as B.C. Island Gets Reprieve from Loggers' Saws," B1.

77 Careless, *To Save the Wild Earth*, 133.

78 "Residents Fear Uranium Will Pollute Lake," *Globe and Mail*, 4 January 1985, 5; "Wollaston Lake Saskatchewan: Residents Oppose Uranium Mining," *Akwesasne Notes*, Summer 1985, 4–5.

79 Michael Goldstick, *Wollaston: People Resisting Genocide* (Montreal: Black Rose, 1987), 214–73; John D O'Neil, Brenda D Elias, and Annalee Yassi, "Situating Resistance in Fields of Resistance: Aboriginal Women and Environmentalism," in *Pragmatic Women and Body Politics*, ed. Margaret Lock and Patricia Kaufert (Cambridge: Cambridge University Press, 1998), 272–81.

80 "Resistance at Wollaston," *Open Road*, Spring 1986, 6–7.

81 Ibid; "Wollaston Lake Saskatchewan: Residents Oppose Uranium Mining", 5.

82 Terry Glavin Interview.

83 A full transcript of the press conference is included in Goldstick, *Wollaston: People Resisting Genocide*, 255; "Residents Fear Uranium Will Pollute Lake," 5.

84 Jamie Kneen. Web Post Regarding Wollaston Lake, Available [Online]: http://sisis.nativeweb.org/clark/nov1198can.html (Accessed 17 November 2016).

85 O'Neil, Elias, and Yassi, "Situating Resistance in Fields of Resistance," 272.

86 Ibid, 266, 81.

87 Sewell, Dearden, and Dumbrell, "Wilderness Decisionmaking and the Role of Environmental Interest Groups," 156.

88 Quoted in David Rossiter, "The Haida Action on Lyell Island," in *Blockades or Breakthroughs?*, 75; Notzke, *Aboriginal Peoples and Natural Resources in Canada*, 97.

89 Gill, *All That We Say Is Ours*, 67–74, 95.

90 Elizabeth May, *Paradise Won: The Struggle for South Moresby* (Toronto: McClelland and Stewart, 1990), 34–36, 42–56.

91 G Bruce Doern and Thomas Conway, *The Greening of Canada: Federal Institutions and Decisions* (Toronto: University of Toronto Press, 1994), 179; May, *Paradise Won*, 50, 104.

92 Doern and Conway, *The Greening of Canada*, 176–77.

93 May, *Paradise Won*, 82–83, 91.

94 Christie McClaren, "South Moresby Logging Plan Sparks Ire," *Globe and Mail*, 21 October 1985, A5.

95 Author correspondence with Guujaw, 2017.

96 Gill, *All That We Say Is Ours*, 120.

97 May, *Paradise Won*, 35, 64–66, 81.

98 "B.C. Haida Block Road," *Globe and Mail*, 31 October 1985, A5.

99 "Haidas Form Human Blockade to Prevent Logging on B.C. Island," *The Citizen*, 15 November 1985, A12; May, *Paradise Won*, 119.

100 Cruickshank, "3 Haidas Charged with Defying Order on Island Logging," A5; Rossiter, "The Haida Action on Lyell Island," 79–80.

101 May, *Paradise Won*, 122.

102 Gill, *All That We Say Is Ours*, 57–63.

103 "Indians Arrested in Logging Dispute," *The Citizen*, 18 November 1985, C18; May, *Paradise Won*, 120–22.

104 "10 More Haida Indians Arrested While Blocking Logging on Queen Charlotte Islands," *The Citizen*, 19 November 1985, A13; "12 More Haida Indians Arrested," *The Citizen*, 21 November 1985, E14.

105 "Haidas Found Guilty of Contempt for Disobeying Court Order," *The Citizen*, 30 November 1985, A20; "9 Haida Given Suspended 5-Month Terms, MP Fined $750," *The Gazette*, 7 December 1985, A2.

106 "28 Haida Are Arrested in B.C. Logging Protest," *Globe and Mail* 26 November 1985, A5.

107 Author correspondence with Guujaw, 2017.

108 "28 Haida Are Arrested in B.C. Logging Protest", A5.

109 Gill, *All That We Say Is Ours*, 134.

110 "Haidas Found Guilty of Contempt for Disobeying Court Order," A20.

111 Jack Danylchuk, "2 B.C. Ministers Meet Haida, Discuss Logging, Land Claims," *Globe and Mail*, 11 December 1985, A8.

112 Terry Glavin, "Tsawout in Front Lines of B.C. Treaties Battle," *Vancouver Sun*, 4 November 1985, B1; "Indians Oppose Marina," *Provincial Citizen*, 26 November 1985, 7.

113 "Indians' Fish Rights Confirmed," *Vancouver Sun*, 10 October 1987, A11.

114 "B.C. Indians Barred from Halting Trains," *Globe and Mail* 9 December 1985, A3; "Indians, CN Reach Deal on Land Claim," *Gazette*, 13 December 1985, C19.

115 John Cruickshank and Jack Danylchuk, "Holds Investment in Forest Company, B.C. Minister Quits," *Globe and Mail*, 18 January 1986, A1; Author correspondence with Guujaw, 2017.

116 Dan Smith, "Temporary Truce Calms Haida-B.C. Dispute," *Toronto Star*, 19 December 1985, A18.

117 Doern and Conway, *The Greening of Canada*, 181.

118 Author correspondence with Guujaw, 2017.

119 "PM, Vander Zalm Sign South Moresby Park Deal," *Ottawa Citizen*, 13 July 1987, A4.

120 Will Horter. The Real Story Behind Gwaii Haanas, Available [Online]: https://dogwoodbc.ca/the-real-story-behind-gwaii-haanas/ (Accessed 1 June 2017); Author correspondence with Guujaw, 2017.

121 Martin Lukacs. From Queen Charlotte to Haida Gwaii, Available [Online]: www.dominionpaper.ca/articles/3248 (Accessed 27 October 2016); May, *Paradise Won*, 309–11; Gill, *All That We Say Is Ours*, 163–64, 98.

122 Wilson, *Talk and Log*, 220–22.

123 Paul George, *Big Trees Not Big Stumps* (Vancouver: Western Canada Wilderness Committee, 2006), 99.

124 Rossiter, "The Haida Action on Lyell Island," 82.

125 This summary is based on an overview of more than 100 articles that appeared in Canadian newspapers throughout the campaign.

126 Gill, *All That We Say Is Ours*, 115; Author correspondence with Guujaw, 2017.

6

"SOMEONE HAD TO STAND UP TO THEM"

Canadian expansion, differentiation, and entrenchment, 1986–1989

This chapter examines the way in which environmental blockading in Canada became fully established as a means of defending biodiverse places. It chronicles, in the context of new court decisions and changing government policies, how First Nation led campaigns continued to combine ODA, litigation, and other strategies, in ways that became increasingly differentiated. Echoing previous developments in Australia and the United States, it also explores how a second body of tactics emerged. These mainly involved non-Indigenous activists who turned to techniques of manufactured vulnerability that involved smaller numbers of activists minimising arrests via tactics such as tree-sits and lock-ons.

1986–1987: A quietening before a new wave begins

During 1986 and 1987 there were only a small number of blockades concerning environmental issues. The first of these involved activists belonging mainly to BC's West Kootenay "back-to-the-land" milieu declaring the region a "pesticide and herbicide free zone" on the basis of health and ecological concerns.[1] Beginning in 1986, Argenta's Nonviolent Action Group and others acted to prevent spraying in the Slocan Valley. Following strict MNS-style nonviolence principles and training drawn from the anti-nuclear movement, protesters blocked roadways, surrounded helicopters, and denied vehicles access to the Forestry Ministry's chemical warehouse. In the midst of concerted lobbying the campaign successfully forced the cancellation of a spraying permit in 1987. Further blockades of roadways and train tracks compelled Canadian Pacific Railways to abandon its use of pesticides in the region by 1989. While these actions did not target large-scale threats to biodiverse areas, experiences would be drawn upon when local activists blockaded logging in the Slocan Valley and elsewhere during the 1990s.[2]

A dispute closer in form and issue to those of the previous two years took place at Wazulis (also known as Deer Island) near Vancouver Island in December 1986. The 68 ha island had been sold to logger Archie Heleta in 1985 for $250,000. He and the government considered that it had been traded into private ownership in an 1851 treaty, but the terms of this were disputed by the Kwakiutl First Nation. In response to clear-cutting an occupation of two beaches began on 1 December 1986 with a ceremony asserting Kwakiutl ownership. Initial blockading, including the lighting of bonfires to discourage logging, forced operations to a different part of the island before up to 100 protesters arrived on 3 December. Faced with such numbers Haleta withdrew to seek an injunction.[3]

This was exactly the outcome the Kwakitul sought, as it not only removed the immediate threat to the area and its sites, but also brought the matter before the courts. Filing a counter-injunction they continued to occupy the island, vowing to meet any resumption of logging with what they described as "non-violent" means.[4] This was not required, as they gained a moratorium on 22 December, in part due to the discovery of a burial site during the occupation. Following the precedent of Meares Island, the BC Supreme Court granted this on the basis of the "irreparable harm" logging would cause before a decision regarding the Kwakiutl's land claim was made.[5] Given the provincial government's resistance to recognising, let alone negotiating such claims, this finding effectively put logging on hold for decades. In terms of the development of ODA-based strategies, it further reinforced the strategy of combining blockading with litigation.[6]

Other than a brief foray covered in the next section, the only environmentally oriented blockade to take place during 1987 was in the province of Alberta, where preparation for gas drilling by Shell was intermittently blockaded for a week in November by activists standing in the way of bulldozers.[7] Along with demonstrating the importance of emotional connections to land, an interview with local outfitter and conservationist Mike Judd indicates that the environmental blockading strategy was by this point well known outside of BC. Nevertheless according to Judd an attempt by members of the Alberta Wilderness Association (AWA) to prevent the clearing of Prairie Bluff – an undeveloped plateau more than 200 km from the provincial capital of Calgary – was largely spontaneous:

> We'd gone through a very lengthy public hearing process and brought in expert witnesses, but the decision makers rejected our argument. We were very disgusted and frustrated so my friend James Tweedy and I said we'd go up and see what they were doing. When we got there they had two bulldozers and were blasting right where the Big Horn sheep trail was.
>
> Some of us in AWA had talked about the possibility of blockading and getting media attention, but no specific plans had been made. Earth First!, Sea Shepherd, Greenpeace and people in BC were out there doing those sort of things, so that indicated that direct action could make gains, but we had no direct contact with anyone involved. What happened was that when we saw what they were doing we looked at each other and said, "It's just about

time we made a stand." So we walked up to the bulldozers and the drivers were shocked when they saw we weren't going to move out of the way. It was quite an experience as the bulldozer blade was above our head and when the driver got right up close they couldn't see us. It was one of those points where you knew you were really putting yourself on the line.[8]

Having realised the pair would not move, the drivers left to call their employer who ordered them to stop work, after which the activists went on further up the slope to disrupt blasting. When the police eventually arrived they used the threat of criminal charges to persuade them to leave. AWA members and others from Calgary arrived the following day and the group continued to bar the way with their bodies until the point at which they were threatened with arrest. As would occur elsewhere, the remote location and lack of police in the area, as well as sympathy on the part of some officers, meant that the time involved in arriving and hiking up the mountain allowed blockaders to halt work for hours at a time. To counter this Shell gained an injunction against anyone other than their employees entering a 485 ha area. Unlike in other Canadian cases, Alberta authorities were confident enough threaten penalties and compensatory costs of up to $100,000. As a result ODA was called off.[9]

While the blockade brought television and newspaper reporters to the remote area it had little impact upon policy. Despite the involvement of leading members in the blockade, AWA had been unable to mobilise more than a small proportion of its hundreds of members, with most actions involving a handful of people at a time. This was attributable to the blockade occurring at late notice, the conservative pro-development character of the province, and AWA's membership being scattered across a massive area.[10]

As in other cases, the introduction of a new repertoire had a major organisational impact. In this case it was a largely negative one. Dissension arose regarding the decision of leading members to blockade before the entire membership had been consulted. Combined with the possibility of having to pay damages arising from ODA, such conflict side-lined the use of radical means, and those advocating them, within AWA. As a result that they would not be employed again in the province by non-Indigenous environmentalists until the 1990s.[11]

1987–1988: The Strathcona Park Dispute

The year 1987 had seen a low level of ODA, but wider developments concerning Aboriginal and environmental causes spurred a spike in activity thereafter. The UN's World Commission on Environment and Development report was released in October 1987 and had a major impact on Canadian publics. The report's call for the level of biodiverse areas under protection globally to be raised from 4 to 12 per cent soon became a core demand for Canadian activists and was incorporated into their arguments.[12]

In the same year Canada's First Ministers, as part of negotiations concerning constitutional change, failed to come to an agreement regarding amendments recommended for Aboriginal self-government. This stimulated a new round of First Nations' activism. A national meeting of chiefs from across Canada endorsed the concept of direct action to assert sovereignty and land rights.[13]

A third motivation came from changes in BC's Social Credit government's policies. During 1987 the Forest Act was amended to allow some areas to be protected from logging through being designated "natural areas". While environmentalists welcomed this, a lack of political support and trained staff meant that the powers would not be used until the second half of 1989, and then only after sustained environmental action. Parallel decisions concerning mineral exploration and mining in Class A provincial parks opened a further front of activity.[14]

Public concern regarding mining within these areas had led the NDP to place a ban on new projects in 1973. Twelve years later the Supreme Court of Canada ruled that the holders of hundreds of mineral exploration licenses would have to be compensated. Citing the need to avoid major pay-outs, the Social Credit government subsequently opted to change the boundaries of six parks, rezoning 110,000 hectares as "recreation areas" subject to mining, hunting, and logging, and removing 50,000 hectares altogether.[15]

The applicability of the original court ruling to all of these areas was questioned by subsequent legal research and findings. Public anger regarding what was seen as a push to erode environmental protection came to focus on a popular hiking spot, Strathcona Park. Located near the centre of Vancouver Island, it was BC's oldest park, sections of which had long been open to resource extraction. Rezoning meant that this would extend into the park's centre, including Canada's highest waterfall. It also generated concerns about effects on fisheries and the nearby residents' water supplies.[16]

Reflecting the resignation of some environmentalists, Warrick Whitehead, president of the Cowichan group of the Sierra Club, told a public meeting in early 1987 that, "We can still end up with a good park, but we will have to compromise … We can't stop the mining companies, they are way too powerful."[17] Demonstrating a growing sense of confidence in the use of confrontational tactics, as well as their deep-felt connection to the park, an initially small group of local residents disagreed.[18] They formed a new group, Friends of Strathcona Park (FOSP), in 1986. Within a year they had recruited 1,000 members and set up chapters in 10 localities.[19]

For the first time in BC, First Nations groups did not play a leading role in a preservation campaign featuring ODA. A study of the campaign by John Dwyer attributes this to the local community mainly being made up of non-Indigenous people, that some Nuu-chah-nulth community members felt aggrieved that they had been left with legal bills and work related to Meares Island, and that although FOSP officially endorsed native title claims made during the campaign, support among members was mixed. Nevertheless, individual First Nations community members took part in protests while local leaders made public declarations and held ceremonies in support.[20]

A scandal regarding ministerial conflicts of interest had played a major role during the Haida Gwaii campaign. Illustrating how wider political incidents can shape decisions within campaigns, another similarly emboldened activists. On 24 February 1987, the BC Minister for the Environment Stephen Rogers was forced to resign after his ownership of shares in Westmin Resources, a company already mining in Strathcona Park, was revealed. Four days later FOSP took advantage of this by holding their first major action, blocking the road leading to Westmin's mine. As it was a Saturday the company claimed few, if any trucks would be travelling that day, but the action gained media attention.[21]

ODA had previously been rejected in 1986 and was limited to this one action in 1987 as it was believed that an ongoing blockade against existing mining would alienate potential supporters on the basis of potential job losses. Despite a number of members being politically conservative, ODA was soon endorsed against new mineral exploration because of clear public opposition to this. For strategic reasons police, politicians, and others were warned well in advance that blockading would ensue if exploration occurred ahead of a public inquiry.[22]

Upheaval would continue to dog BC's environmental bureaucracies, with multiple changes in ministerial configurations during the period of the dispute, but the government stuck to its policy. After Cream Silver Mining received a permit to explore claims on 1 January 1988, FOSP went on standby. Twenty-two days later, they responded to impending exploration by setting up a camp and a full-time blockade made up of parked cars and people sitting or standing in the road.[23]

Company vehicles were regularly turned away before an injunction ordering blockaders to disperse was obtained on 28 January. The company later brought a SLAPP suit against campaigners. Although ostensibly to recoup costs, campaigners interpreted it as an intimidatory act. A minority withdrew but, unlike in Alberta, widespread support and the commitment of a much larger number of local residents meant the threat of damages did not have a decisive effect; the suit was later dropped.[24]

As previous chapters have discussed, while the remote location and rugged geography of many protest sites may pose difficulties for campaigners in terms of mobilization, it can lend them advantages in terms of obstruction. There was only one small access road to the work site and since it was bounded by a lake on one side and a mountain on the other it was easily sealed off. The nearest police station was 90 minutes away and there were only a small number of local officers, most of whom were sympathetic to the campaign. As a result the mining company regularly encountered lengthy delays in having blockaders removed.[25]

In order to minimise media coverage and protester numbers, RCMP officers counter-blockaded a key entrance before arresting three activists on 30 January at 5.30 a.m. On this occasion reinforcements had been brought from outside the area. Underlining the careful and cordial nature of protest policing within BC at the time, protesters were given time to voluntarily move their vehicles and decide who would be arrested.[26]

Following a rally of 100 during the day, police arrested another four protesters at 2 a.m., finally allowing a drilling team to finally enter the area. The continuing inability, or unwillingness, of police to fully enforce the injunction was demonstrated within hours as protesters moved to the mine camp site and disrupted work by surrounding machinery.[27]

By this point ODA had successfully prevented work for ten days, garnering much publicity in the process. Over the next six weeks activists would engage in daily protests and site invasions with arrests taking place on six occasions. At times theatrical pieces were enacted, such as lowering mannequins in lawn chairs next to equipment in place of protesters and attaching hearts, symbolising the excision of the park's centre, to security fencing on Valentine's Day.[28]

Despite freezing weather, between 10 and 30 occupiers were generally present at the protest camp and work site with hundreds attending larger events and rallies. As the park was the first to face the expansion of mining, activists from other affected areas regularly travelled to lend support. Due to this, and the park's regular use by a wide cross-section of the province, the number of blockaders involved from outside the area was higher than during other BC blockades of the period.[29]

In keeping with previous practices, arrestees were generally released after agreeing not to return to within 5 km of the mine site, but one, whose wife's ashes were scattered in the park, refused to do so. Remaining in jail he carried out a ten-day hunger strike until the conditions were dropped in late February. Others would spend time in custody in March after initially refusing to sign bail conditions. Both actions introduced new tactics to environmental blockading in BC.[30]

Members of FOSP cited the Meares Island and Haida Gwaii campaigns as a primary inspiration for their decision to form an organisation and subsequently blockade. However, the way in which they conducted ODA reflected the individual nature of tactical and organisational repertoires within Canadian blockades during the 1980s. Although their approach resembled that used during West Kootenay's anti-pesticide blockades, it was not influenced by them. Instead key activists' previous experiences in anti-nuclear, peace, feminist, and other activism, led them to employ ONV/MNS style templates.[31]

The majority of FOSP's membership had no prior experience of protest. Among those who did were a small group from a back-to-the-land community at nearby Denman Island. These activists were committed to Gandhian influenced non-violent resistance and had set up a community mediation and conflict resolution group prior to the Strathcona Park dispute.[32] Flowing out of non-violence training workshops they organised during 1987, FOSP adopted a strict non-violence code eschewing property destruction, running, resisting arrest, and the possession of drugs, alcohol, or guns.[33]

Alongside Gandhian sources, an oral history of Tasmania's Franklin campaign was a key transnational influence. *The Franklin Blockade* was produced by Tasmanian Wilderness Society (TWS) campaigners Pam Waud and Robin Tindale in 1983. It included stories and discussion regarding tactics and organisation from a variety of viewpoints, although those of activists associated with TWS were most evident.[34]

Interviews with activists involved in Clayoquot Sound campaigns and US Earth First! indicate that the book was readily available in at least some Northern American circles. While these activists had read it, and cited the Franklin as an example that encouraged them to employ ODA, the campaigns they were associated with did not choose to follow the approach to non-violence TWS had. This was either because they rejected such principles, had already developed their own styles, or did not feel ONV was applicable.[35] In contrast, Strathcona activists took the book's stories as validation and used it to shape their overall approach to campaigning. As FOSP co-founder Marlene Smith recalls:

> We had a firm goal in that this was a place that had been set aside by the government to be protected and we were going to hold their feet to the fire until they did so … Once the blockade had begun people from FOCS gave me a copy of the Franklin book and it made me realise we had to be more sophisticated at the political end … It was our bible and handbook on what to do next. I often saw similarities between what was happening here and what had happened at the Franklin blockade and it helped us predict what the government would do. I'm not sure if it is because humans have a tendency to act in a predictable way, especially when it comes to politics, or if it was that we shaped history according to a previous event … They were convinced there was someone in their chambers tipping us off. We didn't, but it unnerved them.[36]

Influenced by Tasmanian and other precedents FOSP sought to ensure that their strict standards of normative behaviour were upheld. Due to the regular influx of new protesters, it was decided that it would be impractical to make training mandatory so organisers, identified by black armbands, outlined principles to newcomers and distributed printed guidelines and legal advice sheets. Protesters who did not follow the code, including those who damaged equipment, were removed.[37]

In keeping with the Franklin campaign, and other MNS-influenced interpretations of openness and safety, organisers regularly met with and briefed police. Unusually for such events, local RCMP officers joined a non-violence training session in the run up to the blockade. Such workshops often include exercises in which attendees act out the roles and behaviour protesters, police officers and workers typically undertake in order to practice conflict de-escalation as well as gain an understanding of other parties' emotions. As local residents many of the officers already shared concerns regarding the impact of mining, but Smith believes involvement in activities in which they role-played as protesters gave them an appreciation of the blockaders' positions and increased sympathy for them.[38]

Blockaders regularly brought operations to a halt by taking advantage of safety regulations that prohibited work whenever members of the public came too close to drilling. Other than on one occasion, when a protester climbed on top of machinery, tactics did not diversify beyond occupying space through standing, sitting, or lying on the ground. This was partially because such tactics were successfully obstructing work, but also reflected the belief of some core activists that tactics such as chaining bodies to equipment invited conflict and were not in the spirit of ethical nonviolence.[39]

Decisions regarding exactly what daily actions would involve were decided by those at the protest camp. Generally ODA would end when the RCMP arrived. If it was felt the campaign would benefit politically and publicity-wise from increased media attention, a select number of protesters, generally decided upon before actions, would refuse to obey police directions and submit to arrest. Those willing to be arrested were assigned a supporter to deal with their legal and other affairs whileothers took witness statements, and in some cases, filmed events.[40]

By mid-February the blockade, combined with pressure from local members of the ruling Social Credit party, forced the government to agree to a moratorium on further exploration as well as to form a Strathcona Park Advisory Committee. Given the province's history of broken promises, that exploration was continuing, and the blockade garnering support, FOSP decided to maintain a disruptive presence to the end.[41] During the blockade police employed lesser charges of mischief rather than contempt of court and, in part due to advice from a local judge opposed to mining, all but 57 of the 64 arrestees had charges dropped on a technicality.[42]

Work was completed on 27 March and FOSP removed its around-the-clock vigil shortly after.[43] Demonstrating the way in which the publicity associated with unconventional protest can increase organisational strength, the group's membership doubled during the course of the blockade and added another 1,000 by June 1987. Although Cream Silver had failed to find any deposits worth mining it applied to drill elsewhere in the park with the result that FOSP set up a new camp in preparation for further ODA.[44]

This would not be required as in September the BC government endorsed report recommendations calling for a ban on new mining and logging in the park. Extractive interests were clearly concerned at this shift and former Forests Minister turned President of the Mining Association of BC, Tom Waterland, told reporters the decision was "in effect, legislation by television, whereby if the people object longly and loudly enough about anything, then they'll get their own way regardless of the legitimate rights of other citizens of the province."[45] Such framing of conservationists as "greedy" and "sectional" would be increasingly deployed by Canadian counter-movements in coming years. At this point it proved unable to affect the debate. With a change of government in 1991 the park's boundaries were upgraded and the mining industry accepted the issue was too controversial, withdrawing from exploration elsewhere.[46]

The success of the FOSP blockade further promoted the blockading template. Its application of the MNS-style approach to non-violence introduced ideas from other movements as well as overseas activists and campaigns. These would be advanced and embraced by other Canadian campaigners in the 1990s.

1988: First Nations' blockades and increased diffusion

During 1988 and 1989 the use of ODA by First Nations to further land claims and secure court victories steadily increased. Actions carried out by the Sekanis of Lake McLeod, BC in June and July 1988 not only had a major influence on the use, breadth and targeting of ODA, but also illustrated the diversity of Aboriginal attitudes towards land use. For while the band sought a moratorium on the logging of traditional forests subject to unfulfilled treaty obligations, they did so on the basis of equity regarding resource exploitation rather than preservation. With the majority of its members unemployed, the band's forestry business had long tried to enter the market but been ruled ineligible by officials or outbid by larger companies. Having exhausted supplies elsewhere these same firms were now moving into claimed territory. With litigation failing and precedents to build on, the band opted for ODA including blockades of logging roads involving barricades and deep trenches. It also closed off access to a popular tourist area in order to impose costs on the government and local non-Aboriginal economy.[47]

Another tactic involved logging 200 truckloads of timber from traditional territory. Although this was sold, its extraction was primarily strategic. As Lake McLeod band manager Harley Chingee told reporters: "We want the government to take us to court. We want them to freeze the area and if they won't, we'll log it."[48] When the activity was subsequently deemed illegal the community responded with a month-long blockade of a key logging road, which effectively shut down operations until the BC Supreme Court granted a moratorium in December. Previous court-imposed moratoriums had concerned comprehensive land claims. This was the first time one had been granted to a BC group based on claims to existing treaty rights, thereby lending Aboriginal activists another legal tool.[49]

Although not concerned with preservation issues, this success formed part of, and influenced, local and national First Nations' militancy during this period. Although each would develop their own approaches, there was correspondence and networking between activists from the Sekani, Nuuchah-nulth, Haida, Gitxsan, and Wet'suwten, Lubicon, and other First Nations.[50] According to journalist Terry Glavin, who covered many blockades during this period, some First Nations communities in BC also employed the same lawyers, providing another channel of diffusion, particularly regarding how ODA could be used to trigger and leverage court action. As he recalls:

> Each major road block had its own unique characteristics and each made ground in law … [some] were about Aboriginal title and consent and forest licenses on specific land, [others] were about duty to consult and who was involved in decision making and [others] were more an environmental victory … [As with fishing] there were different tribes with different interests in different objectives, sometimes doing it with a lot of environmentalist support, sometimes with a lot of network support in Indian country and a lot of legal support … There was a lot of solidarity and tribal leaderships were swapping notes all the time, but they had different priorities and different issues to deal with.[51]

FOCS co-founder Susanne Hare Lawson recalls that "The actions at Meares Island spurred those in Haida Gwaii, but for the most part, people took information and used it for their own region." Similarly, "The injunctions in the Gitxsan area by the chiefs [in 1988] were a great tactic and every step helped the next." Her husband Steve Lawson was the national coordinator of the First Nations Environmental Network in Canada and the couple worked with the Indigenous Environmental Network in the United States, creating opportunities for direct discussion regarding ODA.[52]

Following their own long and unsuccessful campaign to gain land entitlements related to 1899's Treaty Eight, the Lubicon Lake Band of Alberta undertook ODA on 15 October 1988. Declaring themselves a sovereign state exempt from Canadian law, they blocked four roads leading into their territory. Most vehicles were allowed through, but oil and lumber companies were locked out of the area. These had been given prior warning and already shut down operations, possibly in a bid to pressure the province to restart negotiations.[53] The government obtained an injunction and carried out raids involving riot police and dogs five days later, removing barricades and arresting 27 people. Within days it agreed to set aside 245 square kilometres of land and begin negotiations regarding the protection of wildlife habitat and compensation.[54]

A third Aboriginal ODA focused campaign peaked around the same time as the Lubicon and Sekani ones. The Gitxsan and Wet'suwten peoples of northern BC, who had a long shared history and run joint campaigns since the early part of the century, were not subject to an existing treaty, but had received federal government approval to begin negotiations in 1977. With BC blocking progress, they subsequently filed the joint *Delgamuukw* land title case in 1984, asserting ownership of 58,000 km of land. While this slowly worked its way through the courts the Gitxsan–Wet'suwten Tribal Council (GWTC) began to assert sovereignty in 1986 by advising members to ignore Department of Fisheries and Oceans licenses, leading to a series of confrontations with officials. During 1988 and 1989, GWTC's campaign, and those of individual bands, extended to the issue of logging.[55]

While not opposed to all resource extraction, the First Nations' leadership, whose composition had been steadily shifting from elected to hereditary chiefs, were concerned at clear-felling. As would be the case in other parts of BC, this primarily reflected alarm among their communities at effects on habitat, hunting, and trapping, but also encompassed the economic concerns of native and non-native residents regarding how long the resource base for logging could survive. While the immediate aim was to halt the spread of logging in the Skeena Mountains, ODA was tied to the council's broader strategy as GWTC President Don Ryan announced, "The big thing for us is the long-term. All of these things are dogfights. We're going to be building it up and building it up until we de facto own and control the area."[56]

The first blockade occurred on 11 February when loggers at the Oliver Creek watershed were issued a notice to leave by the Gitwangak Frog clan, who announced "permits granted for the extraction and utilisation of these resources are deemed illegal".[57] Wishing to avoid a protracted conflict, and with only a few weeks work left, the contractor withdrew and agreed to engage in negotiations regarding their next site.[58]

One arrest was carried out, that of Gitwangak Fireweed Chief Luulak, for "theft" and "possession of stolen property" after he "confiscated" a front-end loader that workers had failed to remove within a given deadline.[59] The charges were later dropped for lack of evidence. Arrests at blockades would be rare however as Gitxsan–Wet'suwten activists generally conserved their human and other resources while stretching their opponents' by withdrawing whenever they loomed. They then sowed uncertainty and maintained media interest through unannounced blockades at new sites.[60]

In keeping with this, the next blockade, on 29 February, was a one-day effort in which 20 or 30 protesters placed a huge cedar log across a logging road and lit two bonfires in the village of Kispiox. Set up at 4 a.m., with a path made to allow emergency and residents' vehicles past, the barricades prevented more than 60 trucks from travelling to sites in the heavily logged Kispiox Valley. Utilising a new counter-tactic for the province, loggers set up their own blockade to prevent Aboriginal residents from travelling to work but lifted it at police request an hour later. During a long meeting of clan groups some argued the protest was having a negative impact on community members working in the industry while others urged the blockade be extended. In the end it was decided that a feast would be held and the community "take a rest" before determining its next move. The barricades were removed at 6 p.m. When a busload of 30 RCMP officers arrived the following day, they found themselves with nothing to do and left shortly after.[61]

The mobilisation of such numbers of police angered activists but did little to deter them. Following a conference in April they announced plans to oppose the sale of a licence in the Takla–Sustut region allowing 400,000 cubic metres of timber to be cut annually for 20 years.[62] They then targeted the likely winner and largest company in the area, Westar, whose former chief forester was now BC's Forestry Minister.[63]

Action in the Skeena Valley followed on the basis that it would create a corridor into areas threatened by the licence, as well as interfere with traplines. On Thursday, 5 May 1988, a new blockade was set up and Westar's contractors ordered to leave within two days. Their equipment remained, but no logging was carried out until after 18 May, when the blockaders withdrew after being served an injunction. An attempt by Westar to have the court extend this to prevent activists from entering any part of the area failed, as BC Supreme Court Judge Mary Southin ruled that she had no intention of "interfering with free speech".[64]

Tensions during the period gave rise to fistfights in local bars between loggers and their opponents as well as harassment of Gitxsan people working in the industry. Reliance on forestry work also led to tensions within communities themselves. In an interview at the time Ryan acknowledged this stating, "You have husband against wife, brother against brother and father against son, the whole thing with Indian–white relations and all that plays itself out. But it's the kind of discussion we have to have."[65]

While economic need brought internal opposition to blockading, experiences in the industry, as in other First Nation communities, also created support for it on the part of those who had witnessed clear-felling's impact. The sector's dominance, and Gitxsan–Wet'suwten people's subordinate position within it as generally low paid and menial labour, exacerbated anger regarding economic flows and encouraged activists to research alternatives, including GWTC running its own forest technologists program.[66]

Despite the difficulties involved with investing meagre financial resources in court cases, GWTC, and a substantial section of the communities it represented, remained committed to confrontational strategies. Activists also played a role in encouraging and diffusing ODA elsewhere. Speaking in the Stein Valley at a gathering held by Lytton and Mount Currie bands of the Lil'wat nation, Ryan argued: "We have to get out and protect the land … We have to be prepared to go to jail. We have to be prepared to die."[67] Reflecting the widening use of ODA and entrenchment of the strategy, these communities set up a four-month roadblock in 1990 and another blockade in 1991, eventually winning protection for key areas.[68]

Amidst statements such as Ryan's, GWTC decided to make a stand. After setting up a camp in mid-September to prevent Westar from constructing a logging bridge across the Babine River in the Kispiox Valley, activists refused to leave or accept injunction papers. Police withdrew to await instructions while Westar pursued further legal options.[69]

In a November speech to the Union of BC Chiefs, Ryan continued to advocate militancy, arguing that: "We have to create the political environment in this country to get good decisions in the courts and the only way to do that is direct political action." Having argued in favour of the formation of a mobile "peace keeping force", a concept others had raised in regard to enforcing fishing rights, he advised his peers that "You have to be prepared for war." He also revealed that activists at Babine River had made unspecified preparations to "arrest" police and civilians if a raid or attempt to cross the blockade was carried out.[70]

Such plans were not tested, as the RCMP did not return and Westar was ordered not to construct a camp of its own. Following further legal action by both parties, the courts made, and affirmed on appeal, a decision that cut both ways. While Westar would be permitted to build the bridge the company would not be allowed to cross it to carry out logging until the wider land claim was resolved. This brought one dispute to a close but, having effectively prevented logging in one area, more ODA would follow in the new year.[71]

The Sulphur Passage blockade introduces tree-sitting to Canada

In very different ways the Strathcona Park and Gitxsan–Wet'suwten campaigns had both seen a shift towards ODA that minimised arrests. This, along with the most significant amount of technical innovation yet seen, was also evident in a campaign that began in Clayoquot Sound on 13 June 1988.[72]

Local residents would make up the bulk of those involved in the longest con-tinuous blockade the province had yet seen. FOCS announced it was planning a blockade at an emergency meeting held in April after photographer Adrian Dorst discovered landslides in the secluded Sulphur Passage area. Located 27 kilometres from Tofino these had been caused by blasting for logging road construction.[73]

This campaign would diverge greatly from the one that had preceded it at Meares Island. Sulphur Passage was similarly only accessible by boat but was far more remote and involved navigating heavy seas and rugged, sloping country.[74] Aboriginal positions this time were divided. Ahousaht hereditary Chief Earl Maquinna George, the traditional protector of the area, was an ardent opponent of logging and would be arrested during the blockade.[75] Some Ahousaht joined members of other First Nations for a large demonstration at the beginning of the campaign, but many among the band worked in the industry and supported the road. As such blockading would be primarily undertaken by non-Aboriginal activists.[76]

The lead into ODA was also much shorter. Following the failure of negotiations with BC Forest Products, who were in the process of being taken over by New Zealand based firm Fletcher Challenge, a protest camp was established near where work was taking place. Opposition to clear-felling had strengthened around Tofino in the previous four years, as demonstrated by environmentalists taking control of the local Chamber of Commerce and later, following the blockade, the town council. Members of the Ahousaht band applied for an injunction based on the ongoing Nuu-chah-nulth land claim, as did environmentalists on other grounds; both were denied.[77]

Previous Canadian blockades had primarily employed tactics that generated obstruction through barricading and the occupation of space by bodies alone. This one would introduce techniques that amplified manufactured vulnerability to take advantage of safety regulations. FOCS Director Steve Lawson made this explicit at the beginning of the campaign when he told reporters, "Basically we're going to put ourselves in the path of blast debris. Hopefully they will stop before they kill us."[78]

Initially blockaders camped on the road and built a wall across it to ensure they were not run over during the night. They commonly sat or stood openly in the blast zone or, having informed workers of their presence, hid in nearby bush. Later when workers were blasting cliffs blockaders positioned boats and people on the beach and sea beneath it.[79] A basket hanging 10 metres over a blast site was occupied and Canada's first tree-sits employed.[80]

The overall direction and high level of technical innovation stemmed from a var-iety of factors. Although publicity was a goal, the immediate halting of road work was of particular importance to participants as they were concerned it would open up the Megin River valley, a corridor of unlogged habitat running from the centre of Vancouver Island to the coast. According to Aspinall, "the corridor was sacrosanct and we would do anything to save it."[81] This commitment was demonstrated by the adoption of the slogan "No Pasaran", Spanish for "They Shall Not Pass", a phrase originally adopted by anti-fascists during the Spanish civil war and more recently by the left-wing Sandinista government in Nicaragua.[82]

Some residents had recently been involved in the Strathcona Park blockade, including Lawson and Dorst who had been arrested.[83] Demonstrating the diversity of Canadian campaigns, the importance of "tactical tastes" shaped by differing outlooks and collective identities, and the fact that a local cohort had already developed its own way of doing things, FOSP's strategic and organisational approach was not replicated here. Instead planning was ad hoc, spontaneity and experimentation encouraged, and innovation facilitated by a minimal standard of non-violence that prohibited only aggressive behaviour and sabotage. Police and workers were not closely briefed as blockaders sought the element of surprise.[84] As Susanne Hare Lawson recalls:

> It was great when we local people got together as we knew the land and waters and were connected to it deeply. We never did any training or anything like that, everything seemed to be very instinctive and in conjunction with some other greater guiding force ... I think in our case, because there were fewer of us and we were mostly locals and were in charge of the boats and who came and went, we didn't have to be very formalized and we collectively worked together pretty well ... Media wasn't our main priority ever, it was an aside and evolved out of the actions taken to stop work.[85]

Despite connections to other provincial activists, and overseas groups such as Australia's Rainforest Information Centre and US EF!, tactics were primarily generated at a local level.[86] This can be seen in the way that tree-sitting was introduced and developed. Work was regularly being interrupted, but few had anticipated blockading would drag on for months. Although some protest events had drawn hundreds of people, by August numbers at the site had gotten extremely low. These factors provided activists with the time, opportunities, and incentive to experiment. At one point, Dorst had held up work for nine hours by climbing and perching on a branch. Inspired by this, activists then hung a wicker basket off a cliff to interfere with blasting. Activist TK (pseudonym) in turn used this to figure out a way to extend the efficacy of tree-sitting to the point where it stopped work for three days.[87] As newcomer Paul Winstanley recalls:

> [They] suggested hanging a hammock between two trees. Steve Lawson and I figured out how to do it by bear-hugging our way up them. It was on a very steep slope and they were very old, but not wide. After we got up about 60 feet or so I spent two or three hours setting the hammock up thinking that someone else would take over, but it slowly dawned on me that [with so many others arrested] no one was ... While I was up there I stumbled on a climbing technique using ropes tied around the tree as stirrups. While the loggers were gone I'd use the stirrups to drag the hammock higher up one tree and then go across to the other one. The fact they already couldn't get

at me and then would come back and find me higher each time really blew their jets.[88]

Despite Fletcher Challenge receiving the province's most generous injunction yet, in that it banned protesters from coming within five kilometres of construction, the company found legal threats had minimal effect.[89] Blockaders ignored the court ruling and regularly conserved human resources by avoiding arrest and disappearing into the forest.[90] The remote location, difficult geography, and a lack of officers meant that RCMP options were hampered on the occasions when they were able to attend actions. A process server employed by Fletcher Challenge instead issued injunction notices to individuals and police began arresting protesters on the basis of these in Tofino.[91]

Penalties given to the 35 arrested during the three-month blockade varied greatly. Ten charged with contempt of court were exonerated and awarded costs. Although authorities avoided sentencing Ahousaht band members, when it came to non-Aboriginal blockaders, they appear to have been less concerned with the risk of causing a local or broader backlash than during the Strathcona Park and the Gitxsan–Wet'suwten campaigns. As blockading ground on they hardened their stance and arrestees received the heaviest fines and sentences yet levied in the province.

Hinke, who had been dubbed "Spike" for his involvement in tree-spiking during the Meares campaign, was jailed for 37 days for multiple contempt of court charges on 9 August. Ananda Lee Tan later served 45 days, Ron Aspinall 37, and a group of women were imprisoned in a maximum security prison.[92] Reflecting the level of resources authorities were willing to invest, a day after taking a lone stand against blasting Aspinall was arrested, pulled out of work and immediately flown by helicopter from Clayoquot Sound to join an ongoing trial of others.[93]

Where possible supporters visited prisoners and took turns to cover childcare and other commitments. While none welcomed the experience, TK argues the authorities' escalation failed to deter activism as "once you'd been arrested a few times it was no big deal, once you'd been to jail it was no big deal … it was like 'Is that all you've got?' It was warm, dry and kind of a relief to be off the frontline."[94] A number of FOCS members would serve further jail time in 1989 after refusing to make court ordered restitution to Fletcher Challenge.[95]

Blockaders also faced concerted opposition on site. Unlike at Strathcona Park, where workers brought in to do exploration were mainly from outside the area and had little stake in the issue, this campaign took place in the context of long-standing enmities. Tents were slashed and blockaders occasionally roughed up. Aspinall was wildly spun around and almost thrown through the air from a drilling equipment after he climbed it to prevent the operator from deliberately destroying habitat unrelated to construction in revenge for earlier blockading.[96] Blasters regularly flaunted safety and environmental regulations, illegally showering the ocean, spawning grounds, occupiers and those in canoes and boats with debris. Although there were many close calls and a blaster was killed by falling rock, no one among the protesters sustained major injuries, although some would suffer from post-traumatic stress disorder.[97] For first time, protesters such as TK the moral shock associated with these acts had a galvanising effect. As she says:

> I was not a radical by any means, I wasn't even an environmentalist, I was living on my boat and the ocean was my thing, but once I had lived in the forest and talked to the Indigenous people about their experiences I changed … But the thing that really did it, that got my goat, was the blasting. I did a lot of research and it was totally illegal [for them to be sending debris into the ocean]. Had it been legal I might have gone "Oh well" and moved my boat, but it was not fair, corporations should have to abide by the law. I took photos and made reports and nothing changed. I then stayed on my boat because they were not meant to be blasting rock that far out into the bay, but they did it anyway and rocks hit it. That's when it started for me, it just wasn't right, someone had to stand up to them.[98]

Although the blockade itself received media coverage, its remote location, and careful behaviour from workers on the rare occasion journalists were present, meant that few incidents were reported. One that was, involved Winstanley as Canada's free tree-sit also involved the country's first ODA related near-fatality. On 4 August he spotted a worker with a gun. Luckily this only shot pellets, but he was peppered by them while sitting 20 metres above the ground. When this failed to remove him a party of off-duty workers cut one of the trees his hammock was tied to forcing him to slide down it to the ground to avoid freefalling onto the debris-filled slope below. With the equipment now relatively affordable, both protesters and Fletcher Challenge officials had begun carrying video-cameras for safety and media purposes. In this case its deployment by activists, alerted by a distress call put out via radio, caused Winstanley's assailants to flee. Despite the treesitter telling the media, "there were two or three moments when I felt, okay I'm going to die", only one of the party of eight involved in the assault was charged. That man later received 20 hours of community service while Winstanley spent 15 days in prison for breaking the terms of Fletcher Challenge's injunction.[99]

Such risks did not deter activists and although numbers continued to drop, ODA persisted into September. Faced with delays, bad publicity, and expense, the company relented at this point, shifting operations elsewhere.[100] The twin experiences of success and hardship solidified the commitment of many of the activists involved. Diffusion in the mould of that facilitated by travelling activists in Australia and the United States would follow from 1990 onwards as a cohort of FOCS members increasingly lent their time, skills, and experience to other blockades across BC. This included Winstanley refining and innovating a separate set of tree-sitting techniques to those used in the United States and Australia, holding workshops at the University of Victoria and elsewhere regarding them.[101]

1988: The Temagami dispute, a tale of two blockades

By the end of 1988 activists in various parts of BC had launched their longest and most widespread blockades yet, introducing new tactics and approaches and cementing ODA as a strategic option in the process. The first campaign outside of the province to last more than a week would be launched in the same year. Focused on clear-felling in Ontario's Temagami region, this involved hundreds of arrests during two years of rolling blockades. Described by the province's Premier David Petersen as "the most difficult issue" he faced during his tenure, ODA would play a major role in bringing down a government recently elected by landslide.[102]

Ontario is Canada's second biggest province and its most populous, containing over a third of the country's population as well as the national capital, Ottawa. Although not as dominant as in BC, forestry and other extractive industries formed a key source of employment in the less developed and populated north of the province. The timber management regime reflected those elsewhere, with extractive values favoured within the bureaucracy amidst allegations of unsustainable levels of harvesting and lax regulatory frameworks.[103]

Conservation had become a major issue for Ontarians by the early 1980s. During this period, the ruling centre-right Progressive Conservative Party created 155 new provincial parks. This remained contentious as expansion had been whittled down from an original plan for 235. Furthermore, logging was permitted in two parks and mining and hunting in the majority. In 1985 the centrist Liberal Party came to power, in coalition with the centre-left NDP, pledging to create another 51 provincial parks within a year. Although environmentalists had unprecedented access and influence, the powerful, and hitherto largely unchallenged, Ontario Ministry of Natural Resources (OMNR) pushed back. Over the next three years, public opposition to extractive activities grew while government attempts to contain the issue via bureaucratic means floundered.[104]

The issue that brought conflict to a head concerned the construction of new logging roads in and around the 72,000 ha Evelyn-Smoothwater Park, which borders the popular tourist destination of Lake Temagami, 450 km north of the provincial capital Toronto. Created in 1983, the park was poorly funded, lacked a management plan, and omitted many sites of biological significance. OMNR – which oversaw resource extraction and the management of parks, and which was yet to define or recognise "old growth" forests – had recommended an overall doubling of logging in the area during the next decade.[105]

Local opinion was divided along conventional lines. The main regional employer, a local iron mine, would shut in 1989 and the secondary employer, logging company Milne and Sons, was facing bankruptcy.[106] While tourism operators and organisations representing youth camps and other lake users favoured conservation, the majority of non-Aboriginal residents considered forests best served by clear-cutting and replanting, and resented their opponents as "newcomers" and "southerners" privileging leisure over traditional jobs.[107]

Regional opposition to the roads also came from those who had lived in the area the longest, the Teme-Augama Anishnabai community, around 150 of whom lived at Bear Island on Lake Temagami. Their representative organisation, the Teme-Augama Anishnabai Association (TAA), employed a novel approach against plans to build a ski resort and greatly expand resource extraction in 1973 by obtaining a land caution under the Land Title Act. In the context of an ongoing land claim this created legal issues preventing and delaying the sale and alienation of land as well as any development on 98 per cent of a 4,000 km area. The caution did not prevent logging however as this was deemed to "change" rather than "alienate" land.[108] This interpretation intensified grievances since, as then TAA councillor and negotiator Mary Laronde explains:

> The courts were saying we don't know if it's your land yet, but in the meantime, they can do whatever they want to it, because trees aren't part of the land. Which is crazy. The land to us is the whole thing, the soil, the trees, the water and the bees, it's an ecosystem, it's not real estate, it's more than that. We need the land to survive as who we are and what they were doing was not sustainable, it was damaging our way of knowing…[109]

Teme-Augama Anishnabai claims to sovereignty were rejected by courts on a variety of grounds in three cases from 1985 to 1991. While the provincial government acknowledged that the community had not received due compensation for prior losses, they considered this a federal matter and pushed on with plans to expand logging and quash the cautions. For their part TAA, through Chief Gary Potts, asserted, "We have never sold our land. We have never been conquered in war. We have never surrendered our human rights."[110]

Originally planned in the 1970s, the 15 km Red Squirrel Road extension, designed to join Highway 11 with a logging road running through the Evelyn-Smoothwater Park wilderness park, gathered government support in 1986. Much public opposition was organised via a new environmental organisation, the Temagami Wildness Society (TWS-C).[111] Formed by three avid canoeists with long-term ties to the area, it would recruit 15,000 members by 1988.[112]

Sidelined during the creation of the Evelyn-Smoothwater Park, TAA emphatically rejected bids by some environmentalists to have the government reassign land use via park expansion or other means, insisting they be centrally involved in all decisions and their territory remain intact.[113] While they sought a moratorium on road building and strongly opposed clear-felling, they did not oppose logging altogether. As Potts informed reporters, "We're not amenable to the wilderness position of turning the bush into a zoo. Our experience is that the bush changes, and so you can work in the bush and change it."[114]

Differences between non-Aboriginal environmentalists and First Nations had previously been resolved or smoothed over during campaigns in BC. In this case TWS-C's core group of directors were privately split over whether they should prioritise environmental protection or Aboriginal issues. TWS-C publicly supported

Teme-Augama Anishnabai land claims and called for a wilderness reserve in the form of a "land trust to be held for the TAA". Nevertheless, tensions with First Nations activists would grow over time.[115]

Reflecting on events in 2003 David T. McNabb, a Metis historian who worked as an advisor to the government on issues regarding the Teme-Augama Anishnabai, characterised the government's approach to the road extension as an ad hoc one in which their opponents maintained the initiative.[116] As in BC, the government initially resorted to a moratorium and creation of a consultative group to contain the issue. Following a Liberal landslide victory in the 1987 election, it announced the creation of 53 long promised new parks and the phasing out of mining and hydro-electric development as well as hunting in them, except by First Nations people. Regarding the Temagami area promises were made to move logging and roads out of the park and increase forest buffer zones along canoe routes. However, under pressure from OMNR and apparently deigning the issue one in which the authorities had to be seen to prevail, approval was given for the Red Squirrel extension to begin.[117]

With construction looming environmentalists sought an injunction in the courts while TAA opted to begin a blockade. In keeping with the weak diffusion of the 1980s, Laronde recalls:

> There was little discussion. Our backs were against the wall and the only recourse left was to try and physically stop them via peaceful action. We didn't talk about it that much; it was very organic. There had been actions in BC and we knew about Clayoquot Sound and Haida Gwaii, but we didn't really take any cues from anyone. It happened because of the circumstances we were in.[118]

As during the Haida campaign, the first blockade carried out in the province in relation to the logging industry immediately brought TAA's interests to the fore. Citing the Ontario government's "high handed" attitude and "colonial maggot mentality", 100 Teme-Augama Anishnabai erected a "Road Closed" sign and set up a camp in the path of the proposed extension on 1 June 1988.[119] With the band's reserve relatively close by, the occupation was strategically placed to allow for easy resupply.[120]

On the same day the blockade began, another involving armed members of a Mohawk Warrior Society was set up on a major highway and bridge near Montreal. This came in response to a raid on a reserve by 200 RCMP officers seeking to end untaxed cigarette sales. In contrast TAA fully briefed police on their actions, eschewed barricading and sabotage, and pledged to avoid violence and submit to arrest, later stating: "Our actions then and thereafter were entirely peaceful in a cause of peace. They were not actions of war on our part."[121]

The blockade widened internal divisions within the government. No construction was carried out and in late September the government offered TAA one of Canada's first co-management arrangements via membership of a community board. As it fell short of guaranteeing power over land use decisions the deal was repudiated and roadblocks remained in place.[122]

Logging supporters applied their own pressure on 2 September by blockading an entry road to Temagami Lake over the Labor Holiday weekend, demanding Premier Kevin Petersen, who was visiting northern Ontario, meet them regarding construction delays. Such tactics had been previously used in 1987 when road supporters attempted to disrupt a Sierra Club gathering in the area. The disruption of the wider public that the 1988 blockade caused led authorities to use a very different approach to that so far employed with TAA. Police initially allowed 300 people to occupy the area with bodies and cars for two hours before ordering them to disperse. Two hundred police personnel in riot gear then cleared the road. Rather than seeking an injunction, criminal law was immediately used and 51 arrested on charges of obstruction, all of which were dropped five months later.[123]

The following day pro-extension representatives met with government ministers. When this failed to bring their desired outcome, 100 loggers and their families slowed holiday traffic again by marching across the Trans-Canada Highway on 4 September. No arrests occurred this time and the main outcome of ODA appears to be that Milnes were provided with extra forest to log. Ultimately the company's debts were too great and their lenders forced them into receivership on 2 December.[124]

In November TAA released a detailed counter-proposal outlining their vision for the "long term stewardship" of an initially 5,000 km, and later 10,000 km, "motherland that would be a distinct administrative district in its own right".[125] Unwilling to cede this much control, wedged between competing and vocal interests, and facing threats from the TAA to begin blockading a second road, the government shifted tactics, gaining a court injunction on 7 December. TAA complied as, with the next phase of their long running land claim due in January, the court had also banned the government from carrying out construction.[126]

Although aware that a permanent solution was yet to be found, the TAA viewed this finding, and the overall blockade as a success. This was measured in both instrumental and personal outcomes. For many, the experience of ODA during 1988 and 1989 proved a powerful political and cultural experience. As Teme-Augama Anishnabai blockader June Twain later enthused in a chapter for the anthology *Blockades and Resistance*:

> The unfolding commitment was a joy to see. This joy was shared by all the people from our community who were involved with the blockade ... Some of the very positive feelings were caring, commitment, determination, unity and putting aside our differences in feelings. Spiritually we were very strong, joining hands together, praying, sharing sweet grass ceremonies and tobacco offerings, which showed we were a nation of people.[127]

1989: The Temagami dispute widens

The dispute continued into the new year and with it came a new round of blockading. Litigation and lobbying delayed the Red Squirrel Road extension until autumn, but in the meantime various parties targeted other work in the area.[128]

On 28 March, up to 50 members of TAA, and non-Aboriginal supporters, blockaded a road extension being built by the Goulard Lumber Company to open up an area known as the "Wakima Triangle". With heavy slush already preventing work from taking place, this was a largely symbolic action, but reiterated opposition to clear-felling anywhere in Teme-Augama Anishnabai territory.[129]

Goulards' operations were then met on 5 June by a new group made up of environmentalists from Southern Ontario, Voices of Temagami. Carrying out three days of what they described as "non-violent civil disobedience", these activists employed tactics designed to extend the obstructive capacity of their small numbers by hanging hammocks across the road, digging trenches and barricading with stones and trees. In action novel for Canadian preservationists, five activists also chained themselves across an entry gate. In contrast to their approach to TAA, whom they clearly did not want to confront, police did not wait for an injunction and immediately arrested 11 people for "breach of the peace".[130]

While these activists eschewed any connection to other organisations, the blockade was secretly organised by Ontario-based members of TWS-C. These had been inspired by forest activism in Australia and elsewhere and had long discussed employing a blockade. Core members lacked direct experience in environmental blockading itself, but as at Strathcona, drew on experiences regarding ODA in other social movements. They also took advice from long-time activists such as Greenpeace co-founder Robert Hunter, then working as a television journalist in Toronto.[131]

The TAA and Voices of Temagami actions were successful in that the permit for the road was cancelled and the land later preserved. However, TWS-C had not wanted to act in its own name for fear of prejudicing ongoing litigation against the Red Squirrel extension.[132]

In recent years TWS-C had distributed up to 250,000 information sheets at a time, formed local groups, visited schools, and created posters. Regularly briefed by advisers close to cabinet, the organisation helped break major news stories regarding problems with the extension's environmental assessment and other issues. Influenced by materials coming out of BC and the United States, it funded its own scientific research into Temagami's forests and worked hard to popularise the existence and concept of "old growth" forest. Litigation was employed, in the words of TWS-C spokesperson, Brian Back as a means of "stalling, PR and fundraising". By 1989 TWS-C was working with a budget of $1.5 million.[133]

The next round of blockading came in mid-September. By this point Milnes had attracted new financiers and, despite replacing its OMNR minister, the government remained committed to the extension.[134] By mid-year TWS-C had largely exhausted all of its legal options and previous concerns regarding court cases were overcome. Balancing considerations regarding the commencement of work, the time required to set up a camp in an isolated location, and the need to get the protest underway before weather became too inclement, TWS-C announced in late August that it would hold a three-day mid-September "camp-in".[135] In a bid to prevent this from occurring, the government tried and failed to obtain an injunction banning anyone from coming within 500 metres of the extension worksite.[136]

Just before the camp was about to begin Goulards announced that a saw had been damaged by tree-spiking and that they had recently received a letter warning of further sabotage. TWS-C immediately asserted that no one involved in such action would be welcome at their blockade. When sabotage continued alongside the blockade, TWS-C tightened its security under the belief that it had become the target of increased police surveillance and infiltration. Although day-to-day decision making would allow protesters to participate in simple consensus processes, a core group, often meeting separately, had a heavy influence upon the direction of the blockade. This was instituted with a view to maintaining what TWS-C director and local resident Kay Chornook later described as "sanity, safety and security".[137]

Up to 200 people attended the September gathering. With OMNR deciding to begin work at the most inaccessible end of the extension, protesters made their way either by a long car journey on poor roads, a canoe trip of three to six hours, or by boat and plane. Transporting people and supplies for the blockade posed numerous logistical difficulties and over the summer local environmentalists had put much time and resources into establishing a protest camp three kilometres from the work site.[138]

Although influenced by advice from members and others who had previously been involved in ODA, the camp did not contain anyone with extensive experience of forest blockading. TWS-C had access to materials describing blockades undertaken elsewhere but did not closely follow an imported model of organisation and normative protester behaviour. According to Back, "We had to figure things out for ourselves."[139] In this case "non-violence" would entail a rejection of aggressive physical and verbal actions. In order to maintain the initiative, police were not closely briefed on tactics, but it was agreed that the only resistance to arrest would involve going limp. Training was available during events but not mandatory.[140]

The gathering was attended by people varying in age from children to retirees and drew numbers far in excess of what TWS-C had expected. Preferring to keep the focus on its own issues, TAA did not endorse the blockade, although individual community members took part. TWS-C's first action came on Monday, 18 September when up to 130 protesters, including two buried in the ground, occupied the construction site. Police took an hour to clear the way and detained 15, including the Leader of the Opposition, NDP member Bob Rae.[141] Given that the protest involved such a high-profile figure, media coverage was heavy and widespread. While Rae's arrest generated antagonism from rural politicians and unionists within his party, it boosted his environmental credentials. The Temagami dispute would prove to be one of a number of intractable issues which led to the Liberal Party's demise at the 1990 election, bringing the NDP to power in its own right for the first and only time in the province's history.[142]

Whether it was due to Rae's presence or an underestimation of likely numbers, the police chose to transport the politician and others out of the area and release them without charges after they agreed not to return to the site. Over the following two days, a further 33 people were arrested and similarly treated, after which police announced they would begin laying charges of "mischief" and "trespass".[143] The shift may have come from to the realisation that blockading would be ongoing or because, with many protesters beginning to return home after the gathering, a much smaller, and more easily handled, core group of 10 to 20 remained. A number of blockaders later faced court, but heavy penalties were not incurred.[144]

Having studied blockades that had occurred elsewhere, TWS-C was well aware of ODA's potential to gain media attention and expose publics to potentially powerful visual images of the forest. For Back and some core TWS-C members this had been the prime motivation for undertaking action. A group dedicated to staying on site and focused heavily on disrupting work soon coalesced however, and the organisation committed itself to supporting them until the end, regardless of diminishing media coverage. With the remote location and lack of local support precluding large numbers, blockaders chose mainly to employ tactics that maximised obstructive potential and minimised arrests.[145]

Lacking specific tactical information and experience, protesters experimented with a range of means. The blockade became the first Canadian event in which people were buried in the ground and U-locks regularly employed to fasten activists to bridges, vehicles and other equipment.[146] The original source of information regarding the use of the latter remains unclear, but Brian Back, one of the two people buried, recalls two indirect sources of diffusion:

> I'd first seen people buried in the ground as part of media coverage of an Australian blockade. I'd forgotten about it, but Robert Hunter met me for lunch and said "You've got to bury someone in the ground." One person was buried up to their waist for part of the first day and I was buried standing up. As the ground was too hard to dig any further, rocks were piled up to my neck. I wore a rain suit, but I hadn't anticipated how compressed I would be, how the temperature would fluctuate, let alone needing to go to the toilet … At times it was unnerving as a bulldozer came close enough for me to feel the ground moving …[147]

Back was eventually dug out by his fellow blockaders after workers found a way to bypass him. Reflecting the influence of geography and counter-policing upon tactics, the risk of exposure, as well as tighter security, ensured the tactic would not be repeated.[148]

As in the United States, police responses to the use of U-locks rapidly evolved. Initially, with an ambulance standing by, they wrapped one activist's head in wet towels and used a blowtorch to remove a kryptonite lock. Thereafter they employed more straightforward means of countering the tactic. Local tourism operator and TWS-C co-founder Hap Wilson recalls that initially,

> Everybody from the government to the companies to the police were having to deal with this kind of thing for the first time and I think part of why we were so successful was that they were having devise their strategies at the same time as we were.

However, eventually it was worked out that kryptonite locks could be sometimes be cut by special bolt-cutters. Police also began to target activist support people bearing food and water as well as keys to the locks.[149]

Locksmiths were also hired to drill out locks, but innovation allowed continued deployment. In one case a group disguised themselves as construction workers and went unnoticed long enough to lock themselves to drilling equipment. In another, Paul Smith, widely credited with having introduced the use of U-locks to the campaign, cemented a blow-torch proof axle casing into a rock face. According to Back this held work up for three days and necessitated drilling to remove the rock itself. A third instance delayed construction for half a day after three protesters managed to get one lock around all their necks.[150]

Another tactical innovation involved placing four protesters in a wooden structure made from creosote-soaked timber removed from road supplies. It had been anticipated that this would require a great deal of time for the police to dismantle, but instead they opted to use small teams to drag out each occupier individually, and work restarted after two hours.[151]

Following what had become standard US practice, blockaders also dug trenches and placed rocks, slash, and other materials on the road throughout the campaign, which was effective in slowing travel down. Such obstacles at times fuelled hostility from police, many of whom were local residents and logging supporters. To ensure safety and capture footage, video cameras were used both by police and, where practicable, protesters. All parties generally acted with care and relations with construction workers, who Chornook points out "were getting paid regardless", were civil. Those with road supporters varied. While Chornook recalls remaining on good terms with other local residents during and after the protests, other activists cite threatening phone calls and attacks on environmentalist-owned businesses. When a pro-logging rally of 70 was held at the worksite, blockaders withdrew for the day.[152]

With worksites eventually guarded and equipment kept under light at night, protesters found it increasingly difficult to mount actions beyond simple road occupations. After police began using a helicopter to surveil the camp, protesters, upon hearing their approach, took to hiding in their tents so that the authorities would be unable to tell whether sizeable numbers were absent. After holding a few trial runs, the blockade also deployed a tree-sitter for the first and only time in late September. Despite threats to cut her tree down, and an attempt to smoke her out by lighting a fire, which failed as she was 100 feet up, she remained in place for roughly a week. However, as Wilson recalls:

> Because of the time of the year it was getting below freezing overnight. They couldn't get her out, but there was no shelter, so with the exposure she eventually had to come down. If it had been a few months earlier we could have done more tree sitting.[153]

By 9 October 116 protesters had been arrested.[154] Damaged by the leaking of a suppressed report which had recommended conserving old-growth forests in the area, the government brought construction to a halt once more on 18 October.[155] This was also because TAA had applied for a new injunction related to their land claim.[156] After this was denied TAA sent out eviction notices on 31 October to TWS-C, the OMNR and building contractor Carman Construction ordering them to vacate the area.[157]

TWS-C was the only one to respond, immediately closing its base camp. The decision, and TAA commentary surrounding it, was a clear rebuff to a sole focus on environmental rather than sovereignty issues. In comments to the media Back reiterated TWS-C's support for the TAA, publicly welcoming, "the opportunity for them to assert some authority on what is their land".[158] Academic Bruce Hodgins, who owned a campsite in the area that was used by TAA during the blockade that followed, later claimed he and others had tried to maintain a "shaky [TWS-C-TAA] entente which was always far apart on their objectives and also to forge a real alliance and understanding between the Native people and the environmentalists, who included the canoeists."[159]

For her part Chornook recalls a sense of partial relief among those at the TWS-C blockade as "whilst we were determined to continue we were also heading into winter and already spending much of our time just ensuring we kept warm and safe."[160] Back concurs, arguing that media coverage by this time had dwindled with TAA's subsequent actions "breathing new life into the whole issue".[161]

With its decision TAA's leadership placed itself and its goals once more at the centre of the dispute. Further litigation against the road and logging was unlikely to be heard until the new year, so plans for a new blockade were announced on 11 November. Potts stressed that this would be another peaceful event and that protesters would submit to arrest. In line with this he also requested that members of militant First Nation Warrior Societies not take part due to their history of physical and armed resistance. Along with camp meetings briefings were held with the police before and after each action.[162]

Work was now progressing at both ends of the extension. On the first day of the blockade more than 150 Teme-Augama Anishnabai protesters occupied one extremity to hold a Remembrance Day ceremony, joined by the 12 police present.[163] While police were clearly at pains to show good faith, and at previous events had chosen not to arrest TAA blockaders, the following day they employed criminal charges against them for the first time. 46 protesters who had sat in front of heavy machinery were arrested on charges of "mischief" before being transported three hours away by van for processing in Haileybury. In keeping with practices employed throughout the dispute all were then released under their own recognizance.[164]

Over the next month TAA members were joined by members of First Nation bands from across the province as well as sympathetic environmentalists, including Chornook. Work was primarily blocked by people sitting or standing in the way. In 2003 Hodgins recalled a typical action:

> That morning there were about twenty of us actually on the road – Anishnabi, Cree, Oji-Cree, and a few of us non-Native Canadians by the gate at the

northeast end of Lake Temagami. Several people stood showing their support along the road's edges. All had gone through a campfire-centred sweetgrass and tobacco ceremony. "Sergeant do your duty", firmly declared the OPP [Ontario Provincial Police] officer in charge. "In the name of the Queen, make way", cried the sergeant. There I was, a seventh-generation Upper Canadian defying the order of my sovereign. A surge of undefinable emotion passed through me. The Teme-Augama Anishnabai themselves were sitting on their own unsurrendered land. We had been invited to join with them. For me, it was a defining moment. We were then individually carried, politely and carefully, each by four constables, to waiting police vans with their exhaust clearly visible in the clear cold air.[165]

TAA had publicly stated that their goal was to make it "impossible for construction to continue", but work was generally not held up for longer than a few hours at a time.[166] When asked why they did not adopt new and potentially more obstructive methods, Laronde states:

> We weren't at all militant and did not want to interfere with other people's property. The rules of engagement were, don't touch the machinery, don't go near the machinery, don't climb trees ....[167]

By purely blocking the road with bodies she believes the protest was kept disciplined and orderly and good relations maintained with the police. Further to this she argues that in issuing eviction notices and adhering to its own methods TAA clearly distanced itself from the means and aims of both TWS-C and those who had been carrying out sabotage.[168]

One point at which work was delayed for an entire day was on 14 November. The government had received a stronger injunction on the same day barring protesters from coming within 500 metres of construction work. However, it could not be legally enforced, as TAA claimed that some Oji-Cree supporters who were present did not understand English and therefore could not be read the injunction or informed of their rights. Two Oji-Cree elders and Potts were among those arrested the next day, but police elected to employ the lesser charge of "'mischief" rather than contempt of court.[169]

Defying cold conditions TAA maintained its blockade into December, helping inflate total policing costs to around $1 million and construction costs to $5 million, not counting legal fees.[170] Around 360 people were arrested before the road was completed on 10 December, five weeks after its original deadline. Pro-logging locals, including 15 local mayors and reeves, held a ribbon cutting ceremony on 29 January.[171]

No one else was celebrating. After a Spring washout the extension was not repaired or upgraded. It would never be used for logging.[172] A variety of subsequent protest actions, culminated on Earth Day, 22 April 1990, when as part of major events around the world, up to 100,000 Ontarians rallied.[173]

The government had already moved to place 585 ha of the Wakima Triangle under moratorium, begun a process of identifying old-growth forest, and allowed licenses in the Temagami region to expire. On the day after Earth Day 1990, it made its biggest concession yet in signing a memorandum of understanding with TAA.[174] This provided, separate to any future outcome regarding TAA's land claim, for the co-management of around 40,000 ha of Temagami forest as well as the purchase and phasing out of Milne and Sons. The agreement only covered roughly 18 per cent of the region's forests but provided TAA with the power to veto logging in recent sites of contention, resulting in a 70 per cent decrease in logging over the next two years.[175] Disputes within TAA would prevent a final agreement being found and contention over land rights and logging continues. Nevertheless, the actions of Teme-Augama Anishnabai and environmental activists effectively delayed, and in some cases fully prevented, intensive mining and forestry development well into the 2000s.[176] In carrying out the longest series of blockades yet held in the country they also helped popularise the concept and importance of forest conservation and expanded and embedded ODA as a response to environmental issues within Canada.

1989: BC activism continues to expand

During 1989 the first instance of environmentally oriented blockading in BC occurred in June when members of the Ulkatcho First Nation targeted road construction for logging 15 km from Anahim Lake in the sparsely populated and largely undeveloped Cariboo-Chilcoton region. Logging company Carrier, which was carrying out a controversial salvage logging operation in areas previously infested with pine beetle, withdrew after a dozen blockaders occupied their worksite.[177]

The main round of BC activism once more involved the Gitxsan–Wet'suwten. Following protests and threats of ODA by various bands earlier in the year, the GWTC asserted their "right to manage our territories in a more responsible manner" by blockading a bridge on the Suskwa River, approximately 50 km east of Kispiox, on 29 September 1989. This was accompanied by an announcement that: "No fishing, no hunting, no logging and no mushrooms are to be picked [in the area] without our permission."[178]

By 13 October this initial blockade had moved and been supplemented by four others. This was the first time that multiple coordinated blockades were employed in Canada. To varying degrees all used what had become basic road occupation and barricading tactics. Local residents were allowed to pass through the roadblocks, but hunters and fisher people had to apply for permission, and loggers excluded altogether. Fear of triggering further action meant that logging companies complied. As a result, following attendance by up to 200 people on their first day, most blockades were only staffed by a few stalwarts and new tactics or people from outside the region deemed unnecessary.[179]

The site of ODA shifted at the end of October when a new blockade disrupted logging in the Kispiox Valley, turning away dozens of trucks and sending 50 workers home. As during the previous year's dispute, loggers slowed traffic entering the town of Kispiox in a demonstration of their own. Gitxsan–Wet'suwten activists lifted their blockade three days later amidst litigation, brought by the BC government and a new group named the Loggers Defence Alliance, which sought compensation for lost work and a widespread ban on blockading.[180]

Both court cases were rejected as the Supreme Court continued to insist that it would only issue injunctions on a blockade by blockade basis. While the litigation itself signalled hostility on the part of some within the logging industry, relations were not universally negative, as evidenced by contractor Roman Pelltier telling journalists: "The Indians are good people. They were sorry they blocked my work, but it had to happen."[181] Glavin also recalls: "The mayor of Hazelton, Alice Maitland, a white lady, totally supported the chiefs in what they were trying to do. A lot of the working-class white people were also trying to keep the wealth of the resource in the territory."[182]

With around one-third of logging in the region blocked by various means dominant logging company Westar shifted its position. As part of a new forestry panel of timber representatives, unions and Gitxsan–Wet'suwten chiefs committed to discussing issues independently of the government, the company supported opposition to the issuing of new, expanded timber licenses in January 1990.[183] Having already reversed earlier opposition to land claims, the company's vice-president David Mitchell admitted the following month that the provincial government no longer fully controlled the region, arguing that pressure needed to be placed on "reactionary politicians" to begin negotiations.[184] Major change would not come in the short term, but these overtures, and the ODA that prompted them, were evidence of the major shifts in tactics and power relations that had occurred in the province since the Nuuchah-nulth had first undertaken ODA at Meares Island just five years earlier.[185]

Canadian developments, 1986–1989

By the end of 1989 environmental blockading had become fully established as a strategic option within Canada. Repertoires of contention had developed along two core lines. The first followed precedents already established by Indigenous activists regarding tactics associated with soft blockading and the deployment of barricades. The main innovation involved Gitxsan–Wet'suwten activists initially using a "hit and run" strategy focused on conserving resources, minimising arrests, and spreading disruption by deploying multiple blockades and withdrawing when presented with injunctions. From 1984 onwards the size and effectiveness of barricades was increased in some campaigns through the introduction of deep trenches and bonfires.

Although also used by other activists, these tactics dominated Aboriginal-led blockades, which sought to delay work seen as environmentally harmful, support or trigger litigation, and generate media coverage. Most importantly such occupations asserted community sovereignty over traditional lands and publicly enforced it. In the case of the Gitxsan–Wet'suwten campaigns, Indigenous communities not only prevented the movement of logging trucks, but further used barricaded points to assert their sovereignty and control of land use activities through the issuing of fishing and hunting licenses.

Where negotiations or other strategies failed to bring about a resolution, authorities employed police to remove blockaders. They generally preferred to do so on the basis of court injunctions rather than through the immediate application of criminal law. In the majority of cases blockaders submitted to arrest, but at times they undertook strategic withdrawals to create economic uncertainty and preserve resources. First Nations-led blockades employed a loosely defined approach to normative protester behaviour that described itself as "peaceful" and eschewed aggressive behaviour, although this was sometimes for primarily strategic rather than ethical reasons. Blockades were often accompanied by ceremonies and other activities that asserted sovereignty and community connections to land.

Due to the contentious state of law regarding land rights, as well as the high density of Indigenous populations in and near many of the areas being blockaded, authorities tended to act more slowly and carefully when dealing with First Nations led campaigns. The ability of tactics based on barricades and soft blockades to achieve goals likely mitigated against the development of other forms. In the case of the Teme-Augama Anishnabai, campaign leaders further eschewed them due to perceptions that they were associated with activists whom they wished to distance themselves from and the belief that interference with property would increase tensions with loggers and their supporters.

A second strand of Canadian environmental blockading followed the direction of developments in Australia and the United States. Enhanced vulnerability tactics were introduced in 1988 alongside barricades and occupations of sites and machinery by bodies alone and the positioning of people in boats near blasting sites. Means involving tree-sits, burying people in the ground, and locking onto gates, trees and machinery with U-locks and chains were solely employed by non-Indigenous-led campaigns during this period. This was partially attributable to the fact that authorities felt less constrained in arresting non-Indigenous activists and in levying penalties for breaking injunctions. In some cases enhanced vulnerability tactics were innovated on a local basis, in others they were based on precedents from Australia, the United States, and elsewhere.

Notes

1 Rodgers, *Welcome to Resisterville*, 143–49.
2 Ibid, 135–43; Rita Moir, "CP Loses Permit to Spray BC Track with Herbicide," *Globe and Mail*, 23 July 1987, A4.
3 "Kwakiutl Indians Will Block Logging ", *Globe and Mail*, 1 December 1986, A9; "Indians Force Halt to Island Logging," *Globe and Mail*, 3 December 1986, N4.
4 "Indians, Logger Argue over Tiny Island," *Gazette*, 4 December 1986, B8.
5 "Indians Win a Logging Ban on BC Island," *Gazette*, 23 December 1986, B1; "Burial Discovered on Deer Island," *Globe and Mail*, 8 December 1986, A5.

6 Kwakiutl Protest. Available [Online]: www.firstnations.de/forestry/kwakiutl_protest. htm (Accessed 24 November 2016).

7 Sid Marty, *Leaning on the Wind: Under the Spell of the Great Chinook* (Victoria: Heritage House Publishing Co, 2011), 279–86.

8 Mike Judd, Interviewed 24 September 2017.

9 Ibid; Marty, *Leaning on the Wind*, 279–86.

10 Ibid; "Sour-Gas Well Wins OK," *Calgary Herald*, 23 December 1988, A10; Judd Interview.

11 Ibid.

12 Wilson, *Talk and Log*, 240–47.

13 Geoffrey York and Loreen Pindera, *People of the Pines: The Warriors and the Legacy of Oka* (Toronto: Little, Brown, 1991), 279–81.

14 Wilson, *Talk and Log*, 251.

15 Glenn Bohn, "Park Boundary Revision Outlines Mining Rights," *Vancouver Sun*, 30 January 1987, A12.

16 John Dwyer, "Conflicts over Wilderness: Strathcona Provincial Park, British Columbia" (Masters, Simon Fraser University, 1993): 181–85.

17 Nancy Brown, "Pillage of Oldest Park 'Will Continue'," *Times-Colonist*, 11 February 1987, B12.

18 Marlene Smith, Interviewed 24 November 2017.

19 Dwyer, "Conflicts over Wilderness", 169–70.

20 Ibid, 192,97–99; Glenn Bohn, "Mine Road Blockade Organized," *Vancouver Sun*, 28 February 1987, A11; Pauline Martin, "Seven Protesters Arrested," *Campbell River Upper Islander*, 2 February 1988, 1.

21 "Four Vander Zalm Colleagues Facing Conflict Accusations," *Ottawa Citizen*, 25 February 1987, D2; Wilson, *Talk and Log*, 253.

22 Ibid., 68, 75–79, 163.

23 Judy Lindsay, "Separating Strathcona Antagonists Not Easy," *Vancouver Sun*, 18 February 1988, F1.

24 Smith Interview; "Protesters Defy Injunction to Prevent Drilling in Park," *Vancouver Sun*, 30 January 1988, A7; Kim Bolan, "Strathcona Protestors Shrug Off Weekend Arrests," *Vancouver Sun*, 1 February 1988, A1.

25 Ibid.

26 "Protesters Defy Injunction to Prevent Drilling in Park," 30 January 1988, A7; Bolan, "Strathcona Protestors Shrug Off Weekend Arrests," 1 February 1988, A1.

27 Doug Ward, "Showdown a War of Nerves," *Vancouver Sun*, 2 February 1988, A1; Doug Ward, "Arrests Clear Way for Park Drilling," *Vancouver Sun*, 3 February 1988, B1.

28 "Two Arrested at B.C. Protest of Mine in Park," *Gazette*, 8 February 1988, A10; Shelley Browne, "Protester Vows Fast after Arrest," *Vancouver Sun*, 15 February 1988, A3; Gordon Hamilton, " Protesters' Oakalla Transfer Delayed," *Vancouver Sun*, 15 March 1988, B5.

29 Smith Interview.

30 Ellen Saenger, "Jailed Park Protester Vows to Continue Fast," *Vancouver Sun*, 15 February 1988, B4.

31 Dwyer, "Conflicts over Wilderness," 160,74–75; Smith, Interview.

32 Ibid.

33 John Cruickshank, "Disobedience Very Civil at B.C. Rig," *Globe and Mail*, 15 February 1988, A4.

34 Waud and Tindale, *The Franklin Blockade*.

35 Wood Interview; Jakubal, Interview; Adrian Dorst, Interviewed 1 September 2017.

36 Smith, Interview.

37 Dwyer, "Conflicts over Wilderness," 172–76, 86–89, 93, 96.
38 Smith, Interview.
39 Ibid.
40 Dwyer, "Conflicts over Wilderness," 190–94, 205–6.
41 "B.C. Agrees to Public Review of Mining Exploration," *Gazette*, 16 February 1988, B8.
42 Nancy Brown, "Park Activists Go Free," *Times-Colonist* 18 May 1989, 2; Dwyer, "Conflicts over Wilderness," 203, 27–33.
43 "Protest Wraps up as Drilling Ends in Park," *Vancouver Sun*, 28 March 1988, A1.
44 Bohn, "Environment Group Expects Government to Side with Miners," 5 July 1988, A10.
45 Mark Hume, "End to Mining, Logging in Park Wins Plaudits," *Vancouver Sun*, 2 September 1988, A1.
46 Wilson, *Talk and Log*, 256.
47 "Road Blocked by Indians," *Vancouver Sun*, 23 June 1988, B6; Ben Parfitt, "Indians' Trenches Shut Road to Park," *Vancouver Sun*, 13 July 1987, B12.
48 "Band Told to Stop Logging Crown Land," *Vancouver Sun*, 14 September 1988, B7.
49 Terry Glavin, "B.C. Native Land Claim Freezes Vast Forest Area," *Vancouver Sun*, 17 December 1988, A1; Ben Parfitt, "Blockade Observers at Loggerheads over Danger Posed by Man with Axe:," *Vancouver Sun*, 6 October 1988, A13.
50 Terry Glavin, "Injunction Won't Curb Us, Native Vows," *Vancouver Sun*, 27 June 1987, A12; "Ontario Natives Set up Blockade, Lubicon Promise Similar Tactic," *Gazette* 2 June 1988, B4; Glavin, "B.C. Blockade Resembles Lubicon Confrontation," D15.
51 Glavin Interview.
52 Lawson Interview.
53 "Lubicon Lake Cree: Oppressed People's Struggle To Take Back Power," *Akwesasne Notes*, Fall 1988, 22; Smokey Bruvere, "Lubicon Lake Land Entitlement Breakdown in Negotiations," *Akwesasne Notes*, Autumn 1986, 23.
54 Thomas Walkom, "Action Speaks Louder Than Words," *Globe and Mail*, 26 October 1988, A7; "27 Lubicon Supporters Freed to Clear Way for Land Talks," *Vancouver Sun*, 21 October 1988, A14.
55 "Zig Zag," *BC Native Blockades and Direct Action*, 2006, 6–9.
56 Terry Glavin, "Blockade Marks Bid by Indians to Exercise 'Ownership' of Land," *Vancouver Sun*, 18 February 1988, A11.
57 Mark Hume, "Blockade Stops Logging: Indians to Pick Another Target," *Vancouver Sun*, 16 February 1988, E8.
58 Ibid.
59 Glavin, "Blockade Marks Bid by Indians to Exercise 'Ownership' of Land," A11.
60 Gitxsan History of Resistance. Available [Online]: www.gitxsan.com/culture/culture-history/gitxsan-history-of-resistance/ (Accessed 20 November 2016).
61 Terry Glavin, "Feast Ends Logging Truck Blockade," *Vancouver Sun*, 1 March 1988, B1; Terry Glavin, "Show of Force by RCMP Angers Kispiox Area Chief," *Vancouver Sun*, 2 March 1988, B1.
62 Terry Glavin, "Indians Plan to Fight Sale of 20-Year Forest Licence," *Vancouver Sun*, 2 May 1988, A8.
63 Terry Glavin, "Indians Hear Call to Battle Logging," *Vancouver Sun*, 12 April 1988, B5.
64 Mike Bocking, "Natives to Fight Logging but Won't Block Trucks," *Vancouver Sun*, 19 May 1988, B8; Terry Glavin, "Indians Block Logging Road," *Vancouver Sun*, 6 May 1988, B6.
65 Terry Glavin, "Natives Say Wilderness Now a Rotting Heritage," *Vancouver Sun*, 14 June 1988, A1.

66 Ibid.

67 "A Sacred Place ... A Working Place," *Globe and Mail*, 7 May 1988, D2.

68 A Story Written on the Land. Available [Online]: lilwat.ca/wearelilwat7yul/history/ (Accessed 3 September 2017).

69 Terry Glavin, "Indians Fighting against Road Stand Firm Despite Injunctions," *Vancouver Sun*, 6 October 1988, A13.

70 Terry Glavin, "Direct Action by Indians Urged," *Vancouver Sun*, 16 November 1988, A1.

71 "Logging Injunction War on Hold," *Vancouver Sun*, 8 October 1988, C16; "Judgment Aims to Neutralize Loggers' Dispute with Indians," *Vancouver Sun*, 21 October 1988, A2.

72 Joanna Streetly. No Pasaran: The Fight for Sulphur Passage, Available [Online]: www.tofino-bc.com/geography/clayoquot-protests-sulphur-passage.php (Accessed 22 August 2016).

73 Ibid.

74 Hinke Interview.

75 Earl Maquinna George, *Living on the Edge: Nuu-Chah-Nulth History from an Ahousaht Chief's Perspective* (Winlaw: Sono Nis Press, 2003), 117–18.

76 Hinke Interview; Lawson Interview.

77 "Police to 'Keep Peace' in Log Dispute," *Vancouver Sun*, 6 July 1988, A2; Horsfield and Kennedy, *Tofino and Clayoquot Sound*, 491.

78 "Blockade Planned for Road Builders," *Vancouver Sun*, 13 June 1988, B8.

79 Terry Glavin, "Tofino Protesters Halt Work on Logging Road," *Vancouver Sun*, 14 June 1988, F7.

80 "Environmentalists, Loggers Meet," *Vancouver Sun*, 2.1 June 1988, A8; C.J. Hinke, "The Prison Experience of a Tree Protector," *Journal of Prisoners on Prisons* 2, no. 2 (1990): 4.

81 Aspinall Interview.

82 Horsfield and Kennedy, *Tofino and Clayoquot Sound*, 491–93.

83 Ibid.

84 Dorst Interview; Aspinall Interview; TK (pseudonym), Interviewed 17 June 2017.

85 Lawson Interview.

86 Hinke Interview; Dorst Interview.

87 Dorst Interview; TK Interview.

88 Paul Winstanley, Interviewed 14 November 2017.

89 "Protesters Fined for Defying Ruling," *Vancouver Sun*, 9 September 1988, A12.

90 Ben Parfitt, "Tofino Police under Order to Nab Logging Protesters," *Vancouver Sun*, 28 June 1988, B7.

91 Mark Hume, "5 Arrests Expected in Logging Protest," *Vancouver Sun*, 12 August 1988, A1.

92 Hinke, "The Prison Experience of a Tree Protector," 4; Jean Kavanagh, "Jail Fails to Fall Logging Protest," *Vancouver Sun*, 10 August 1988, B5.

93 Aspinall Interview; C. J. Hinke. Available [Online]: www.memorybc.ca/hinke-c-j-carl-john-1950 (Accessed 7 December 2016).

94 TK Interview.

95 Ibid; Aspinall Interview.

96 Aspinall Interview.

97 TK Interview.

98 Ibid.

99 Karen Gram, "Environmentalist Says Life Threatened," *Vancouver Sun*, 5 August 1988, A1; Winstanley Interview.

100 Bob Bossin. The Clayoquot Women, Available [Online]: http://www3.telus.net/oldfolk/women.htm (Accessed 6 December 2016); Aspinall Interview.

101 Winstanley Interview; TK Interview.
102 Gerald Killan, "The Development of a Wilderness Park System in Ontario, 1967–1990: Temagami in Context," in *Temagami: A Debate on Wilderness (Toronto, 1990)*, ed. Mark Bray and Ashley Tomson (Toronto: Dundurn Press, 1990): 85.
103 Mark Winfield, *Blue-Green Province: The Environment and the Political Economy of Ontario* (Vancouver: UBC Press, 2012), 1–2; Craig McInnes, "Hearing to Assess All Aspects of Forest Industry in Ontario," *Globe and Mail*, 10 May 1988, A15.
104 Rosemary Speirs, "Parks Issue Catches Peterson in a Crossfire," *Toronto Star*, 16 January 1988, D1.
105 Jamie Benidickson, "Temagami Old Growth: Pine, Politics and Public Policy," *Politics and Public Policy* 23 (1996): 41–50.
106 John Temple, "Tourism, Lumber Industries Clash over Plan to Build Logging Road in Northern Ontario Wilderness," *Toronto Star*, 5 May 1988, A24.
107 Bruce W Hodgins, "Contexts of the Temagami Predicament," in *Temagami: A Debate on Wilderness*, 128–30.
108 James Lawson, "Space, Strategy and Surprise," in *Blockades and Resistance: Studies in Actions of Peace and the Temagami Blockades of 1988-1989*, ed. Bruce W Hodgins, Ute Lischke, and David T McNab (Ontario: Wilfrid Laurier Univ. Press, 2003): 167–77.
109 Mary Laronde, Interviewed 20 September 2017.
110 Darcy Henton, "Band Members Making Last Stand Today in Bid to Halt Forest Logging Road," *Toronto Star*, 11 November 1989, D5.
111 The abbreviation TWS-C has been used to distinguish the Canadian organisation from the Australian TWS.
112 David Israelson, "Just Plain Folks: The Environment's New Champions," *Toronto Star*, 15 November 1987.
113 Lawson, "Space, Strategy and Surprise," 180.
114 Thomas Walkom, "At Loggerheads over Logging," *Globe and Mail* 28 September 1987, A7.
115 Brian Back, "Temagami: An Environmentalist's Perspective," in *Temagami: A Debate on Wilderness*, 141–42; Kay Chornook, Interviewed 14 February 2017.
116 David T McNab, "Remembering an Intellectual Wilderness," in *Blockades and Resistance*, 43.
117 Craig McInnes, "53 Parks to Be Created in Ontario," *Globe and Mail*, 18 May 1988, A1.
118 Laronde Interview.
119 Don Dutton, "Ontario Indians Block Logging Road," *Toronto Star*, 3 June 1988, A22.
120 Lawson, "Space, Strategy and Surprise," 164; Rudy Platiel, "Ontario Native Band Blocks Planned Road into Disputed Region," *Globe and Mail*, 2 June 1988, N12.
121 Ute Lischke and David T McNab, "Actions of Peace," in *Blockades and Resistance*, 4; "Rifle-Toting Mohawks Set up Road Blockades," *Windsor Star*, 2 June 1988, A1.
122 Christie McLaren, "Plan Shares Logging Control with Indians," *Globe and Mail* 28 September 1988, A5.
123 "Protesters Charged after Logging Road Blocked," *Gazette*, 3 September 1988, B7; "Judge Clears 51 Protesters," *Globe and Mail*, 15 February 1989, A3.
124 "Loggers Block Road Again Despite Meeting Ministers," *Ottawa Citizen*, 6 September 1988, A13; "Lumber Company Declares Bankruptcy," *Ottawa Citizen*, 3 December 1988, A4.
125 Christie McLaren, "New Conflict Looms on Logging Roads," *Globe and Mail* 22 November 1988, A12.

126 Christie McLaren, "Ontario Going to Court in Indian Dispute," *Globe and Mail*, 30 November 1988, A9; Christie McLaren, "Indians Agree to End Blockade after Appeal Court Decision," *Globe and Mail*, 8 December 1988, A9.

127 June Twain, "The Joy of Unfolding Commitment," in *Blockades and Resistance*, ed. Bruce W Hodgins, Ute Lischke, and David T McNab (Ontario: Wilfrid Laurier University Press, 2003): 17.

128 Christie McLaren, "Ontario Government Starts Cutting Trees in Temagami Area," *Globe and Mail*, 21 March 1989, A5; Thomas Claridge, "Wilderness Society Wins First Round," *Globe and Mail* 24 March 1989, A5.

129 Christie McLaren, "Temagami Indians Set to Block Logging Road," *Globe and Mail*, 24 March 1989, A5; Chornook Interview.

130 Darcy Henton, "Police Arrest 5 as Group Blocks Logging Road," *Toronto Star*, 5 June 1989, A3; Deborah Wilson, "Police Break up Blockade of Temagami Logging Road," *Globe and Mail* 6 June 1989, A18.

131 Brian Back, Interviewed 15 September 2017.

132 Ibid.

133 Ibid; Darcy Henton, "Northerners Blast Toronto Activists for Meddling in Temagami Wilderness," *Toronto Star.*, 6 June 1989, A10.

134 "Roynat Loan to Let Milne Resume Lumber Production," *Globe and Mail* 10 June 1989, B3.

135 "Temagami Blockade," *Globe and Mail*, 26 August 1989, A12.

136 "Ontario Can't Stop Environmentalists' Protest in Temagami: Court," *Gazette*, 15 September 1989, B1.

137 Chornook Interview; "Loggers Warned of Spikes," *Edmonton Journal*, 15 September 1989, B7; Hap Wilson, "Interviewed 17 January 2017.

138 Darcy Henton, "Group Endures Trip through Rugged Bush in Bid to Save Forest," *Toronto Star*, 17 September 1989, A8; Back Interview.

139 Ibid.

140 Rachel Leaney, "200 Environmentalists Given Training to Resist Temagami Logging," *Globe and Mail*, 18 September 1989, A12.

141 Timothy Appleby and Richard Mackie, "Rae among 16 Arrested at Temagami Protest," *Globe and Mail*, 19 September 1989, A1.

142 Lischke and McNab, "Actions of Peace," 7.

143 "Logging Road Work Proceeds as Protest Ebbs," *Ottawa Citizen*, 21 September 1989, A17.

144 Kelly Egan, "Environmentalists Battle Loggers in Temagami," *Ottawa Citizen*, 10 October 1989, A1; Wilson Interview.

145 Chornook Interview; Wilson Interview.

146 Kelly Egan, "Anti-Logging Road Activists Score Only Small Victories," *Ottawa Citizen*, 10 October 1989, A2; Wilson Interview.

147 Back Interview.

148 Chornook Interview.

149 Wilson Interview.

150 Ibid; Back Interview.

151 "Police Arrest, Release 31 at Temagami Logging Site," *Ottawa Citizen*, 20 September 1989, A5.

152 Chornook Interview; "Residents Support Logging; Protesters Have Road to Themselves," *Ottawa Citizen* 24 September 1989, A7; Back Interview.

153 Wilson Interview.

154 "7 Protesters Face Mischief Charges," *Ottawa Citizen* 10 October 1989, A2.
155 "Temagami Trees at Risk Study Finds," *Toronto Star* 13 October 1989, A12.
156 "Logging Road into Temagami Will Be Suspended, Ontario Says," *Gazette*, 20 October 1989, B1.
157 Darcy Henton, "Band Evicts Environmentalists from Logging Road Ministry, Construction Company Also Told to Leave," *Toronto Star*, 1 November 1989, A12.
158 Ibid.
159 Bruce W Hodgins, "The Tamagami Blockades of 1989," in *Blockades and Resistance*, 26–27.
160 Chornook Interview.
161 Back Interview.
162 Donald Grant, "Natives to Resume Logging Road Blockades," *Globe and Mail* 3 November 1989, A13; Lana Michelin, "Temagami Protesters Urged to Go," *Globe and Mail*, 6 November 1989, A12; Laronde Interview.
163 Darcy Henton, "Native Demonstrators Celebrate as Temagami Road Work Halted," *Toronto Star* 12 November 1989, A2.
164 "Police Arrest Protesters," *Edmonton Journal* 13 November 1989, A4.
165 Hodgins, "The Tamagami Blockades of 1989," 23–24.
166 Grant, "Natives to Resume Logging Road Blockades," 3 November 1989, A1.
167 Laronde Interview.
168 Ibid.
169 Darcy Henton, "Police Charge 3 Indian Chiefs as Temagami Fight Continues," *Toronto Star*, 16 November 1989, A15; "Nineteen More Arrests at Temagami," *Ottawa Citizen*, 20 November 1989, A4.
170 Darcy Henton, "OPP Bill Almost $1 Million in Temagami Logging Fight," *Toronto Star*, 4 December 1989, A10.
171 "Supporters Mark Road's Opening," *Gazette*, 29 January 1989, A7.
172 Bruce Hodgins, "Gary G. Potts," in *Blockades and Resistance*, 20.
173 David Israelson, "Thousands March in Metro to Celebrate Earth Day," *Toronto Star*, 23 April 1990, A1.
174 David Israelson, "Province yet to Rule on Temagami Plan," *Toronto Star*, April 5 1990, A4.
175 Suzanne Steel, "Logging Ended in Prime Temagami Area," *Gazette*, 24 April 1990, B1.
176 Native Forest Network. Temagami Chronology, Available [Online]: http://temagami.nativeweb.org/temagami-chronology.htm (Accessed 26 December 2016).
177 Keith Baldrey, "Ulkatcho Band Vows to Continue Blockade," *Vancouver Sun*, 3 August 1989, B5.
178 Justine Hunter, "Indians Set up Road Blockade near Hazelton," *Vancouver Sun*, 29 September 1989, F10.
179 Terry Glavin, "New Native Road Blockade Joins Four in Skeena Valley," *Vancouver Sun* 14 October 1989, C16; Glavin, "Interview," 29 August 2017.
180 Suzanne Fournier, "Natives Blockade Loggers," *Province*, 31 October 1989, 1; Suzanne Fournier, "Logging Drivers Reciprocate on Blockade," *Province*, 1 November 1989, 4; Suzanne Fournier, "Court Imposes Deadline: Indians, Loggers Get 14 Days," *Province*, 3 November 1989, 43.
181 Glavin, "New Native Road Blockade Joins Four in Skeena Valley," C16; Suzanne Fournier, "Gitksan Chiefs Jubilant over Court's Action," *Province*, 17 November 1989, 24.
182 Glavin Interview.
183 "New Alliance Adds Voice to Forest Protest," *Vancouver Sun*, 20 January 1989, H15; Terry Glavin, "To the Barricades," *Vancouver Sun*, 21 July 1989, B1.

184 Terry Glavin, "Westar Joins Northwest Timber Protest: Tired of Government Stance, Firm Reveals," *Vancouver Sun*, 23 February 1990, B3; Terry Glavin, "Province Must Alter Policy, Westar Head Tells Meeting," *Vancouver Sun* 27 October 1989, A16.

185 Notzke, *Aboriginal Peoples and Natural Resources in Canada*, 102–3.

7

"WE REFUSE TO BEQUEATH A DYING PLANET TO FUTURE GENERATIONS BY FAILING TO ACT NOW"

Developments in environmental blockading and ODA since 1990

With environmental blockading embedded as a strategic option in Australia, Canada, and the United States the repertoire was deepened and diffused nationally and internationally during the 1990s. This period saw a rapid expansion in tactical forms, the majority of which originated in the three countries covered by this book as well as in the United Kingdom, where widespread ODA against road construction began in 1992. Tactical diversity peaked in the late 1990s, the point by which detailed environmental blockading manuals had been released. Although innovation has continued to occur, the main development since this time has been in the diffusion of tactics first developed in the four countries to other places and movements.

The presence of strong, ODA-oriented environmental preservationist movements in these countries supported tactical development in large part because activists were not regularly subject to major repression. Blockades in the Philippines, Russia, Nigeria, and elsewhere during the period were curtailed by police and military crackdowns, long jail sentences and preventative detentions, as well as the killing and maiming of activists. Such action obviously undermined the logic behind techniques of manufactured vulnerability.[1]

Despite facing similar repression, blockades in Brazil and Malaysia were able to build into protest waves which commanded much international publicity and support. The dozens of *empates* in Brazil were canvassed in the introduction. In Malaysia Penan, Kelabit and other Indigenous communities in the state of Sarawak carried out dozens of blockades between during the 1980s and 1990s against rainforest logging. These were generally in the form of roadblocks, which included community members, sometimes armed with blowpipes, occupying key entrances and roadways with their bodies as well as constructing villages across them. These campaigns did not extend the blockade as a form of action beyond one or two national regions and the range of tactics employed were limited and generally unsuited to other countries.[2] As such while coverage of these events in international publications and films served as inspiration to environmentalists internationally few specific elements of their repertoires were diffused at this time.

The following sections chronicle the origins of many of the techniques that developed between 1988 and 1997 as well as how they were connected to and reflected developments and dynamics that had occurred during earlier periods of embedding. They canvas specific campaigns and blockades that were key to the continuing expansion of the repertoire along with how increased direct and indirect diffusion was facilitated via international travellers, publications, and the internet. Following this, the chapter explores the continuing impact of environmental blockading upon a variety of movements today.

Australia, 1988–1997 developments

Between 1985 and 1989 ODA at the point of destruction underwent a lull in Australia and blockading received little support from most mainstream environmental organisations. For their part radicals shifted their focus to ODA against rainforest timber imports, visits by US nuclear warships and other issues. Tasmania, where political closure reigned due to both centre-left and centre-right parties staunchly supporting logging, saw intermittent blockades, sometimes organized by TWS, during 1984 and from 1986 onwards. This included some tactical development, as discussed in Chapter 3, as activists adapted and extended tree-sitting means from the United States. Political closure, due to the state's arch-conservative government, similarly led activists in Queensland to plug blasting holes and occupy caves for six weeks in 1987 and 1988 to prevent the destruction of bat habitat at Mt Etna through limestone mining. Although the level of blockading steadily increased across the country thereafter, the main sites of repertoire development occurred during the South East Forest Alliance (SEFA) campaign of 1989–1990 and a series of blockades carried out by the North East Forest Alliance (NEFA) from 1989 onwards.[3]

South East Forest Alliance (SEFA), 1989–1990

Despite concerns about rates of logging, prior to 1989 mainstream organisations such as the Total Environment Centre (TEC) had ruled out blockading in NSW for fears of jeopardising their ALP allies hold on power.[4] With the state ALP voted out and the federal ALP government opting to issue a 15-year woodchip license for areas environmentally significant enough to be on its "National Estate" register, this strategy was abandoned. Local campaigners and state and national organisations such as TWS, ACF, and TEC came together as SEFA and in February 1989 launched a campaign of "peaceful resistance" to achieve the "cessation of all logging activity".[5]

Despite campaign materials citing "the obstruction of rainforest logging at Terania Creek", the SEFA campaign was initially based on the model used by TWS during the Franklin Dam.[6] This combined an MNS/ONV style approach to normative protester behaviour with the outcomes of consensus decision making largely limited to the practicalities of undertaking actions within predetermined parameters and strategies set by key SEFA decision makers. Nonviolence training was carried out in the city and at protest camps by TWS and others and police briefed about actions ahead of time. Although some actions involved blocking access points with bodies, civil disobedience, typically involving the limited trespass of prearranged numbers of protesters within exclusion zones, was prioritised over the disruption of work. Up to 1,200 people were eventually arrested and this initially generated a high level of publicity, thus meeting organiser's goals on the basis of a belief that they could "change the positions of politicians … through the power of public opinion".[7]

What the strategy failed to do was have any significant immediate or medium-term impact on the level of logging itself. In keeping with other campaigns described in this book the nature of living in the forest and witnessing its daily destruction, coupled with frustration at state and federal government intransigence and a sense that repetitive tactics were fast losing media appeal, led some to embrace more disruptive means. Having been attracted to the campaign by goals such as "slowing down or halting logging and roading activity"[8], a significant minority of activists lost faith in the official campaign strategy. If they were to be arrested, face fines and court, and suffer the enmity of local residents who supported logging, then they wanted to feel that their sacrifices were having a greater impact.[9]

This group began to act independently of the rules set down by SEFA. Focusing on the direct disruption of logging operations they adopted streamlined means of consensus decision making and secrecy to deploy tactics designed to surprise their opponents and disrupt operations. The combination of a concentration on minimising arrests with wanting to prevail on a regular basis over loggers and a special police squad assembled for "Operation Red Gum" resulted in a high level of tactical innovation.[10]

Mark Blecher and other innovators were responsible for a series of inventions which extended disruption through enhanced vulnerability. These included designing durable tree-sit platforms with which activists were able to put themselves out of the reach of cherry picker cranes. As a result, one woman held up logging and attracted widespread media attention by remaining in a tree for 56 days. In another case three brothers engaged in tree-sits together. The campaign marked the most intensive use of the tactics thus far in Australia with 50–100 protesters eventually taking part in tree-sitting actions. Their efficacy, as at the Daintree and elsewhere, was enhanced by tying adjoining trees together, on this occasion with fencing materials. For the first time tactics previously employed interstate and internationally, such as using U-locks to attach activists to logging equipment and people supergluing hands their hands to equipment and each other, were used in a NSW forest campaign.[11]

Most importantly for the development of national and international ODA, Blecher built the first devices designed to allow blockaders to shut down entry points by locking their wrists and arms into steel plates and cylindrical cannisters buried and concreted into the road or ground. Known variously as the sleeping dragon, wog wog, or Blecher device the difficulty of safely removing activists meant that over the next decade its use extended widely to blockades around the globe, spread directly by travelling activists and indirectly via manuals and mainstream and alternative media coverage.[12]

A third innovation, also widely adopted internationally and still in use today, was the tripod. Designed by construction worker Peter Vaughn, with advice from an engineer, the version created at this time had three metal scaffolding tubes staked to the ground, which were clamped and bolted at the top to form a pyramidical structure. A dentist's chair was then suspended from the apex for an activist to sit in. Later versions would use wooden poles lashed together with a person sitting on top. These devices replicated the vulnerability associated with tree-sits and forced authorities to remove activists and structures located on roads or over equipment.[13]

The effectiveness of tactics based on manufactured vulnerability is directly related to the willingness of authorities to avoid injuring activists. Unfortunately, the first deployment of a tripod also saw the first person fall out of one. Despite warnings from blockaders an attempt by police and loggers to lower the tripod saw the otherwise stable structure tip over. Despite falling eight metres activist Dave Burgess avoided major injury. The negative publicity associated with this, and other incidents, forced authorities to take greater care, especially when journalists or video cameras were present.[14]

As elsewhere, rapid tactical innovation wrongfooted opponents and the "catch up" factor required by authorities increased the ability of campaigners to directly close down worksites. At the same time violence increased against blockaders. Loggers were authorised to make "citizens arrests" and police accused of standing by during incidents, which included a car with an activist chained beneath being forced off a road and workers brandishing rifles and other weapons.[15] Despite operating in this hostile environment, and lacking a clear strategy beyond fighting coupe by coupe, regular action was sustained into 1990. The news media and public were largely unaware of the differences between factions, but key mainstream organisations began to withdraw resources and momentum was lost. The combined efforts of all campaigners forced both the federal and state governments to negotiate, but concessions were minimal.[16] Nevertheless the techniques developed fed into a much more focused and successful movement in another part of the state.

North East Forest Alliance, 1989–1997

In 1989 NEFA launched the first of several blockades in Northern NSW. Defining itself as "nonviolent" the group differed from the official SEFA approach by employing a looser version of normative protester behaviour. With only monkeywrenching and physical aggression ruled out, this was conducive to obstructive innovation in allowing for secrecy and by concentrating efforts on directly disrupting logging operations from the beginning. Innovation was constantly encouraged and at one point NEFA ran a public, and partially self-satirising,

"Blockade Design Competition". NEFA's approach was adopted in part due to the involvement and influence of Nomadic Action Group members and others who had brought a similar approach to ODA during the environmental blockades of 1979–1984.[17]

During the SEFA campaign the cost of policing had run close to A$120 million with the result that the state government eventually withdrew police and shifted back to the use of local officers. There had been attempts by campaigners to use litigation, but this had only brought about short-term delays.[18] Noting the successes and failures of the SEFA campaign, NEFA adopted a flexible strategy through which it would successfully use ODA to generate media coverage, inflict political and financial costs, and delay work in the short term. At the same time, a longer term political and legal strategy, influenced by the Nightcap campaign and led by lawyer John Corkill, was designed to force the NSW Forest Commission to comply with existing environmental planning laws.[19]

The campaign further differed from SEFA in that while blockades would some-times be held in hostile areas, proximity to Northern NSW's alternative commu-nities provided a vital political base and source of support. Added to this was the influx of a new cohort of blockaders who became known as "ferals". Drawn from city-based anarcho-punk and squatter scenes these activists embraced both ODA and living in the forest and brought additional creativity, as well as an aggressive edge, to proceedings.[20]

NEFA's first blockade occurred at North Washpool in 1989 with the support of members of the local Aboriginal community. Cars were used to block the road and 14 people arrested before the NSW Forest Commission was directed by the courts to suspend roadbuilding due to the presence of sacred sites. With logging looming again in 1990 a blockade was set up, including the use of a CB system for communications and tripods made from wooden poles drawn from the bush. Before it could be tested a further injunction was gained due to illegal logging and faults with the EIS. The success of these events fully established NEFA's strategy of combining litigation with ODA.[21]

A major step forward came with the Chaelundi campaign. In March 1990 NEFA had gained an injunction on logging in this forest. This had been won in the context of a blockade involving lock-ons to cars and a logging truck. ODA had also involved the first use of lock tubes, metal cannisters into which activists' arms were encased and locked together, which could be placed around equipment, gates, and other points.[22]

Having carried out the requisite EIS, the FC announced plans to log in July 1991, but soon found itself stymied by a five-month long blockade, which included three camp sites. Having declared the Chaelundi People's Wilderness Park/Chaelundi Free State in April, blockaders, including an engineer and riggers, went all out in constructing barricades and devices. The results demonstrated the interplay of geography, beliefs about normative protester behaviour, emotions, resources drawn from the local community and environment, and other factors in creating a protest ecology favourable to effective ingenuity and obstruction.[23]

Some tactics were new. Building supplies intended for a new road were scavenged from the area and concrete culvert pipes used either as barricades or devices with activists locked inside and on top. In addition, some blockaders were further anchored around the wrist and neck by chains connected to a concrete slab. With wooden poles supplied by a local alternative community, variations on tripods were designed including placing activists upon two pole bipods and monopods, the latter featuring a single pole attached to a "Star of David" assemblage of timber poles or a tripod. In another innovation a "web of life" was created by connecting netting between trees to allow activists to hover and move above ground. These were deployed alongside updated versions of tactics from Australia and overseas, such as people locking onto equipment, devices and each other, tripods, tree-sits, road barricades, car bodies, soft blockades, a dam, bonfires, heated rocks, and activists buried in the ground and chained to concrete slabs or locked to vehicles placed above them. Action was coordinated by an efficient communications system and combined in a series of obstructions. When police moved in, they found themselves spending entire days dismantling obstructions only for a new set to be in place by the time they returned the next morning.[24]

After more than two weeks and 200 arrests authorities were only able to regain to take control of the area for a day before NEFA gained a court injunction against logging. The group subsequently secured a decision protecting the area on the basis of threats to endangered fauna. A failed appeal from the NSW conservative government led to regulatory attempts to exempt logging from environmental protections but this in turn was blocked after a government MP resigned over the issue, costing conservatives their parliamentary majority. Legislation toughening protections for fauna was subsequently passed, but continued ODA and litigation was required to test its provisions and ensure the FC met its requirements.[25]

Amidst occasional losses NEFA's strategy continued to secure victories with the FC being forced to withdraw from areas either through court action or by deciding to do so unilaterally due to obstruction, cost, and attendant controversy. All of this was caused by or came in the context of regular blockades as well as sporadic "hit and run" ODA in the forests, for instance in the lead up to elections in 1995, and actions at FC offices and elsewhere. Although loggers sometimes intervened, the costs and difficulties imposed by remote locations and tactics meant that police were either rarely seen or largely reduced to an observation and peace-keeping role.[26]

During this time the tactics described above were deployed in various combinations and new ones added to the Australian repertoire, including linking multiple tripods together with a platform on top and placing activists on a 12-foot-log suspended above a road by steel chain. Logging cable was strung across a road to create a "witchy web weaving" through netting and bush materials that could be hidden within and locked onto. Cantilevers, devices using a pole affixed to a rock wall and extended across a road or track, were introduced from the United States, after long-time Northern NSW activist Dailan Pugh read about them in the *EF! Journal*. The tactic was adapted through the addition of a blockader being hung off the end.[27]

Beyond setting an example shared through various forms of media, tactics innovated and adapted by NEFA were also directly diffused nationally and internationally by travelling activists, correspondence, and coverage in alternative media. In the early 1990s NEFA produced and widely distributed its *Intercontinental Deluxe Guide to Blockading*. Although handbooks had been produced for blockades and campaigns previously, this was the first manual to go beyond outlines of expected approaches to normative protester behaviour and legal and other practical advice. It provided diagrams showing how to construct and use tripods, tree-sits, cantilevers, and other devices and techniques, as well as advice on site infiltration and strategy. It would eventually be shared on the internet, but as most activists had limited access to such means in the early 1990s the booklet was initially emailed and posted before being copied and distributed in hard copy form overseas by US and UK EF! groups.[28]

US developments during the 1990s

Throughout the 1990s localised campaigns continued across the United States. Even where campaigners opted to operate under other names, to avoid association with monkeywrenching, the EF! network remained a key link between them. It also continued to be the prime vehicle for tactical and strategic diffusion via the now familiar channels of members who initiated actions and travelled between them, the *EF! Journal*, songs, and gatherings, as well as the production of a detailed blockading manual in 1997. Following Redwood Summer, a Forest Action Workshop group was formed, which toured the United States and Canada in 1991 holding 23 workshops including a slide presentation on ODA and training regarding action planning.[29]

Diffusion between Australia and the United States occurred in similar ways with articles regularly covering blockades and developments in each country carried in the *EFJ*, *World Rainforest Report* and other national and international publications. US activists such as Orin Langelle and Jake Jagoff took part in Australian gatherings and blockades while John Seed continued to tour the United States regularly.[30]

The successful use of litigation and appeals related to court findings protecting endangered species led to a downturn in anti-logging ODA across the Pacific Northwest (PNW) and California in the first half of the 1990s. Sporadic forest blockades occurred in the region and Illinois alongside new campaigns in Colorado and Florida and the first use of the environmental blockading template in Wisconsin, Maine, and Vermont. These generally adopted the now standard EF! approach to normative protester behaviour with secrecy employed and a, sometimes shaky, division maintained between monkeywrenching and forest blockading.[31]

For the most part blockades during the early 1990s combined barricading, soft blockading, tree-sitting, and lock-on tactics developed in previous years. A new form of barricading/soft sabotage was introduced during the eight-week Californian Albion blockade in 1992 where brightly coloured yarn was woven between trees. Due to its ability to stop chainsaws from working, it added another layer of obstruction that needed to be removed before work could commence. The first of what would be six years of blockading in Idaho's Cove/Mallard area used tripods used for the first time in the United States in 1992. The technique was employed in a fresh round of blockading at Mount Graham in Arizona in 1993 and soon became a regular feature of American ODA.[32]

A new protest wave, triggered by political responses to recent wildfires in Idaho and Washington, began in 1995. Having gained control of both federal legislatures, Republicans passed a series of bills which rolled back numerous environmental regulations. These included what came to be dubbed as the "Salvage Rider" and "Logging Without Laws" legislation. Similar provisions, in which changes were attached to unrelated appropriations, had been previously used by federal politicians in 1989 to enable logging. In this case reforms were ostensibly designed to allow access to timber from areas damaged by fire or infestation as well as to thin forests deemed to be at risk from them. Environmentalists rejected arguments that forests would benefit from such action, but what truly outraged them was that during 1995 and 1996 the measures effectively removed environmental and wildlife protections from all forests. Federal authorities were also ordered to maximise logging and open up timber sales that had been previously halted, and the ability for activists to make administrative appeals was removed.[33]

A heavy lobbying campaign immediately followed. After exercising an initial veto, the Clinton administration signed through the provisions. With legal challenges limited, and logging companies moving rapidly to take advantage of the situation, blockades were soon mounted. Rapid political and legal closure, as well as outrage at the blanket opening of areas previously considered saved, meant that these were supported by a wider range of mainstream organisations than in the past and involved a new flood of activists.[34]

Between 1995 and 1997 approximately 30 forest blockades occurred alongside other protest activities involving thousands of arrests. Many were sustained and some spread over years. Alongside multiple blockades in Oregon and California were those in Washington, Idaho, Illinois, Colorado, Montana, and for the first time Minnesota and Virginia. The central role of radical environmentalists demonstrated that the EF! network, and allied groups, such as the newly formed Ruckus Society, initiated by long time EFers, remained the central vehicle for carrying out and training activists in ODA.[35]

Tactics employing U-locks, sleeping dragons, car bodies, and sleeves were introduced, adapted, and extended. As police became more adept at removing them, activists introduced innovations using barrels and moulds filled with concrete and fitted with sleeves, which were then locked into. During the long-running Californian Headwaters campaign one "superbox" proved resistant enough that a pair of women were taken to jail and court still locked into it. Another 1996 blockade within Oregon's Olympic National Forest involved the "American Family" affinity group locking on to a couch that was filled with cement, reinforced with rebar, and locked to concrete in the road, with a living room set up around it. Echoing a trend towards increased scale, density, and complexity across the repertoire, combinations of boxes, barrels, car bodies, and other tactics were regularly employed.[36]

Tree-sitting tactics were also extended via processes of diffusion and innovation. Based on construction, netting and other techniques primarily developed during road protests in the UK (discussed below) the first "tree villages" allowing multiple sitters to live in high canopy locations for extended periods were introduced to the United States in 1996 during the Headwaters campaign. In 1997 the blockade added the "Love Pod", a circular wrap-around tree-sit which could accommodate six people, to the repertoire. These appeared alongside tree villages perched 100 to 200 feet in the air with tipi coverage and space for lock-boxes, water, stoves, supplies, and solar-powered communications equipment. Such improvements made evictions much more difficult and heightened media novelty, most famously allowing activist Julia Butterfly to occupy a 1,500-year-old California Redwood tree for 738 days.[37]

A series of "Free States" set up during these years served as key sites of development. In a similar fashion to NEFA's 1991 "Chaelundi Free State" these occupied large areas of roads and forest with multiple assemblages of lock-ons, barricades, and other obstructions. The most effective of these, and the one largely credited as key point for the refinement and diffusion of tactics, was set up in Oregon during September 1995. Litigation had initially prevented salvage logging in Oregon's Warner Creek, but a court ruling influenced by the new Salvage Rider reopened it. EFers, who had been heavily involved in organizing lobbying, rallies, forest walks and EIS submissions, immediately launched the "Cascadia Free State". Its landscape was dominated by "Fort Warner", which featured a timber wall across a road supplemented by a watchtower, cat walk, moat, and drawbridge. In addition to this deep trenches and multiple slash and rock barricades covered long sections of road and culverts were removed to ensure flooding. Barrels, sleeping dragons, including one covered with a steel door, car bodies, a crawlspace, and other points were set up for locks-ons alongside tipis, tripods, bipods, and other pole-based structures. The area was held for 343 days before forest salvage operations were cancelled.[38]

Responses from police and other authorities during this period largely followed a trend of increasing repression. Exclusion areas were widened, up to 50 miles in the case of the 1996 Oregon Sugarloaf campaign, armed security patrols deployed, and protest camps regularly harassed. Fines and jail sentences were increased in various states. In California activists were charged with resisting arrest when they refused to unlock themselves from devices. Interfering with logging operations was made a felony offence in Idaho. Some of these charges and penalties proved illegal or unworkable and legal penalties failed to curtail blockading.[39]

More concerning was a shift towards increased violence. In Colorado, California, and Oregon loggers drove through picket lines. Threats and attacks were regularly levied against protesters, forcing one blockade in the Blue Mountains to disband. In 1998 David "Gypsy" Chain was killed by a tree felled in the direction of blockaders during ongoing actions in the Headwaters area.[40]

Responsibility for countering blockading shifted from local Forest Officers to a specialised and highly resourced Forest Service Law Enforcement and Investigations branch in 1994. While these FS officers and police developed more sophisticated removal techniques involving climbers, cherry pickers and diamond grinders they also, deliberately or through exercising minimal care, tipped over tripods and other structures. Most controversially they began to use pain compliance holds and in a small number of cases applied pepper spray to activist's eyes with cotton swabs during office occupations and other actions to force them to release themselves from lock-on devices. Dogs were also used to intimidate blockaders and tree-sitters were targeted with CS spray. Despite media representatives often being locked out of logging areas the increased availability of portable video equipment allowed activists to film some incidents. The subsequent release of footage generally served to increase public sympathy as well as shift police behaviour. At the same time increasing violence towards activists, as well as rider itself, was partially credited for a new upswing in monkeywrenching, primarily away from blockade sites and much of it under the banner of the Earth Liberation Front.[41]

Between 1995 and 1997 blockading slowed or prevented logging in various areas while providing media coverage that mainly supported campaigning strategies carried out by a spectrum of environmentalists. Although each dispute involving a blockade had its own particularities all, including long running campaigns concerning the Cove/Mallard and Headwaters areas, benefited from the overall rise in contention. The broad movement had a major political impact with the result the Clinton administration moved to roll back anti-environmental provisions.[42]

Beyond directly developing and diffusing the environmental blockading model and tactics, and demonstrating their effectiveness, the protest wave also led to the compilation of the *EF! Direct Action Manual* by a collective in Oregon. Released in 1997, at 152 pages this was a major expansion on the NEFA booklet and included detailed instructions and diagrams regarding dozens of techniques as well as information about security, organisation, and legal issues. In keeping with standard US blockading practices of the time, it advocated a loose definition of normative protester behaviour with the majority of tactics relying upon secrecy. Similarly, although its sample nonviolence code only stated that campaigns "will not condone senseless acts of sabotage", the authors did not advocate or include information about monkeywrenching. Sections on training primarily focused on practical issues to do with conflict de-escalation, safety, and scouting. Up to 30 minutes out of the suggested model for a three-and-a-half hour training session were allotted for a section regarding the history and practice of nonviolence and the majority of this involved discussions in which participants were expected to define the meaning of the term for themselves.[43]

Copies of the manual were distributed via the *EFJ* and EF! chapters around the world. Updated and expanded editions were produced in 2002 and 2015. Echoing the fact that most innovation occurred during the 1980s and 1990s, they have mainly included extra advice, commentary, and details, with few genuinely new tactics added.[44]

Canadian developments, 1989–1995

By the end of 1989 environmental blockading had become fully established as a strategic option within Canada. Repertoires of contention had developed along two core lines, both of which intensified in the years to come. Precedents established by Indigenous activists regarding tactics associated with soft blockading and the deployment of barricades continued to dominate many First Nation-led blockades, which sought to delay work seen as environmentally harmful, support or trigger litigation, generate media coverage, and enforce community sovereignty over traditional lands.

The entrenchment of the repertoire became clear in its rapid expansion and diffusion in response to a paramilitary raid on an armed Mohawk blockade against golf course development on sacred land in the town of Oka, Quebec, in July 1990. This incident, which led to the death of a policeman and the subsequent deployment of soldiers, triggered dozens of solidarity blockades across Canada, providing a clear focus for ongoing grievances regarding sovereignty.[45] The majority of these were unarmed. Roadblocks were largely spontaneous and locally organised but had major economic outcomes with some entire regions closed off for periods. They mainly targeted major roads and rail lines to maximise pressure on the government and non-Indigenous population, but also formed part of ongoing campaigns against logging, dam construction and low-level flights by the military.[46]

British Columbia was the main site of action with around 30 blockades. Some held by Nisga'a, Chilcoton, Fountain, and Gitxsan–Wet'suwet'en communities focused on logging. One was led by a group of 14-year-olds who told the media "We're not allowed to argue with the loggers, we must be very polite."[47] Widespread disruption led the conservative BC government to seriously engage with the issue of Aboriginal sovereignty for the first time. In another first the government also deployed major force against some First Nations blockades.[48]

Tactical innovation was limited, but expansion and diffusion of the existing repertoire was high throughout the 1990s, leading to numerous inquiries into policing and First Nations rights. Although the number of such blockades has ebbed and flowed since this time, many disputes regarding sovereignty, land use, and racism remain deadlocked and the repertoire remains a key strategic response.[49] In 2020 a new wave of road blocks, occupations, and rail closures were held in the provinces of Ontario, Quebec, and BC following the eviction of Wet'suwet'en camps preventing the construction of an internally contested gas pipeline through 190 kilometres of territory.[50]

The second strand of Canadian environmental blockading followed a similar direction to developments in Australia and the United States in that the number, combination and diversity of enhanced vulnerability tactics increased. Although these tactics extended across Canada most of the adaptation and innovation continued to occur in BC. Clayoquot Sound would remain a key node as members of the Sulphur Passage campaign developed into a cohort of travelling activists responsible for further innovation and diffusion through workshops and participation in blockades. Focused on obstruction and employing secrecy these activists, those who picked up on their skills and approach, and later groups such as Forest Action Network, were responsible for extending the efficacy of tree-sitting and introducing various forms of lock-ons, tripods, and other tactics at places such as Bulson Creek, Walbran Valley, Fog Creek/ISTA, and elsewhere.[51]

Authorities and companies experimented with a range of new responses including charging non-violent activists with endangering police officers, employing special removal teams and spotters, and cutting large swathes of forests pre-emptively if a blockade was imminent. In response some activists improved radio communications, infiltration techniques, clothing, climbing equipment and platforms to move higher and remain longer within canopies. Some of these responses were innovated locally and others learnt and shared directly with activists from United States and Australia via visits to one another's countries.[52]

There was an increasing use of such tactics in campaigns involving or initiated by members of First Nations communities. These included campaigns regarding the Tsitika Valley, Lillooett Creek, ISTA, and Carmanah Valley. As with First Nations campaigns wholly predicated on barricading and soft blockading, not all in the community were against logging or development and those that were sometimes came under heavy pressure from relatives and others. Although the situation would change in Australia and the United States in later years, the leadership and involvement of First Nations people in blockading overall remained higher in Canada during the 1990s.[53]

Some in the Clayoquot Sound area, frustrated at government prevarication and continued clear-felling, began to employ sabotage more extensively. Tree-spiking had already been used during the 1980s and members of Sea Shepherd, whose boat was moored in the area for a period in the 1990s, also had an influence. In 1991 a major logging bridge was burnt down, shutting down access to sites and putting 240 people out of work.[54]

Although the NDP's election in 1991 saw some degree of reform and concession in other parts of the province, its decision to continue logging 74 per cent of Clayoquot Sound's old-growth forest shocked locals and environmentalists. An attempt was made to burn another bridge, but the primary response, led by a new group within Friends of Clayoquot Sound, was to hold Canada's biggest blockade in 1993.[55]

Despite the name "Clayoquot Summer" citing the Californian Redwood Summer campaign, tactics of enhanced vulnerability at the point of destruction were sidelined in favour of soft blockades at entry points. Practices developed during Strathcona Park, Franklin Dam, and other blockades were drawn upon in that Clayoquot Summer employed a strict MNS/ONV style "Peaceful Direct Action Code". This approach to normative protester behaviour, which was reinforced through regular training sessions, posters, and speeches, combined a strand of eco-feminism with a commitment to full "openness" with authorities, and a strict definition of and ban on sabotage. This led to an almost three-month daily pattern of brief early morning occupations in which police were informed of the number of people willing to be arrested before an injunction was read out and those choosing to remain on the road removed. With their own negotiations in train and differences over how to best campaign for an end to clearcutting the Nuu-chah-nulth community largely sat the event out.[56]

Highly media focused, the blockade caused very little obstruction. However, with more than 10,000 participants, 932 arrests, and mass trials of activists that ended with some spending months in jail for "criminal contempt of court", it commanded widespread national and international attention. This was subsequently parlayed into an international boycott campaign that successfully cut into profits and generated, alongside continued Aboriginal land claims and more obstructive blockades, major economic uncertainty. By the mid-1990s this was enough to force companies and the government to agree to some reductions in logging and the formation of joint timber ventures with traditional owners.[57]

UK developments during the 1990s

Social movements in the UK regularly employed blockades and ODA as part of campaigns regarding nuclear weapons, urban squatting, and animal rights during the 1980s, but it was not until the 1990s that a major protest wave concerning the protection of sites on the basis of their environmental value began. This movement was initially focused on the issue of preventing clearing for road construction and brought together grievances regarding pollution, car use, community input into land use decisions, and the protection of both urban spaces and woods and meadows. The latter had often experienced exploitation at some point, and were generally small, if not tiny, compared to the areas subject to blockades in the United States, Australia, and Canada. Nevertheless such "green space" had either been regenerated or preserved to a point where it generated similar grievances to those regarding the protection of lightly or undeveloped ecosystems overseas. In a densely populated and highly developed country such sites were highly valued enough to trigger dozens of blockades against threats from road and airport construction, mining, and suburban development from 1992 onwards.[58]

Inspired by direct experiences within radical environmental movements in Australia and the United States, as well as correspondence and information about them, a handful of activists launched UK Earth First! in 1991. Initially focused on issues such as the importation of rainforest timber the nascent network, alongside allied ones such as Alarm UK and Road Alert, rapidly grew in response to the "Roads for Prosperity" construction program. Launched by the UK government in 1989 this involved a £23 billion investment over ten years to build 24,000 miles of new road.[59]

The first major opposition came with a blockade at Twyford Downs in 1992. Inspired by Australian forest blockades and national anti-nuclear ones, such as Greenham Common, it brought together local residents with radicals from across the UK to set up a series of protest camps from February 1992. Blockaders took part in site occupations, climbed on equipment, and flooded the site with water.

Techniques of enhanced vulnerability diffused from overseas and other national movements, such as tripods, tree-sits and lock-ons, were also used. Although the main camp was evicted over three days in December, regular ODA continued into 1993 after which intermittent, but large-scale trespass actions involving up to 1,500 people were held.[60]

As previously discussed, trends in blockading overseas had led to long-term occupations in which protest camps combining a series of barricades and other tactics had become obstructions in themselves through being placed in the path of work. Alongside intermittent site occupations such camps were employed in many of the blockades that followed Twyford Downs. UK activists did not generally enjoy the advantage of remote locations subject to minimal policing. However, laws treating trespass as a civil issue in England and Wales lent them other advantages as police could not unilaterally and immediately evict occupiers, at least legally. Authorities generally had to had to make a court application to remove protest camps and blockades and in some cases activists used squatting rights to further delay evictions. Combined, these conditions allowed activists to spend up to months putting in place increasingly complex obstructive assemblages.[61]

As trespass was not a criminal offence the ability of authorities to quickly use court fines and bail conditions to curb protest was also lowered. Injunctions against protest and other activities however could be taken out against blockaders and seven activists at Twyford Downs were imprisoned for breaking one. As in overseas cases, suits for damages could also be brought and the Department of Transport (DOT) attempted to sue protesters for £1.9 million over losses accrued during the Twyford Downs blockade. Mounting such cases however used up resources and, as in overseas situations, had relatively little impact on mobilisation. As a result this one, like many, was dropped. Activists had recourse to the courts themselves and other than seeking injunctions against projects, which were more rarely granted than overseas, could seek their own damages. In the case of Twyford Downs ten blockaders successfully sued the Hampshire police for £50,000 over unlawful arrest and detention.[62]

Property and trespass laws, as well as limits on local police budgets, meant that private security was generally employed to block and remove trespassers, further driving up the cost of delayed projects. During Twyford Downs and later blockades protesters were assaulted by security guards and police. Riot police were deployed during some evictions and activists closely surveilled by state and private agencies. Over time chainsaw operators, climbers, and cavers were privately contracted to assist with evictions and specialist teams of police formed.[63]

In 1994 the UK government passed the Criminal Justice Act (CJA) targeting hunt sabotage, dance parties, squatters, road activists, nomadic travel, and other practices. While this provided authorities with new eviction and criminal powers, it also had the effect of bringing a range of groups together into a popular opposition movement, which increased the level of blockading, among other practices, in the short to medium term.[64]

Between 1994 and 1997 more than 50 blockades were launched in defence of sites designated as environmentally important with a large number of allied actions targeting urban locations set for demolition as well as company and government offices. Some campaigns involved years of ODA and multiple protest camps. At times these appeared simultaneously, in others new camps were set up following one eviction after another. In Scotland, where trespassing was a criminal offence throughout the period, there were fewer campaigns, but ODA at Glasgow's anti-roads Pollok Free State lasted for more than two years.[65]

The decentralised UK EF! network served as a unifying symbol and coordinating network via gatherings and the national *EF! Action Update* newsletter. By 1993, 45 chapters had been established across the UK. Allied publications and theoretical publications such as *Do Or Die!* and national and local newsletters, broadsheets, and bodies, which grew out of the broad anti-CJA movement, such as *Schnews* and local Free Information Networks, further promoted specific blockades, diffused tactics, and debated strategy and approaches. Fluid cohorts of blockaders, many broadly identifying with EF! and some part of long existing traveller convoys, moved from blockade to blockade sharing skills and discussion.[66]

Although many blockades officially designated themselves as "nonviolent", significant sections of the movement did not. In terms of normative protester behaviour, the key internal movement distinction was between "fluffies" and "spikies". Those on the "fluffy" end of the spectrum eschewed aggression towards opponents and sabotage, but the adoption of strict MNS/ONV type rules was rarely advocated in the UK at this time. Partially influenced by and echoing trends that had occurred within anarchist, autonomist, squatting, animal rights, and anti-nuclear movements in the UK and Europe during previous decades, and responding to increased state repression, the "spiky" approach included aggressive behaviour, physically resisting arrest and eviction, and the use of monkeywrenching.[67]

An early EF! gathering had elected to separate monkeywrenching, or "pixieing" as it came to be known as in the UK, and blockading, with the former to be carried out under the name of the Earth Liberation Front and the latter as EF!. The delineation was not sustained for long and major property damage was commonly employed during blockades with fencing torn down and equipment wrecked. During one occupation opposing mine expansion at the Whatley Quarry in 1995 £250,000 of property was destroyed and operations closed for a week due to sabotage.[68]

The rate of innovation and diffusion of tactics was rapid. In addition to the factors already discussed were those such as opponents' countermoves, terrain, and ready access to materials that could be scavenged from road construction sites. While the dangers and specialisation involved in increasingly complex techniques reduced the number of activists able to employ them, mass site occupations involving up to thousands of people maintained broader involvement and remained a regular part of blockades throughout the 1990s.[69]

Tree-sits had been used in 1992 at Twyford Downs and during a 70-day site occupation in Bristol against clearing for a supermarket. Longer term sits were linked with nets and combined with site occupations and sabotage at Jesmond Dene in 1993. UK activists tended to use tree houses rather than platforms, and blockades in 1994 at Solsbury Hill and against the M65 near Manchester introduced high canopy tree villages to the global repertoire. The latter involved 40 treehouses connected by walkways which took authorities a week to evict.[70]

During the 1995 to 1997 Newbury blockade tree houses accommodating up to 14 people and including kitchens and other amenities were constructed. Stand-alone towers were constructed during blockades, such as those against mining at Brenhenllys Opencast Mine, South Wales, and the construction of the M11 Bypass in East London and the M66 near Manchester, with the latter incorporating a concrete barrel lock-on device. ODA against clearing for the construction of a second runway in Manchester during 1996 and 1997 involved multiple tree villages and a "Battlestar" treehouse. Surrounded by high walls the tree was covered in grease, corrugated metal, empty gas bottles, and barbed wire to deter climbers. During its five-day eviction some activists placed their necks in nooses at the top of branches to make removal unsafe. At some blockades sitters threw urine and other liquids at opponents and repelled chainsaws and climbers with staves.[71]

Another major global innovation introduced during this wave was the use of tunnels. Due to the danger of them collapsing on activists these delayed the movement of machinery for construction. They also aided in eviction resistance by stalling the introduction of cherrypicker cranes to remove tree-sitters. Extensive tunnelling, combined with lock-ons, was popularised during ODA targeting construction of the A30 motorway near Devon. During the eviction of the Fairmile camp in 1997 bunker-like fortifications meant that it took police two days to remove a single door. Another camp combining 30 treehouses alongside ground level obstructions at Lyminge forest, Canterbury included tunnels containing computer equipment, CB radios, lighting, and air pumps. By the time of the Manchester runway campaign's month-long eviction, tunnels had been dug up to 40 feet deep with one activist able to remain underground for 17 days.[72] Tunnelling not only extended disruption, but also proved particularly appealing to the media. One activist, known as "Swampy", who had spent long periods in tunnels at Fairmile and Manchester, briefly became a national celebrity and was given his own column in the *Sunday Mirror* newspaper.[73]

Do or Die and other publications regularly included instructions and diagrams regarding different tactics and in 1997 the Road Alert network published their own manual *Road Raging: Top Tips for Wrecking Roadbuilding*. At 214 pages it had a broader campaigning focus than Australian and US guides and included advice on organising groups, recruiting, security, publicity, and research. Much of the book was dedicated to blockading and UK-specific laws were covered alongside instructions regarding building tree villages incorporating ladders, lockable trapdoors for platforms, walkway seats, suspended platforms, poles, and ladders. Guides to barricading roads and buildings, digging tunnels and dams, and invading worksites were included along with information on eviction tactics typically used by opponents.[74]

In relation to normative protester behaviour, the manual briefly discussed non-violence and broadly characterised the tactics it covered as such. However, it did not attempt to define the term or to identify ideal behaviour, noting that " the right to defend oneself against violence is a controversial point" and that "the people who wrote and edited this themselves differ in opinion on these issues". Various vehicles and types of equipment, which could be targeted by either lock-ons or monkeywrenching, were described. Different points of view regarding sabotage were briefly canvassed, but beyond some basic tips concerning security and surveillance, detailed instructions were not provided.[75]

Although early on in the protest wave activists had hoped to halt existing projects the core strategy of the movement soon developed into one of "noisy defeats, quiet victories" with one activist surmising that "anti-road direct action is very unlikely to stop [a] particular road, but creates a climate of opinion where other road schemes are more likely to be defeated before they start."[76] In terms of causing delays and ramping up costs the movement was highly successful with the three-year Newbury campaign alone involving up to 1,600 security guards at a time, accruing more than £29 million pounds in security and policing costs on top of project overruns of more than £30 million. With blockaders directly impeding projects while drawing media attention, and acting as a radical flank, financial and political costs combined to eventually force local and central governments to cancel or put on hold close to 500 out of 600 road building schemes. By 1998 *Construction News* concluded that "the major road-building programme has virtually been destroyed".[77] A number of opencast mines and suburban developments were similarly cancelled. While activists continued to oppose clearing projects, the focus of environmental ODA in the late 1990s and into the 2000s shifted to genetically modified crops and global justice issues[78]

Diffusion within and beyond environmental movements

Diffusion and debate regarding tactics, strategy, and other ideas and practices constantly occurred within radical environmental movements, and beyond them, throughout the periods discussed thus far. Within environmental campaigns ODA was regularly employed at sites related to the exploitation of biodiverse places including offices, timber mills, processing plants, transport routes, warehouses, and retail outlets. Tactics first developed at the point of destruction were often used during protests at such places. For example, a tree-sit was held outside British Columbia's parliament and tripods and barricades employed to blockade entrances to timber mills and company offices in Australia and the United States.[79]

In acting as a radical flank, blockaders shifted some existing non-ODA based environmental organisations in a more ambitious and assertive direction, as was discussed earlier in relation to Canada and Australia. Radical action also brought new organisations aggressively using litigation, education and lobbying onto the scene, such as the large number of Grassroots Biodiversity Groups in the United States.

As has been discussed, in addition to environmentalism blockaders commonly focused on other political issues. Many also identified themselves as part of broad social justice, anarchist, and leftist movements. Approaches regarding normative protester behaviour, strategy, media relations, and other elements of environmental blockades were often influenced by, if not wholly based on, those from peace, feminist, civil rights and other movements, particularly during the initial stages of protest waves. Developments regarding these ideas also flowed from blockades into associated movements.

Inter-movement diffusion regarding tactical forms regularly occurred during the 1980s and 1990s. Within two years of the tactic first being trialled during the SEFA campaign tripods were deployed in an urban setting during a two-week blockade of the 1991 AIDEX military arms fair. US EF! brought ODA tactics into solidarity campaigns with steelworkers and other unionists against a common enemy, the Maxxam Corporation. Tripods and other techniques originating from forest campaigns were also used during the Seattle blockade of the World Economic Forum in 1999.[80] From the first Reclaim The Streets event, held in London during 1995, onwards tripods were employed both as a symbol of resistance and to hamper the eviction of anti-car dance parties from city thoroughfares.[81]

Lock-on tactics have proved to be some of the most modular. As noted in Chapter 3 the use of U-locks moved from anti-nuclear to forest campaigns in the 1980s and then broadened out. Lock-on boxes, sleeves, barrels filled with concrete, and other extensions first developed during environmental blockades lent themselves to office, factory, and other occupations regarding issues ranging from student fees to refugee rights and live animal exports. As previously discussed their use was not limited to progressive movements as anti-abortion groups also employed U-locks during blockades of medical clinics in the United States in the 1980s and developed a series of innovations in parallel to those used by EF! and others.

The continuing deployment and relevance of environmental blockading

From the 2000s onwards the primary development in environmental blockading has been further national and international diffusion and cross-fertilisation with other movements. Despite improvements in eviction methods, including police producing their own anti-ODA guides, blockading remains a strategic option that is used in many countries.[82] In the birthplace of the template, site occupations, lock-ons, and other forms of ODA were undertaken on multiple occasions across four states in Australia against logging, mining, and the clearing of Indigenous sites of significance during 2020. One tree-sit, occupied by a woman who had previously spent more than 18 months in a tree during a blockade, was made up of a bed frame and mattress hung in the forest canopy.[83]

These actions, in part due to the COVID pandemic, were relatively short lived but long-term blockades regarding biodiverse places and associated issues have been mounted globally on a regular basis since the 1990s, as demonstrated by the following two examples. From 2012 onwards a series of encampments in Germany's Hambach forest delayed clearing for the expansion of Germany's largest open-cut coal mine. Combining grievances regarding climate change, forest protection and local input into land use decision-making, three blockades were mounted for six to nine months at a time between 2012 and 2014, with the most recent one remaining in place from 2015 to 2018.[84]

German communities had mounted a series of occupations of nuclear power and airport runway sites in the 1970s and 1980s that included long-term hut villages and barricades. These were combined at Hambach with tactics developed in the decades since. The final eviction in September 2018 began with police raids to remove road barricades and activists locked to the top of tripods and other structures, resulting in hundreds of arrests and some activists spending months in pretrial detention. Eventually tree villages comprised of more than 70 huts formed the final line of defence. Despite the deployment of massive numbers of police, it took close to a month to remove the last of them. Illustrating the continuing dangers involved in ODA, police temporarily suspended operations after activist and documentary maker Steffen Meyn fell 15 metres to his death. "Sunday strolls", regular site invasions, were another tactic pioneered in the 1970s and one of these during the eviction overwhelmed police with 7,000 protesters.[85]

Continuing resistance, including rebuilding treehouses, disrupting mining operations and bringing 50,000 people to a demonstration near the site, led the German government to preserve the remaining parts of the forest in early 2020. With some local villages still to be demolished, and the coal industry to survive until 2038, residents and activists have maintained ODA.[86]

Another major blockade, which illustrated the persistence of the repertoire in contesting issues intertwining Indigenous sovereignty and land protection, was the 2016–2017 Standing Rock No Dakota Access Pipeline protest in North Dakota, USA. During this campaign thousands of protesters opposed to the construction of a pipeline linking the Bakken oilfields to Illinois joined members of local youth organisation RezPect Our Water and other Water Protectors at the Standing Rock Indian Reservation. A series of protest camps rapidly expanded to include members of over 300 First Nations from across the United States, as well as non-Indigenous environmentalists and other supporters. The camps became the focus of general opposition to the abrogation of First Nations sovereignty as well as the environmental and cultural destruction the project would bring, both through clearing, threats to local water supplies and sacred sites, and by accelerating climate change. Occupying sites for over six months, hundreds of those who refused to move out of the way of construction were arrested. Blockaders were fired upon with tear gas and rubber bullets, and endured strip searches alongside other forms of violence from security guards, police, and the National Guard. Information about the blockade was heavily featured in activist, Native American and broader social media networks with footage of attack dogs savaging protesters who had entered a work site on 3 September 2016 going viral. Illustrating the contemporary interconnection of online activism and ODA over 1.5 million people "checked in" to Standing Rock on Facebook as part of a campaign to deter and confuse law enforcement surveillance as well as show support for those on the ground.[87]

Extinction Rebellion and continuing urban connections to the environmental blockading repertoire of contention

Tactics, strategies, and approaches pioneered, developed, tested and popularised via environmental blockades from 1979 onwards also remain in regular use during occupations and protests in urban locations. In recent years the best-known exponent of ODA in this regard has been Extinction Rebellion (XR). In terms of its repertoire of contention connections were extant in the examples of early XR actions which opened this book as these included the use of tripods, tree-sits, and lock-on devices. These tactics have reappeared in numerous actions carried out by over 1,100 groups in 70 countries since 2018.[88]

XR's use of disruption has had tactical and strategic similarities to blockades at the point of destruction in terms of expressing alarm and opposition, acting as a radical flank and seeking to draw public and media attention to environmental issues. As with many activists involved in the defence of biodiverse places XR members have often turned to ODA due to moral shocks associated with the urgency of threats to the environment, political closure regarding solutions to them, and dissatisfaction with mainstream environmental organisations' strategies.[89]

From its foundation the movement's primary focus has been to raise broad consciousness regarding climate change via generalized disruption of the public. Its central strategy, based on highly contested research by Erica Chenoweth regarding civil resistance movements, was initially to mobilise "3.5% of the population to achieve system change". This is based on a belief that "no regime in the 20th century managed to stand against an uprising which had the active participation" of that proportion of the population.[90] Mass mobilisations involving increasing number of arrests were envisaged as building to the point where legal and political systems would be overwhelmed. In such a situation it is held that governments will acquiesce to movement demands and introduce deep reforms, including their replacement with citizens assemblies. Although broad change remains the movements' raison d'être, and its principles explicitly oppose the "blaming and shaming" of individuals, XR activists have increasingly also engaged in the use of ODA at points of decision making, consumption and production to pressure opponents to abandon specific policies, judgements and choices.[91]

The environmental blockades detailed in this book have often included activists with revolutionary agendas and a focus on the broader economic and social roots of environmental destruction, some of whom were active in first warning of the dangers of climate change. However, the nature and focus of their activities was, and generally continues to be, preventing the immediate destruction of specific sites. For some radicals this has been viewed as a stop-gap or partial measure while broader societal changes, ranging from apocalyptic collapse to peaceful or violent revolution, unfold. Localised, individual victories have also been regarded as a key part of publicising broader issues and building confidence and the capacity to initiate or develop wider movements.[92]

Various approaches to normative protester behaviour have been canvassed in earlier sections. Although some blockades at the point of destruction have adopted specific rules and formulations most have tended towards looser, fluid definitions. As part of extended debates during the global justice blockades of peak economic decision-making bodies in urban areas a new approach, often dubbed "diversity of tactics", was adopted by some activists from 1999 onwards. This arrangement designated specific "zones" for different tactical tendencies to operate in, ranging from those favouring basic rallies and soft blockading through to those employing violent offensives to break police lines. According to sociologist Amory Starr, such zoning was intended to maintain "a united front of solidarity among activists and organisations with divergent beliefs about tactics while also isolating (geographically and temporally) within a protest day actions involving different levels of 'risk'".[93]

XR's decentralised network form and use of consensus decision making and working groups has built on contemporary practices developed by radical environmentalists, anarchists, feminists, and others. Choices are bounded however within the parameters of a set of basic principles and values to which all who identify with XR are expected to adhere. Among those outlined in XR's founding policy documents was a commitment to "nonviolence" that included being "polite and respectful" to police officers and security guards "even when provoked". This was adopted on the basis of both ethics and the goal of conversion. Media and opponents would be briefed on "the general escalation plan of the rebellion" but "the specific operational details of actions will only be known to those taking part and not be shared to the extent that it will undermine their ability to carry them out".[94]

In concert with its rapid growth XR attracted critiques concerning its approaches to strategy and tactics as well in relation to informal hierarchies, diversity, and the movement's articulated and implicit positions and practices regarding class, racism and other oppressions. Debates regarding racism, capitalism, and gender, particularly in the United States, and Indigenous viewpoints and leadership, particularly in Canada, had been a feature of radical environmentalism during the 1980s. By the late 2010s such issues were being regularly debated across a range of progressive movements with the result that discussion arose rapidly within XR. Just as the diffusion of XR's repertoire and news of its actions had been facilitated globally via the internet at a pace unachievable in earlier decades, so has contention regarding its ideas and practices.[95]

In response core sections of the movement have moved away from XR's founders' "Beyond Politics" position to emphasise opposition to capitalism, patriarchy, colonialism, and other oppressive beliefs and structures, as well as the need for awareness regarding the ways in which they manifest internally and externally. These changes have been expressed in movement goals, published materials, and daily practices.[96]

There has also been modification and clarification regarding the ways in which XR's key principles are interpreted in relation to normative protester behaviour. For instance, the website for XR's founding UK chapter included a section in 2020 recognizing "that many people and movements in the world face death, displacement and abuse in defending what is theirs". As a result the authors stated "We stand in solidarity with those whom have no such privilege to protect them and therefore must protect themselves through violent means; this does not mean we condone all violence, just that we understand in some cases it may be justified." Similarly while stating that "our network" will not "undertake significant property damage because of risks to other participants by association" the statement of principles did "not condemn other social and environmental movements that choose to damage property in order to protect themselves and nature."[97]

Guiding principles and rules are difficult to enforce unless the majority of a campaign or movement agrees with them. Even then sections of movements may act unilaterally or move in other directions, as did members of XR in France when they sabotaged 3,600 electric scooters in late 2019. Nevertheless, despite its breadth and size most actions carried out by XR's many chapters, up to the time of publication, have largely adhered to its central strategy and principles.[98] As demonstrated by the changes that have already taken place in the movement, and by the course of the other protest waves covered in this book, changing political conditions, responses from authorities, internal debates and other factors are likely to result in further shifts in strategy and practice.

Notes

1 McIntyre, Environmental Blockading Timeline.
2 Ibid.
3 Ibid.
4 Angel, "Letter from Total Environment Centre Executive Director to John Seed ", 1.
5 South East Forest Campaign Strategy Paper, (SEFA, 1989), 1–3; *South-East Forests Campaign Handbook*, (SEFA, 1988), 3.
6 Ibid, 2, 12–20.
7 Ibid; Cohen, *Green Fire*, 172–75.
8 *South-East Forests Campaign Handbook*, 12.
9 Jacob Grech, Interviewed 12 May 2006; Author correspondence with David Burgess, 22 October 2020.
10 Ibid.
11 Cohen, *Green Fire*, 174–80; "SEFA Activity Sheet: 9–14 August 1989," (SEFA, 1989), 1.
12 "Press Release 20 March 1989," (South East Conservation Working Group, 1989), 1; Branagan, *Global Warming, Militarism and Nonviolence*, 86–87.
13 Peter Vaughn, "Police Interview Transcript: 5 January," (1990), 1–5; David Burgess, "Police Interview Transcript: 13 November," (1989), 1–7.
14 Ibid; "Media Release: 18 September," (South East Conservation Working Group, 1989), 1.
15 "Police OK Citizens Arrests," *Bega District News*, 12 September 1989, 1.
16 Daniel Lunney, "The Eden Woodchip Debate: Part II (1987-2004)" (paper presented at the 6th National Conference of the Australian Forest History Society, 2005), 286–320; Cohen, *Green Fire*, 174–80.
17 Nicole Rogers, "Law, Order and Green Extremists," in *Green Paradigms and the Law*, ed. Nicole Rogers (Lismore: Southern Cross University, 1998), 143; "Who Are NEFA?," (NEFA, 1993), 3–7.
18 "Arrests and Legal Report," *South East Forest Report* 7 (1989): 1–2.
19 "NEFA Meeting Minutes: 29 September," (NEFA, 1991), 1–3; Aiden Ricketts, "Om Gaia Dudes: The North East Forest Alliance's Old Growth Forest Campaign," in *Belonging in*

the Rainbow Region: Cultural Perspectives on the N.S.W. North Coast, ed. Helen Wilson (Lismore: Southern Cross University Press, 2003), 121–25.

20 Cohen, *Green Fire*, 183–184; Seed Interview.

21 "Ten Year Fight Achieves Hope for North Washpool," *World Rainforest Report*, no. 19 (1991): 6–7; "Who Are NEFA?," 8–14.

22 Ibid; Author communication with Aidan Ricketts, 16 October 2020.

23 Ibid; Branagan, *Global Warming, Militarism and Nonviolence*, 41–45.

24 "Green Activism," in *Green Paradigms and the Law*, 173–75; Cohen, *Green Fire*, 181–202.

25 Rogers, "Law, Order and Green Extremists," 145–48; "Who Are NEFA?," 12–13.

26 Branagan, *Global Warming, Militarism and Nonviolence*, 46–73.

27 Ibid; Personal correspondence with Dailan Pugh, 16 November 2020; "Green Activism," 175–76; "NEFA Report: May", (NEFA, 1992), 1–8.

28 NEFA, *Intercontinental Deluxe Guide to Blockading*; Vanessa Bible, "Aquarius Rising", 61.

29 Kelpie Willsin and George Shook, "Forest Action Workshop," *EFJ* 11, no. 6 (1991): 25; Bari, *Timber Wars*, 200–1; Coulter Interview.

30 Seed Interview; Langelle Interview.

31 McIntyre, Environmental Blockading Timeline.

32 DAM Collective, *Earth First! Direct Action Manual*; Roselle, *Tree Spiker*, 156–58; Randall Restless, "First American Tripod is Constructed in Cove/Mallard," *EFJ* 21, no. 1 (2000): 56–57.

33 Durbin, *Tree Huggers*, 233–63; Patti Goldman and Kristen Boyles, "Forsaking the Rule of Law: The 1995 Logging without Laws Rider and Its Legacy," *Environmental Law* 27, no. 4 (1997): 1035–96.

34 Ibid, 265–289; Anthony Silvaggio, "The Forest Defense Movement", 151–153,183; Roselle, *Tree Spiker*, 212–19.

35 Ibid; McIntyre, Environmental Blockading Timeline.

36 Ibid; DAM Collective, *Earth First! Direct Action Manual*, 56, 68–69.

37 Sprig, "Our Kind Live in Forests to Save Forests from Our Kind," *EFJ* November– December (2000): 28–29.

38 Tahoma (pseudonym), "Cascadia Rising!," *EFJ* November-December: 14–15, 73; DAM Collective, *Earth First! Direct Action Manual*, 119–20; Silvaggio, "The Forest Defense Movement", 176–195.

39 Ibid, 154–162; Silvaggio, "The Forest Defense Movement", 139–140; McIntyre, Environmental Blockading Timeline.

40 Darren Speece, "Defending Giants: The Battle over Headwaters Forest and the Transformation of American Environmental Politics, 1850 to 1999" (PhD thesis, University of Maryland, 2010), 311.

41 Ibid., 279–280. Silvaggio, "The Forest Defense Movement", 154–62.

42 Speece, "Defending Giants", 261–304; Durbin, *Tree Huggers*, 280–289. Silvaggio, "The Forest Defense Movement", 163–167.

43 DAM Collective, *Earth First! Direct Action Manual*.

44 Ibid; DAM Collective, *Earth First! Direct Action Manual* (*EFJ*, 2015).

45 York and Pindera, *People of the Pines: The Warriors and the Legacy of Oka*, 271–81.

46 Ibid.

47 Zig Zag, "BC Native Blockades and Direct Action,", 8–9.

48 Ibid; Nicholas Blomley, "Shut the Province Down," 9–35.

49 Don Clairmont and Jim Potts, *For the Nonce: Policing and Aboriginal Occupations and Protests* (Ontario: Ipperwash Inquiry, 2006), 19–23; Rima Wilkes and Tamara Ibrahim, "Direct Action over Forests and Beyond," in *Aboriginal Peoples and Forest Lands in Canada*, ed. D.B. Tindall, Ronald Trosper, and Pamela Perreault (Vancouver: BCM Press, 2013), 74–83.

50 Tracey Lindeman, "'Revolution Is Alive': Canada Protests Spawn Climate and Indigenous Rights Movement," *Guardian*, 28 February 2020, Available [Online]: www.theguardian.com/world/2020/feb/28/canada-pipeline-protests-climate-indigenous-rights (Accessed 1 November 2020).

51 Bondi Interview; Winstanley Interview; William Lawrence, Nuxalk People Obstruct Logging of ISTA Old-Growth Forest, 1995–1998, Available [Online]: https://nvdatabase.swarthmore.edu/content/nuxalk-people-obstruct-logging-itsa-old-growth-forest-1995-1998 (Accessed 1 November 2020).

52 Winstanley Interview; Aspinall Interview; Seed Interview; Coulter Interview, Hinke Interview.

53 Winstanley Interview, Blomley, "Shut the Province Down", 32–35; Lawrence, Nuxalk People Obstruct Logging of ISTA Old-Growth Forest; Jacinda Mack, *Remembering ISTA: Nuxalk Perspectives on Sovereignty and Social Change* (York University, 2006), 5–6, 27–34.

54 Horsfield and Kennedy, *Tofino and Clayoquot Sound: A History*, 511–12; Bondi Interview.

55 Ibid; Horsfield and Kennedy, *Tofino and Clayoquot Sound: A History*, 513–29.

56 Ibid; Michaela Mann, *Clearcut Conflict: Clayoquot Sound Campaign and the Moral Imagination* (Ottawa: Saint Paul University, 2013), 36–42, 55–65.

57 Kim Goldberg, "Clayoquot Crackdown," in *Witness to Wilderness: The Clayoquot Sound Anthology*, ed. et al Howard Breen-Needham (Vancouver: Arsenal Press, 1994), 197–201; Hoberg and Morawski, "Policy Change through Sector Intersection", 388–400.

58 Giorel Curran, *21st Century Dissent: Anarchism, Anti-Globalization and Environmentalism* (Hampshire: Palgrave MacMillan, 2007), 214–17; Derek Wall, *Earth First! and the Anti-Roads Movement: Radical Environmentalism and Comparative Social Movements* (London, New York: Routledge, 1999), 149–50.

59 Department of Transport, "Roads for Prosperity," (London: Department of Transport, 1989), 1–23; Benjamin Seel, Alexandra Plows, and Brian Doherty, "Direct Action in British Environmentalism," in *Direct Action in British Environmentalism*, 7–8; Wall, *Earth First! and the Anti-Roads Movement*, 45–64.

60 "Down with the Empire! Up with the Spring!," *Do or Die*, no. 10 (2003): 6–10; Wall, *Earth First! and the Anti-Roads Movement*, 65–74.

61 Ibid; Road Alert!, *Road Raging*.

62 Ibid.; Doherty, "Manufactured Vulnerability", 68–69; Wall, *Earth First! and the Anti-Roads Movement*, 73.

63 Ibid., 125–131.

64 Ibid., 137–139.

65 Ibid., 65–86; "Down with the Empire!", 6–25.

66 Ibid; Wall, *Earth First! and the Anti-Roads Movement*, 88–91,150–155.

67 Wall, *Earth First! and the Anti-Roads Movement*, 69–71, 81–82, 130–31, 155–58; Alexandra Plows, "Praxis and Practice: The 'What, How and Why' of the U.K. Environmental Direct Action (E.D.A.) Movement in the 1990s" (PhD thesis, University of Wales, 2002), 276–88.

68 "Down with the Empire!", 8–35.

69 Ibid.

70 Ibid.; Wall, *Earth First! and the Anti-Roads Movement*, 65–87; McIntyre, Environmental Blockading Timeline.

71 Ibid; "Life on the Battlestar," *Do Or Die*, no. 6 (1997): 82–86; Doherty, "Manufactured Vulnerability," 65–66.

72 Ibid, 69-70; "Farewell Fairmile," *Do Or Die*, no. 6 (1997): 48–52; McIntyre, Environmental Blockading Timeline.

73 Matthew Paterson, "Swampy Fever: Media Constructions and Direct Action Politics," in *Direct Action in British Environmentalism*, 151–56.
74 Road Alert!, *Road Raging*.
75 Ibid.
76 "Direct Action: Six Years Down the Road," *Do Or Die* 7 (1998): 22.
77 Quoted in "Down with the Empire!," 23; Seel and Plows, "Coming Live and Direct", 115–28; "Direct Action: Six Years Down the Road," 1–4.
78 "Down with the Empire!," 25–34.
79 Winstanley Interview, Seed Interview, Coulter Interview.
80 Iain McIntyre, *Always Look on the Bright Side of Life: The AIDEX '91 Story* (Melbourne: Breakdown Press, 2008), 30; Silvaggio, "The Forest Defense Movement", 223–25.
81 Wall, *Earth First! and the Anti-Roads Movement*, 87.
82 National Policing Improvement Agency, *ACPO Manual of Guidance on Dealing with the Removal of Protestors: Faslane 365 2006–2007* (Bedfordshire: NPIA, 2007).
83 Southeast Forest Rescue, Available [Online]: https://zetblogs.com.au/394016-sefr-southeast-forest-rescue/ (Accessed 8 November 2020); Logging Operations Shut Down Across Victoria, Available [Online]: www.ecoshout.org.au/blog/logging-shut-down-across-victoria-seven-protests-one-day (Accessed 8 November 2020); Updates, Available [Online]: https://dwembassy.com/334-2/ (Accessed 8 November 2020); Takayna 2020, Available [Online]: www.bobbrown.org.au/takayna_2020 (Accessed 8 November 2020); Protesters Blockade Adani Minesite, Available [Online]: www.thebigsmoke.com.au/2020/08/24/protestors-blockade-adani-mine-site-as-traditional-owners-issue-eviction-notice (Accessed 8 November 2020).
84 News Ticker, Available [Online]: https://hambachforest.org/blog/2018/09/25/ticker-from-september-25 (Accessed 31 July 2020).
85 Ibid; McIntyre, Environmental Blockading Timeline.
86 Karin Jager and Gero Rueter, "Hambach Forest: Germany's Sluggish Coal Phaseout Sparks Anger," *Deutsche Welle*, 19 January 2020, Available [Online]: www.dw.com/en/hambach-forest-germanys-sluggish-coal-phaseout-sparks-anger/a-52059845 (Accessed 31 July 2020).
87 Johnson, Hayley. "# NoDAPL: Social Media, Empowerment, and Civic Participation at Standing Rock." *Library Trends* 66, no. 2 (2017): 155–175.
88 Why Rebel?, Available [Online]: https://rebellion.global/why-rebel/. (Accessed 31 July 2020); Knights, "The Story So Far", 16–21; Taylor, "Extinction Rebellion Activists Claim Victory in HS2 Tree Protest; Events," Available [Online]: https://extinctionrebellion.uk/act-now/events/ (Accessed 2 November 2020).
89 Extinction Rebellion Occupy Greenpeace Offices; Extinction Rebellion UK, About Us, Available [Online]: https://extinctionrebellion.uk/the-truth/about-us/ (Accessed 1 November 2020).
90 Ibid.
91 Ibid; Andrew Charles, "Extinction Rebellion: A Short Critical Guide," *Overland*, October 2019, Available [Online]: https://overland.org.au/2019/10/extinction-rebellion-a-short-critical-guide/ (Accessed 2 November 2020); Matthew Taylor, "The Evolution of Extinction Rebellion," *Guardian*, 4 August 2020; Available [Online]: www.theguardian.com/environment/2020/aug/04/evolution-of-extinction-rebellion-climate-emergency-protest-coronavirus-pandemic (Accessed 1 November 2020); Oscar Berglund and Daniel Schmidt, *Extinction Rebellion and Climate Change Activism* (Cham, Switzerland: Palgrave Macmillan, 2020), 87–92.
92 "Down with the Empire!," 28–101; Christopher Manes, *Green Rage*, 225–34. Winstanley Interview; Coulter Interview; Seed Interview.

93 Amory Starr, "'… (Excepting Barricades Erected to Prevent Us from Peacefully Assembling)': So-Called 'Violence' in the Global North Alterglobalization Movement," *Social Movement Studies* 5, no. 1 (2006): 66.

94 "Whole Rebellion Action/Policy Document," (London: Extinction Rebellion, 2018), 2–32; Berglund and Schmidt, *Extinction Rebellion and Climate Change Activism*, 27–34, 87–92.

95 Ibid, 35–36, 48–50, 99–100; Charles, "Extinction Rebellion"; Taylor, "The Evolution of Extinction Rebellion"; Imogen Watson, "After a Soul Searching Lockdown Extinction Rebellion Plots Its Next Move," *The Drum*, 20 July 2020, Available [Online]: www.thedrum.com/news/2020/07/20/after-soul-searching-lockdown-extinction-rebellion-plots-its-next-move (Accessed 1 November 2020).

96 Ibid; *Power Together* (London: Extinction Rebellion UK, 2020), 3–6; Taylor, "The Evolution of Extinction Rebellion."

97 "About Us."

98 "Activists Sabotage 'Ecologically Catastrophic' E-Scooters in France," *France 24*, 12 May 2019, Available [Online]: www.france24.com/en/20191205-activists-sabotage-ecologically-catastrophic-e-scooters-in-france (Accessed 1 November 2020); Taylor, "The Evolution of Extinction Rebellion."

CONCLUSION

During the 1980s the protection of biodiverse places became a major global issue, one whose importance would grow in the decades to come. In part this resulted from efforts by Indigenous people in a variety of countries to protect and reclaim territories that had come under the ownership and exploitation of others via colonial dispossession. Challenges to dominant practices also came from non-Indigenous conservationists as well as alternative "back-to-the-land" communities and others who had settled in rural areas and formed connections to land. Contention regarding logging, mining, and other activities reflected and fed into a widening ecological consciousness, as broader communities turned their attention to the plight of forests, rivers, and other places within their own countries as well as overseas.

A series of environmental blockades that were launched in Australia, the United States, and Canada from 1979 onwards played a significant role in capturing and shifting these publics' awareness. In providing a national and comparative history of these blockades this study has detailed and analysed how the environmental blockading repertoire was initially developed and embedded in each country. It has established that the sustained, close, and intense level of protest and associated work and living, and the engagement with the biodiverse environments involved, made environmental blockades key points of tactical innovation for obstructive direct action more generally. In doing so it has contributed to a fuller understanding of the nature and development of environmental movements in these nations as well as their transnational dimensions.

Each national case study analysed how and why the environmental blockading template first emerged and was subsequently embedded through its use by different communities, milieux, and networks. Despite some differences in political institutions, economics and patterns of resource extraction, common themes driving grievances existed across the three. Following colonisation each country had experienced long periods of intensifying and unsustainable exploitation of land and resources. By the 1970s and 1980s this was having noticeably deleterious effects as well as threatening areas previously considered protected by their remote location or importance to local communities. Although each country, and regions within them, had administrative regimes that allowed for varying degrees of public input, dissenters faced political closure. When combined with the deep emotional, spiritual, and cultural connections to biodiverse places that people experienced on a personal and community level, the inability to effectively contribute to or challenge decisions regarding land use was a decisive factor in almost all campaigners' turn to ODA.

In Australia EB first appeared at Terania Creek in 1979 as a locally generated and largely spontaneous response to impending logging. A number of tactics and organisational and ethical approaches, loosely influenced by practices from other social movements and local "New Settler" communities, were improvised and then carried forward into later campaigns during the 1980s. In the United States the adoption of EB from 1982 onwards formed part of the emerging Earth First! network's explicit rejection of "insider" political strategies and conscious search for fresh responses to the nation's rightward turn and attendant escalating attacks on environmental gains. It was partially inspired by Australian examples and those from other domestic movements, with EF! soon developing its own tactical and organisational repertoires.

The emergence of EB in Canada followed a different trajectory. While all three countries are settler nations and featured varying levels of contestation by Indigenous communities over conservation, this was particularly extant in Canada due to demographic and historical factors. Here a mixture of issues regarding sovereignty and the destruction of habitat led to the application of existing First Nation blockading traditions to the protection of Clayoquot Sound in 1984. In the campaigns that followed over the next five years the majority were led by or involved alliances with Aboriginal communities. Relationships between Indigenous and non-Indigenous activists varied and were often conditioned by the degree to which the former's rights and leadership were recognised and incorporated. EB in Canada was also influenced by the scale of cutting's frequent impact on oceans and rivers in British Columbia, which posed major threats to fishing, hunting, and emergent tourist industries. These drove economic grievances to a larger degree than elsewhere.

Interviews conducted and other primary sources consulted for this thesis indicate that all these early campaigns featured people with strong emotional and spiritual connections to the biodiverse places involved. These not only motivated them to directly hinder destruction and overcome counter-tactics and costs, but also led to radicalisation through exposure to the way in which ecosystems was treated. During the blockades that followed, connections and moral shocks intensified and affected new participants, creating long-term campaigners and new activist milieux. All of the early blockades were launched in response to decisions by opponents, and those that followed would often react as threats arose. Over time some activists began to use ODA pre-emptively and became more strategic in terms of when, where, and how it was deployed.

Repertoires of contention

A key contribution of this study has been to develop understandings of repertoires of contention. It has introduced new terms such as "obstructive direct action" and "normative protester behaviour" in order to better define and specify forms of contention and approaches to them. The use of original terms has also been used to avoid the often loaded and polysemous terminology that surrounds concepts such as "nonviolence" and "direct action".

The coverage of numerous campaigns has explored why certain tactics, strategies, forms of organisation, and approaches were chosen and adapted from existing repertoires, why some endured, and what shaped the rate and direction of innovation. Broad cultural and contextual factors as well as incidents, dynamics, and geographies particular to specific events have been identified as key drivers of development, with EBs and their associated protest camps acting as "submerged networks" and "laboratories of insurrectionary imagination".

Histories regarding tactical emergence within and between each country have been provided. Some common patterns have been identified in relation to four main tactical categories: soft blockades, barricades, enhanced vulnerability, and sabotage. In all three countries the first set of tactics to be employed belonged to the soft blockade category and involved people standing, lying, or sitting on the ground in order to occupy work sites or prevent the movement of opponents. These means were often combined with forms of barricading, which included placing rocks, logs and debris, and digging trenches.

While some cohorts only ever employed these tactics, many moved in the direction of increasing activist risk to stop or slow opponents' movement and work. In some cases, they extended soft blockading tactics to include hiding and running through work sites. The use of enhanced vulnerability tactics further allowed activists to minimise arrests and maximise obstruction via means such as tree-sitting, burying people, putting in place constructions such as tripods, and locking on to trees, machinery, concrete-filled boxes and barrels, and entry gates.

Beyond damage to roads and fences incurred through barricading and entering sites, major sabotage was rarely used in concert with other tactics during environmental blockades prior to the 1990s. When it was, organisers usually publicly denounced its use. Because of its popularity in other contexts among members of the EF! network, this was less the case in the United States but the two repertoires were still generally separated. A specific sabotage technique, that of tree-spiking, became increasingly controversial within EF! during the 1980s with some activists denouncing its use in any circumstances on the basis of ethical and political concerns. Underlining the role of differing local and national factors major sabotage was regularly used as a part of blockading in the UK during the 1990s.

Generally, forms of barricading, soft blockading, and enhanced vulnerability were firstly aimed at preventing workers and equipment from accessing a site and then subsequently at interfering with ongoing work. Differentiation and the rate and direction of innovation in each country, and regarding each category of tactics, stemmed from a wide range of factors. These included evolving and emerging collective identities, emotional responses, and previous experiences and traditions of activism. Such influences were combined with cognitive aspects regarding how campaigners best thought they could achieve their objectives, as well as what those objectives should be. Over time a trend was seen in many campaigns towards using protest camps themselves as a form of obstruction and to combine tactics together to create a multiplier effect.

Chapters 5 and 6 established that Canadian patterns were far more localised and varied, and examined how the involvement of Indigenous communities, as well as initially lower costs of action, helped facilitate larger and more sustained blockades than in the United States. In 1984 existing First Nation strategies involving occupying land and setting up roadblocks to protest government policies and assert sovereignty were extended to include the defence of forests and other biodiverse areas. Over the next five years two core repertoires emerged. One, primarily used by Aboriginal-led campaigns, involved soft blockading and/or barricades to occupy land and roads. The other, developed by non-Indigenous environmentalists, followed similar patterns to Australia and the United States in adding techniques of enhanced vulnerability, such as tree-sits and lock-ons. As in the other countries the activists who did the most to develop such tactics were those focused on maximising disruption and minimising arrests. They also favoured approaches to normative protester behaviour, which allowed for spontaneity, secrecy, and minor property damage. From 1990 onwards the two strands began to merge to a degree although the stand-alone roadblock remained the preserve of First Nations protests.

As discussed in the Introduction, the development and deployment of differing tactics and repertoires were related to complex ecologies of protest action. The influence of national and regional political and economic contexts, structures, and institutions in generating grievances, defining goals, opening opportunities, and creating costs at a variety of levels has been discussed. Organisational and decision-making forms, protest personnel, and the influences of often meagre financial and human assets available to campaigns also had effects on types of innovation and experimentation.

The way in which activists framed and perceived grievances and goals was another key theme. Rates, forms, and definitions of success were seen to have further impacts upon tactical choices and in encouraging or discouraging innovation. Although universally committed to taking a stand few campaigners believed they could bring a permanent halt to destructive activities through obstruction alone. Instead, confirming experiences and observations regarding other social movements, obstruction was more typically viewed as a means of attaining other outcomes. These included imposing coercive financial and political costs on authorities and businesses, generating publicity, and delaying work while litigation, lobbying, and other strategies unfolded. The histories provided in this book confirm the argument made in its Introduction that the degree to which campaigners prioritise and balance these ends with others, the importance they assign to blockading within their overall strategy, and their views concerning the nature and role of the mainstream media, all contribute to tactical direction.

Journalistic coverage was very important to activists in all the campaigns covered. Although many distrusted reporters and developed their own media, particularly in the United States, very few radical environmentalists at this point cleaved to what Maddison and Scalmer describe as the "separate camp" position of avoidance. Most fitted into a spectrum running from the pair's "modernising accommodation" category, in explicitly tailoring actions and modifying behaviour in search of sympathetic coverage, to "expressive militancy", in terms of seeking coverage but not changing tactics, messages or appearance in light of it.[1] Within these, activists employed a variety of means of ensuring their frames cut through. These reflected past practices and included tactics such as hanging banners with slogans in the heart of ODA or holding actions which juxtaposed clear-cuts with old-growth forest.

Common tensions also emerged regarding the degree to which effective and sustained obstruction should be prioritised. Campaigners dedicated to the pursuit of positive media coverage were often divided as to what kind of actions best drew, held, and shaped public attention. In seeking to control events and limiting means to soft blockading tactics they believed would foster a benign image, some organisers curtailed obstruction. By attempting to minimise conflict, in order to keep the focus on key messages rather than debates regarding protester or opponents' behaviour, they ran the risk of their actions being deemed inauthentic and ineffectual "publicity stunts" by journalists as well as the activists they relied upon to carry out actions. For other activists the novelty and controversy associated with barricading and enhanced vulnerability tactics was seen as a surer means of ensuring attention.

Other key strategic and tactical influences identified included the degree to which campaigns and groups within them were emotionally and strategically invested in prevailing during a particular incident or blockade, or saw such events as primarily contributing to a longer term strategy of success. The physical nature and geographical location of the places and work involved had a range of impacts in terms of what kind of soft blockades, barricades, and enhanced vulnerability tactics could be used. While planning was important, emotional and spontaneous responses to situations as they unfolded were also sources of innovation and success.

Another key influence upon the shape and direction of blockading was differing approaches to normative protester behaviour. Ideas regarding what constituted "violence", "sabotage", "secrecy", and "openness", and whether they were efficacious and ethical, were flexible and loosely defined at the beginning of each protest wave, and remained so during many of the blockades that followed. Even when campaigns and cohorts imported or adapted a defined model, and organisers attempted to strictly apply it, processes of evolution, contention, and adaption unfolded.

By 1984 Australian activists had formed key lines of approach to normative protester behaviour, appropriate tactics, and means of change. One adapted the strict application of Orthodox Nonviolence principles developed by the US Movement for New Society and other Gandhian-influenced activists. The other involved more loosely and minimally defined guidelines. In allowing for secrecy and minor property damage these encouraged the use and development of tactics involving barricades and enhanced vulnerability. Both approaches would reappear in later Australian conservation protest waves, often sparking debates similar to those that had occurred during the first half of the 1980s.

In the United States, following some initial debate, divisions were less about guidelines for public protests and ODA than about the role and relationship of sabotage and other tactical forms to each other, as well as within wider movement strategies. As with other characteristics, Canada showed the greatest variation. With minimal crossover of personnel, particularly in comparison to the United States and Australia, approaches were primarily generated and adopted on a campaign by campaign basis.

The means of decision making and organisation were often linked to ideas of normative protester behaviour. Although many campaigns featured informal hierarchies, and a minority formal ones, the use of egalitarian forms of consensus decision making emerged as a common element across all three countries. Most methods were relatively informal and unstructured, evolving as campaigns and protests waves unfolded. Campaigns led by Canadian First Nations generally employed long-standing forms already in common use. Those with strict rules regarding nonviolent behaviour were more likely to import and adapt formalised and highly structured methods as well as place boundaries on the range of possible outcomes. In campaigns that featured training workshops group decisions were further influenced by the content and style of formal training processes and whether participation in them was considered mandatory.

The level of input and flexibility each process allowed, the speed at which decisions were made, the form and amount of communication between different parties, and whether key decisions were made at the point of blockading or elsewhere, had an influence on the evolution and use of repertoires. Events and tactics during campaigns were further affected by the degree to which participants felt obligated to follow group decisions, whether spontaneity was encouraged or discouraged, and by the perceived or actual need to provide immediate responses to unfolding events. Overall, activists focused on creating obstruction and minimising arrests while employing a loose code of normative protester behaviour were the primary drivers of technical innovation and the main users of enhanced vulnerability techniques across all three countries.

The role of tactical interaction with authorities and the evolution of patterns of policing was another key theme. Although the diffusion of counter-tactics between different regions and the pace of innovation were generally much lower than among protesters, a similar range of variables regarding authorities' repertoires influenced individual choices and wider patterns. The repertoires available to all parties involved were further influenced by existing laws and traditions as well as by changing political contexts. Responses to ODA by authorities, the criminal justice system and workers affected blockading repertoires and values regarding normative protester behaviour, and in turn were affected by them.

In Australia and the United States, authorities tended to use criminal law provisions to arrest and remove protesters. Bureaucratic regulations placing restrictions on anyone other than workers entering certain areas were also regularly employed. US companies increasingly deployed private security guards and technology and worked with authorities, including paramilitary officers, to prevent activists from accessing such areas. The use of climbers to remove tree-sitters was also experimented with. Such an approach would be extended to all three countries and the UK in the 1990s as specialist teams, often already tasked with rescue work, were used to remove activists from devices and positions that enhanced vulnerability.

Although simple soft blockading and barricading tactics were more commonly used in Canada between 1984 and 1989, authorities typically took much longer to remove protesters. This was related to the recourse to more formalised civil law processes involving injunctions and contempt of court proceedings. Policing and sentencing patterns also tended to be more affected by local affiliations and calculations regarding political ramifications, particularly where First Nations were concerned.

In all three countries bail conditions were often imposed that prevented protesters from returning to the places where they had been arrested. Campaigns dealt with this in differing ways, some by minimising the numbers subject to arrest, others by making the conditions and other elements of the criminal justice system a civil rights issue. On a day-to-day level police responses to specific tactics, such as the use of U-locks, were highly varied, especially when a new tactic had been introduced and all parties were experimenting.

In the United States and Canada companies and authorities employed the use of lawsuits and restitution orders requiring individuals and groups to pay compensation for lost profits. This counter-tactic at times effectively tied up activist time and resources, but rarely had a nullifying effect on ODA.

Counter-movements were launched in all three countries. Their differing forms, repertoires, growth, and impact have been canvassed and evaluated in light of blockaders' strategies and practices. The source and level of local community and worker support or hostility varied in relation to culture and lifestyles as well as the changing political economy of various industries and regions. Personal relationships and day-to-day conflicts between workers and activists were affected by the broader context and also fed into it.

Diffusion

Among environmental blockaders processes of information sharing and translation were active and contextual. Despite some means and ideas emerging sequentially, diffusion was far from automatic. Tactics and organisational forms were often created in parallel, particularly when the tactic involved was relatively simple and open to being innovated under similar geographic and campaign conditions.

All countries experienced indirect national and transnational diffusion via mainstream media coverage of blockades. In terms of national and regional activist circles, the United States had a dominant vector of direct diffusion via the EF! network and its associated members, songs, gatherings, workshops, and publications. In Australia brokerage was primarily carried out by NAG's roving blockaders, but ONV principles were also disseminated by nonviolence activists to TWS. Following adaption during the Franklin dispute, TWS campaigners continued to train participants during Tasmanian campaigns and attempted to convince campaigners to adopt ONV style principles at the Daintree and elsewhere. All three groupings diffused tactics, strategies, and approaches to other networks and campaigns, including issues beyond the protection of biodiverse places. At times they continued to clash over similar concerns to those that had caused division during the Franklin blockade.

Local activists, often in concert with brokers, recontextualised existing concepts and tactics, translating and transforming them to suit their conditions, incorporate critiques, and respond to experience. This was particularly the case in Canada, where campaigns had little crossover in terms of personnel, and information about blockading in other parts of the country served as inspiration rather than guidance.

The Canadian situation was largely replicated in regard to cross-national ties and influences during the 1980s. These played less of an influence than they would during the 1990s when activists would share detailed information via ODA handbooks as well as connect more regularly via travel and the internet. Transnational links during the 1970s and 1980s were weak in comparison to regional and national ones. While some activists in each country were aware of what was happening elsewhere, they generally lacked detailed information about already existing techniques and strategies or rejected them in favour of locally developed knowledge. Nevertheless, Australian campaigns were cited as a key impetus for American blockades and played an important role in inspiring US activists to expand their activities. In the case of Canada's Temagami and Strathcona Park campaigns environmental activists directly based a portion of their tactical and/or organisational forms on examples from Australia and the United States.

The outcomes of environmental blockading

Outside of the UK in the 1990s there were few cases where ODA directly saved areas by completely shutting work down and imposing, or threatening, sufficient immediate financial and policing costs to cause companies and governments to withdraw. Typically in Australia, and to a lesser degree Canada, the mobilisation of wider publics via media coverage of environmental destruction, and resistance to it, fed into pressure that forced governments to impose moratoriums, legislate regulatory reform, and rezone endangered areas as national parks. In Australia blockading was at times explicitly combined with leveraging political support via electoral campaigning while in the United States it was unofficially tied in with ballot measures during Redwood Summer.

Although the strategy was used in all three countries, during the 1980s it was primarily in the United States that combinations of litigation and blockading were employed to directly compel bureaucracies to apply, enforce, and extend existing environmental protection laws and policies. This strategy also became central to NEFA's campaigns in NSW during the 1990s. In Canada court cases arising out of civil injunctions brought by or against blockaders helped build up a body of findings that recognised First Nations' rights and by extension their ability to enact conservation.

In each country the publicity associated with blockading helped popularise scientific knowledge and terminology regarding ecology, rainforests, old-growth forest, and biodiversity. In the process, existing policy regimes were de-legitimised and understandings of harm and crime reshaped. Within environmental and other social movements protest waves promoted discussion and debate regarding concepts such as "nonviolence", "deep ecology", and "monkeywrenching".

On a personal level blockading expressed individual and group outrage and commitment to biodiverse places. Many blockaders were only concerned with protecting a single place, but for others ODA fitted into wider strategies regarding environmental and social change. Blockading also played a key role in promoting, creating, and expressing collective and personal identities.

ODA often had unpredictable outcomes. Sometimes militant action built support for radical solutions and organisations, broke through public and media indifference, and had a radical flank effect, which boosted the position of moderates. At other points it tarnished reputations, diminished influence, and triggered or galvanised harmful counter-movements. Internally the emotional, political, and cultural experience of living and blockading within biodiverse places could lead to burnout, grief, and paralysing internal dissent or bond groups together and inspire long-term activism.

A paramount finding of this study is that these periods of contention ultimately established the environmental blockading action template as an enduring strategic option for conservationists and created a repertoire of contention drawn upon by many movements. Existing scholarship concerning campaigns during the 1990s and since has gone some way to answering the kind of questions this study has raised in regard to tactical forms and the patterns and processes of innovation, adaption, and diffusion. However, patterns of learning amongst activists, authorities, counter-movements, and business interests deserve further research, particularly in regards to the ways these did, or didn't, build on experiences from the 1980s. Connected to this is a need to recognise and research the transnational and cross-movement importance of campaign handbooks from the 1980s onwards, including guides to ODA published by environmental blockaders during the 1990s. The impact of increasing familiarity with repertoires in relation to media coverage and other campaign outcomes is another area deserving of further longitudinal research.

In relation to the initial protest waves covered by this book, a great deal remains to be explored. This study has included a close analysis of a select number of campaigns, but the many others touched upon merit more detailed analysis. The history and role of music, banner making, and other cultural expressions during the period has been cited, but similarly deserves the level of attention they have received in regard to other periods and movements. The increasing use of photography, video, and film technology during the 1980s, its role in diffusion, and the degree to which campaigners were able to influence publicity by providing media outlets with footage are other potentially illuminating areas. Finally, while this study has concerned itself with the three countries in which the environmental blockading prospered, comparative investigation considering why it failed to do so in places where ODA was only briefly deployed would provide further understandings regarding both the period and the nature of repertoires of contention more generally.

Note

1 Maddison and Scalmer, *Dissent Events*, 215–21.

BIBLIOGRAPHY

"3 Nest in Humboldt Redwoods to Protest Building of Road." *San Francisco Chronicle*, 21 May 1988, A2.

"5 Who Chained Themselves to Logging Trucks Are Arrested." *San Francisco Chronicle*, 14 February 1990, A20.

"6 Stage Treetop Sit-in, Protest Forest Harvest." *Houston Chronicle*, 3 July 1985.

"7 Protesters Face Mischief Charges." *Ottawa Citizen*, 10 October 1989, A2.

"9 Haida Given Suspended 5-Month Terms, MP Fined $750." *The Gazette*, 7 December 1985, A2.

"10 More Haida Indians Arrested While Blocking Logging on Queen Charlotte Islands." *The Citizen*, 19 November 1985, A13.

"10 Protesters Praised, Convicted." *The Bulletin*, 28 October 1984, A9.

"12 Arrests Made During Two Logging Protests." *Ukiah Daily Journal*, 28 October 1988, 3.

"12 More Haida Indians Arrested." *The Citizen*, 21 November 1985, E14.

"17 Protesters Arrested at Logging Roads." *San Francisco Chronicle*, 25 July 1990, A4.

"18 Arrested in Three Actions in North Kalmiopsis." *Earth First! Journal* 7, no. 6 (June 1987): 6.

"22 Things to Do as an Earth Firster!" *Earth First!* 2, no. 6 (June 1982): 4.

"27 Lubicon Supporters Freed to Clear Way for Land Talks." *Vancouver Sun*, 21 October 1988, A14.

"28 Haida Are Arrested in B.C. Logging Protest." *Globe and Mail*, 26 November 1985, A5.

"44 Arrested at L-P Mill." *Earth First! Journal* 10, no. 7 (August 1990): 1.

"700 Arrested in Australia." *Earth First! Journal* 3, no. 3 (March 1983): 1, 6.

"Aboriginals Stop Anti-Road Group." *Courier Mail*, 18 August 1984, 13.

"Activists Sabotage 'Ecologically Catastrophic' E-Scooters in France," France 24, 12 May 2019, Available [Online]: www.france24.com/en/20191205-activists-sabotage-ecologically-catastrophic-e-scooters-in-france (Accessed 1 November 2020).

"Anti-Loggers Call for Ban." *Northern Star*, 10 September 1982, 1.

"Arrests and Legal Report." *South East Forest Report* 7 (1989): 1–2.

"B.C. Agrees to Public Review of Mining Exploration." *Gazette*, 16 February 1988.

"B.C. Haida Block Road." *Globe and Mail*, 31 October 1985, A5.

"B.C. Indians Barred from Halting Trains." *Globe and Mail*, 9 December 1985, A3.

"Bald Mountain Road Stopped!!" *Earth First! Journal* August (1983): 1.

"Ban on Plywood." *Canberra Times*, 17 October 1989, 4.

"Bar against Logging Set to Expire Today." *St. Louis Post-Dispatch*, 19 December 1990, 7.

"Blockade." *Groundswell: Newsletter for a New Society* 1, no. 2 (January 1983): 5.

"Blockade Planned for Road Builders." *Vancouver Sun*, 13 June 1988, B8.

"Blockade Updates." *Earth First! Journal* March (1983): 1, 5.

"Burial Discovered on Deer Island." *Globe and Mail*, 8 December 1986, A5.

"Buses Stuck on Daintree Road." *Courier Mail*, 8 October 1984, 1.

"Cabinet Probe on Protesters." *Courier Mail*, 14 August 1984, 10.

"Call for Logging Ban on All Rainforests." *Sydney Morning Herald*, 18 September 1979, 3.

"Conservationists Arrested at Errinundra." *Canberra Times*, 4 February 1984, 7.

"The Continuing Story of up the Creek without a Study." *Nimbin News*, 27 August 1979, 6–7.

"Cover." *Groundswell: Newsletter for a New Society* 1, no. 2 (January 1983): 1.

"Daintree Lull Ends: Police Clash with Daintree Protesters." *Canberra Times*, 13 August 1984.

"Daintree Protesters Dig In." *Canberra Times*, 7 August 1984, 3.

"Daintree Update." *Nimbin News* (August 1984): 3–4.

"Daintree Update 2." *Nimbin News* (September 1984): 5–6.

"Damn Hetchy Dam!" *Earth First! Journal* 4, no. 5 (May 1984): 19.

"Date Set to Hear Mt Nardi Dispute." *Northern Star*, 21 September 1982, 2.

"Developers Attack Vireo." *Earth First! Journal* 8, no. 8 (September 1988): 6–7.

"A Diary from the Festival for the Forest." *Nimbin News*, 20 August 1979, 5.

"Direct Action: Six Years Down the Road," *Do Or Die* 7 (1998): 22.

"Down with the Empire! Up with the Spring!," *Do or Die*, no. 10 (2003): 1–101.

"Earth First! Founder Turns to Persuasion." *Arizona Republic*, 1 June 1991, D3.

"Earth First! Link Cut to Ballot Issue." *Ukiah Daily Journal*, 24 July 1990, 1.

"Earth First! Protesters in Sequoia." *Santa Cruz Sentinel*, 31 July 1990, 26.

"Editorial." *Earth First!* 1, no. 1 (November 1980): 1.

"Editorial: Logging War Heats Up … Still the Government Is Silent." *Northern Star*, 1 October 1982, 2.

"Editorial: Two Wrongs Don't Make a 'Riot'." *Register-Guard*, 5 August 1984, 16 A.

"Editorial: Working within the System Gets RACE Results." *Southern Illinoisan*, 6 September 1990, 8.

"EF Local Groups and Contacts." *Earth First! Journal* 3, no. 3 (March 1983): 8.

"Environment Groups to Support Protesters." *Northern Star*, 22 July 1982, 3.

"Environmentalists, Loggers Meet." *Vancouver Sun*, 21 June 1988, A8.

"Eugene Woman Pleads Innocent to Trespassing on Native Forest." *Register-Guard*, 2 May 1985, 14C.

Extinction Rebellion – First time in living memory central London's bridges blocked by protest group, Available [Online]: https://extinctionrebellion.uk/2018/11/17/update-extinction-rebellion-first-time-in-living-memory-central-londons-bridges-blocked-by-protest-group/ (Accessed 1 September 2020).

Extinction Rebellion protests block London bridges, Available [Online]: www.bbc.com/news/uk-england-london-46292819 (Accessed 1 September 2020).

Extinction Rebellion Occupy Greenpeace Offices, Available [Online]: https://risingup.org.uk/extinction-rebellion-occupy-greenpeace-offices (Accessed 1 September 2020).

Extinction Rebellion says 'We're Fucked", block access to Downing Street, Available [Online]: https://extinctionrebellion.uk/2018/11/14/breaking-london-extinction-rebellion-says-were-fucked-block-access-to-downing-street/ (Accessed 1 September 2020).

"'Fairly Extensive' Tree Spiking Found." *Register-Guard*, 12 May 1985.

"The Fanfare and the Tumult Gone." *Macleay Argus*, 9 August 1980, 1.

"Farewell Fairmile." *Do Or Die*, no. 6 (1997): 48–52.

First Tribal Park in BC/Indigenous Relations, Meares Island, Turns 30 Years Old and Is Expanded. Available [Online]: http://fnbc.info/news/first-tribal-park-bcindigenous-relations-meares-island-turns-30-years-old-and-expanded (Accessed 22 August 2016).

"Foes of Road Win Delay after Chaining Selves to Earthmovers." *Los Angeles Times*, 6 June 1989, A3.

"Forest Service Says It Won't Push Bald Mountain Road." *Register-Guard*, 30 January 1985, 10B.

"Forestry Offices Occupied by 200 Conservationists: No Arrests Made." *Northern Star*, 22 July 1982.

"Former Minister amongst 42 Arrests." *Canberra Times*, 11 January 1983, 3.

"Four More Arrests at Mt Nardi." *Northern Star*, 1 October 1982, 2.

"Four Vander Zalm Colleagues Facing Conflict Accusations." *Ottawa Citizen*, 25 February 1987, D2.

"Fourth Day of Elk Logging Blockade." *Ukiah Daily Journal*, 22 August 1990, 1.

"Franklin-Lower Gordon Rivers Campaign." *Wilderness News* (July 1983): 2.

"Greenies Retreat to Fight Another Day." *Courier Mail*, 17 August 1984.

"Gun-Toting Photo a Joke, Friends Claim." *Ukiah Daily Journal*, 12 June 1990, 1.

"Haidas Form Human Blockade to Prevent Logging on B.C. Island." *The Citizen*, 15 November 1985, A12.

"Haidas Found Guilty of Contempt for Disobeying Court Order." *The Citizen*, 30 November 1985, A20.

"Indian Tribe's Protest Halts Logging Operation." *San Francisco Chronicle*, 29 October 1988, A15.

"Indians Arrested in Logging Dispute." *The Citizen*, 18 November 1985, C18.

"Indians' Fish Rights Confirmed." *Vancouver Sun*, 10 October 1987, A11.

"Indians Force Halt to Island Logging." *Globe and Mail*, 3 December 1986, N4.

"Indians, Logger Argue over Tiny Island," *Gazette*, 4 December 1986, B8.

"Indians Oppose Marina," *Provincial Citizen*, 26 November 1985, 7.

"Indians Take Stand on Logging in Park." *Globe and Mail*, 22 October 1984, 4.

"Indians Win a Logging Ban on B.C. Island." *Gazette*, 23 December 1986, B1.

"Indians Win Appeal as B.C. Island Gets Reprieve from Loggers' Saws." *The Gazette*, 28 March 1984, B1.

"Indians, CN Reach Deal on Land Claim." *Gazette*, 13 December 1985, C19.

"Interspecies Action," (Unknown, 1985), 1.

"Jeers Greet 1500 Logging Protesters at 'Redwood Summer'." *The Pantagraph*, 23 July 1990, 4.

"Judge Clears 51 Protesters." *Globe and Mail*, 15 February 1989, A3.

"Judge Won't Ban Activists from Island Logging Site." *The Globe and Mail*, 18 December 1984, 5.

"Judgment Aims to Neutralize Loggers' Dispute with Indians." *Vancouver Sun*, 21 October 1988, A2.

"Kwakiutl Indians Will Block Logging ". *Globe and Mail*, 1 December 1986, A9.

Kwakiutl Protest. Available [Online]: www.firstnations.de/forestry/kwakiutl_protest.htm [Date Accessed 24 November 2016].

"Lawmakers Seeks to Save Wilds of Shawnee Forest." *St. Louis Post-Dispatch*, 25 June 1990, 24.

"Life on the Battlestar," *Do Or Die*, no. 6 (1997): 82–86

"Loggers Block Road Again Despite Meeting Ministers." *Ottawa Citizen*, 6 September 1988, A13.

"Loggers Fight Has Just Begun." *The Sun*, 4 February 1984.

"Loggers Warned of Spikes." *Edmonton Journal*, 15 September 1989, B7.

"Logging Firm Gets Injunction against Meares Protesters." *Globe and Mail*, 4 December 1984, 4.

"Logging Injunction War on Hold." *Vancouver Sun*, 8 October 1988, C16.

Logging Operations Shut Down Across Victoria, Available [Online]: www.ecoshout. org.au/blog/logging-shut-down-across-victoria-seven-protests-one-day (Accessed 8 November 2020).

"Logging Protest Gives Alec a 'Birds-Eye' View of the World." *Mercury*, 28 February 1986, 1.

"Logging Protest: Twelve Arrested." *Sydney Morning Herald*, 23 August 1979, 3.

"Logging Road into Temagami Will Be Suspended, Ontario Says." *Gazette*, 20 October 1989, B1.

"Logging Road Work Proceeds as Protest Ebbs." *Ottawa Citizen*, 21 September 1989, A17.

"Logging War Flares." *Northern Star*, 30 September 1982, 1, 3.

"Lubicon Lake Cree: Oppressed People's Struggle To Take Back Power," *Akwesasne Notes* Fall (1988): 22.

"Lumber Company Declares Bankruptcy." *Ottawa Citizen*, 3 December 1988, A4.

"Middle Hd Protest Widens." *Macleay Argus*, 9 September 1980, 1,2.

"Mining: 24 Hours Warning." *Nambucca Guardian News*, 27 June 1980, 27.

"The Mining Front ... Fence Goes up; Man Arrested." *Nambucca Guardian News*, 9 July 1980, 2.

"New Alliance Adds Voice to Forest Protest." *Vancouver Sun*, 20 January 1989, H15.

"New Group in Mining Protest." *Nambucca Guardian News*, 8 October 1980, 1.

News Ticker, Available [Online]: https://hambachforest.org/blog/2018/09/25/ticker-from-september-25 (Accessed 31 July 2020).

"Nightcap." *Colong Committee Newsletter*, no. 75 (November 1982): 5.

"Nineteen More Arrests at Temagami." *Ottawa Citizen*, 20 November 1989, A4.

"No G-O Road." *Earth First! Journal* June (1983): 15.

"NSW Cabinet Saves Rainforests from the Axe." *Canberra Times*, 27 October 1982, 3.

"Old-Growth Backers Chain Themselves Up." *Spokane Chronicle*, 31 May 1990, B4.

"Old-Growth Protest." *Seattle Times*, 30 May 1990, E1.

"Ontario Can't Stop Environmentalists' Protest in Temagami: Court." *Gazette*, 15 September 1989, B1.

"Ontario Natives Set up Blockade, Lubicon Promise Similar Tactic." *Gazette* 2 June 1988, B4.

"Orbost Angry with Conservationists." *Canberra Times*, 6 February 1984, 7.

Over 1000 people block Parliament Square, Available [Online]: https://extinctionrebellion. uk/2018/10/31/over-1000-people-block-parliament-sq-to-launch-mass-civil-disobedience-campaign-demanding-action-on-climate-emergency/ (Accessed 1 September 2020).

"Peace Protests Roll Call." *New Statesmen* 109(1985): 6.

"PM, Vander Zalm Sign South Moresby Park Deal." *Ottawa Citizen*, 13 July 1987, A4.

"Police Arrest Protesters." *Edmonton Journal* 13 November 1989, A4.

"Police Arrest, Release 31 at Temagami Logging Site." *Ottawa Citizen*, 20 September 1989, A6.

"Police Clash with Forest Protesters." *Canberra Times*, 13 August 1984, 3.

"Police OK Citizens Arrests," *Bega District News*, 12 September 1989, 1.

"Police Ride Shotgun as Logs Are Taken." *Daily Mail*, 30 August 1979, 3.

"Police Thwart Dam Protest." *Canberra Times*, 13 January 1983, 3.

"Police to 'Keep Peace' in Log Dispute." *Vancouver Sun*, 6 July 1988, A2.

"Post Rendezvous Action Shuts Downtimber Sale." *Earth First! Journal* 9, no. 7 (August 1989): 1, 19.

"Power Together," London: Extinction Rebellion UK, 2020.

"Protest in Victorian Rainforest." *Canberra Times*, 12 January 1984, 7.

"Protest over Logging in Rainforest." *The Age*, 18 August 1979, 5.

"Protest Plan Vigil in Wilderness." *Canberra Times*, 19 December 1982, 3.

"Protest Wraps up as Drilling Ends in Park." *Vancouver Sun*, 28 March 1988, A1.

Protesters Blockade Adani Minesite, Available [Online]: www.thebigsmoke.com.au/2020/08/24/protestors-blockade-adani-mine-site-as-traditional-owners-issue-eviction-notice (Accessed 8 November 2020).

"Protesters Charged after Logging Road Blocked." *Gazette*, 3 September 1988, B7.

"Protesters Defy Injunction to Prevent Drilling in Park." *Vancouver Sun*, 30 January 1988.

"Protesters Fined for Defying Ruling." *Vancouver Sun*, 9 September 1988, A12.

"Protesters Try to Block Route but Loggers Reach Mount Graham Site." *Mohave Daily Miner*, 3 October 1990, 2.

"Rainforest Blockades Are Dropped." *Courier Mail*, 29 August 1984, 9.

"Rainforest Row Turns to Violence." *Courier Mail*, 13 August 1984, 1.

"Redwood Summer Activists Assaulted by Loggers." *Earth First! Journal* 10, no. 7 (August 1990): 7.

"Redwood Summer Activists Harassed by Police." *Earth First! Journal* 10, no. 7 (August 1990): 7.

"Redwood Summer and Beyond." *Polemicist* 2, no. 2 (November 1990): 13.

"Redwood Summer Chronology." *Earth First! Journal* 11, no. 1 (November 1990): 7.

"Residents Fear Uranium Will Pollute Lake." *Globe and Mail*, 4 January 1985, 5.

"Residents Support Logging; Protesters Have Road to Themselves." *Ottawa Citizen*, 24 September 1989, A7.

"Resistance at Wollaston." *Open Road*, Spring 1986, 6–7.

"Resistance Gets Hairy at Bald Mountain." *Earth First! Journal* November–December (2000): 11.

"Road Blocked by Indians." *Vancouver Sun*, 23 June 1988, B6.

"Round River Rendezvous." *Earth First! Journal* 3, no. 6 (1983): 1, 4–5.

"Roundup Message Clear." *Eugene Register-Guard*, 18 September 1987, 22A.

"Roynat Loan to Let Milne Resume Lumber Production." *Globe and Mail* 10 June 1989, B3.

"A Sacred Place...A Working Place." *Globe and Mail*, 7 May 1988, D2.

"Salt Creek Arrests." *Earth First! Journal* March (1983): 1,5.

"Saving Sugar Bear," *Seattle Times*, 31 August 1990, 2.

"Sawmillers Call Temporary Halt at Terania." *Sydney Morning Herald*, 4 September 1979, 1.

"Shawnee." *Southern Illinoisan*, 5 September 1990, 8.

"Sour-Gas Well Wins Ok." *Calgary Herald*, 23 December 1988, A10.

Southeast Forest Rescue, Available [Online]: https://zetblogs.com.au/394016-sefr-southeast-forest-rescue/ (Accessed 8 November 2020).

"Spiked Trees Found in Oregon – Cutting Perils a Protester." *San Francisco Chronicle*, 23 June 1987, 8.

A Story Written on the Land. Available [Online]: lilwat.ca/wearelilwat7yul/history/ (Accessed 3 September 2017).

"Supporters Mark Road's Opening." *Gazette*, 29 January 1989, A7.

Takayna 2020, Available [Online]: www.bobbrown.org.au/takayna_2020 (Accessed 8 November 2020).

"Temagami Blockade." *Globe and Mail*, 26 August 1989, A12.

"Temagami Trees at Risk Study Finds." *Toronto Star* 13 October 1989, A12.

"Ten Year Fight Achieves Hope for North Washpool." *World Rainforest Report*, no. 19 (1991): 6–7

"Texas EF Fights Freddie Godzilla." *Earth First! Journal* 7, no. 1 (November 1986): 1, 11.

"Timber Group, Activists Clash on North Coast." *Ukiah Daily Journal*, 4 September 1990, 1.

"Togetherness." *Newsweek*, 4 June 1990, 4.

"Tree Spiking Renounced Behind Redwood Curtain." *Earth First! Journal* 10, no. 5 (May 1990): 12.

"Troubled Road at the Cape." *Courier Mail*, 3 December 1983, 1.

"Two Arrested at B.C. Protest of Mine in Park." *Gazette*, 8 February 1988, A10.

"TWS Blockade Feedback." *Groundswell: Newsletter for a New Society* 1, no. 4 (1983): 3.

Updates, Available [Online]: https://dwembassy.com/334–2/ (Accessed 8 November 2020).

"Violence after Gippsland Meeting: Inquiry Announced into Victoria's Timber Industry." *Canberra Times*, 14 January 1984, 3.

"Where Were You When We Cracked Glen Canyon Dam?" *Earth First!* 1, no. 4 (March 1981): 1–2.

"Whole Rebellion Action/Policy Document." (London: Extinction Rebellion, 2018).

Why Rebel? Available [Online]: https://rebellion.global/why-rebel/. (Accessed 31 July 2020).

"Wilderness Society Turnover $1m." *Canberra Times*, 12 February 1983, 7.

"Wilderness War in Oregon." *Earth First! Journal* 3, no. 5 (June 1983): 1, 4.

"Wollaston Lake Saskatchewan: Residents Oppose Uranium Mining." *Akwesasne Notes* Summer (1985): 4–5.

"Work Disrupted by Forest Protesters." *Northern Star*, 5 August 1982, 3.

"Workers Hear the Bad News: Sand Miner Quits Macleay." *Macleay Argus*, 13 December 1980, 1, 5.

A Nony Moose (pseudonym). "Shooting Cows: A Novel Idea." *Earth First! Journal* 11, no. 8 (September 1991).

Abbey, Edward. "Letter ". *Earth First! Journal* 8, no. 2 (December 1987): 3.

— *The Monkey Wrench Gang*. Philadelphia: Lippincott, 1975.

Acosta, Ana. "Quakers, the Origins of the Peace Testimony and Resistance to War Taxes," In *War and Peace: Essays on Religion and Violence*, edited by Bryan Turner. 101–120. London: Anthem Press, 2013.

Ahmed, Safir. "Ban on All Tree Cutting in Shawnee Sought." *St. Louis Post-Dispatch*, 3 October 1990, 14.

Angel, Jeff. "Letter from TEC Executive Director to John Seed Concerning Blockading." Undated, 1988, 2.

Anderson, Christopher. "Aborigines and Conservationism: The Daintree-Bloomfield Road." *Australian Journal of Social Issues* 24, no. 3 (August 1989): 214–26.

Anderson, Patrick, Telephone Interview, 9 June 2015, Sydney, Australia. 1 hour, 12 minutes.

Aspinall, Ron. Telephone Interview, 22 October 2017, Coral Harbour, Canada. 2 hours, 4 minutes.

Associated Press. "Environmentalists Protest Fiddler Timber Sale." Available [Online]: http://tdn.com/news/state-and-regional/environmentalists-protest-fiddler-timber-sale/article_990c41cf-3b37-5f13-87de-b733f5db5c91.html (Accessed 7 November 2015).

— "Old Growth Logging Foes Sit in Trees." *Spokane Chronicle*, 7 June 1990, C8.

— "Protest Continues as Old Firs Cut." *Spokesman-Review*, 1 April 1986, A10.

— "Protesters Told to Pay Damages." *Register-Guard*, 4 September 1985, 5A.

— "Redwood Summer' Rolls Along." *Orange County Register*, 31 July 1990, A3.

— "Seven Logging Protesters Arrested." *Register-Guard*, 5 May 1987, 3C.

— "Some Earth Firsters Shed Spikes." *Spokane Chronicle*, 11 July 1990, A10.

Appleby, Timothy, and Richard Mackie. "Rae among 16 Arrested at Temagami Protest." *Globe and Mail*, 19 September 1989, A1.

Aries. "Go Climb a Tree!". *Earth First! Journal* 5, no. 6 (June 1985): 7.

Asplund, Ilse. "Evan Mecham Eco Tea-Sippers International Conspiracy." *Earth First! Journal* 31, no. 2 (February 2011).

Australopithicus. "Nemesis News Net." *Earth First! Journal* 5, no. 7 (August 1985): 12.

Back, Brian. Telephone Interview, 15 September 2017, Toronto, Canada. 2 hours, 6 minutes.

— "Temagami: An Environmentalist's Perspective." In *Temagami: A Debate on Wilderness*, edited by Mark Bray and Ashley Thomson. 141–47. Toronto: Dundurn Press, 1990.

Bailey, Ric. "Bald Mountain in Restrospect." *Earth First! Journal* 4, no. 1 (November 1983): 6–7.

Baird-Windle, Patricia, and Eleanor J Bader. *Targets of Hatred: Anti-Abortion Terrorism.* New York: Palgrave Macmillan, 2001.

Baldrey, Keith. "Ulkatcho Band Vows to Continue Blockade." *Vancouver Sun*, 3 August 1989, B5.

Bantjes, Rod. *Social Movements in a Global Context: Canadian Perspectives.* Toronto: Canadian Scholars' Press, 2007.

Barbosa, Luis. *The Brazilian Amazon Rainforest: Global Ecopolitics, Development and Democracy.* Lanham, MD: University Press of America, 2000.

Bari, Judi. "California Rendezvous." *Earth First!* 9, no. 1 (November 1988): 4.

— "Californians Start a New Fad: Tree-Sitting Becomes a Pastime." *Earth First! Journal* 9, no. 8 (September 1989): 4.

— "Expand Earth First!". *Earth First! Journal* 10, no. 8 (September 1990): 5–6.

— *Timber Wars.* Monroe: Common Courage Press, 1994.

Barnes, Trevor, and Roger Hayter. "Introduction." In *Troubles in the Rainforest: British Columbia's Forest Economy in Transition*, edited by Trevor Barnes and Roger Hayter. 1–14. Victoria: Western Geographical Press, 1997.

Barnes, Trevor J, Roger Hayter, and Elizabeth Hay. "Stormy Weather: Cyclones, Harold Innis, and Port Alberni, BC." *Environment and Planning A* 33, no. 12 (2001): 2127-47.

Barron, David. "CD Begins Anew in Kalmiopsis." *Earth First! Journal* 7, no. 5 (May 1987): 5.

Batio, Christopher. "Shawnee Battle Reaches State Capital," *Southern Illinoisan*, 27 February 1990, 1.

BC Treaty Commission. *What's the Deal with Treaties?* Vancouver: BC Treaty Commission, 2007.

Beder, Sharon. "SLAPPS: Strategic Lawsuits against Public Participation: Coming to a Controversy Near You." *Current Affairs Bulletin* 72, no. 3 (1995): 22.

Belanger, Yale D., and P. Whitney Lackenbauer. "Introduction." In *Blockades or Breakthroughs?: Aboriginal Peoples Confront the Canadian State*, edited by Yale D Belanger and P Whitney Lackenbauer. 3–50. Montreal: McGill-Queen's Press, 2014.

Benidickson, Jamie. "Temagami Old Growth: Pine, Politics and Public Policy." *Politics and Public Policy* 23 (1996): 41–50.

Berglund, Oscar and Daniel Schmidt, *Extinction Rebellion and Climate Change Activism.* Chaim: Palgrave MacMillan, 2020.

Bernstein, Steven, and Benjamin Cashore. "The International-Domestic Nexus: The Effects of International Trade and Environmental Politics on the Canadian Forest Sector." In *Canadian Forest Policy: Adapting to Change*, edited by Michael Howlett. 65–93. Toronto: University of Toronto Press, 2001.

Bevington, Douglas. *The Rebirth of Environmentalism: Grassroots Activism from the Spotted Owl to the Polar Bear.* Washington, DC: Island Press, 2009.

Bevington, Douglas, and Chris Dixon. "Movement-Relevant Theory: Rethinking Social Movement Scholarship and Activism." *Social Movement Studies* 4, no. 3 (December 2005): 185–208.

Bible, Vanessa. "Aquarius Rising: Terania Creek and the Australian Forest Protest Movement." Honours, University of New England, 2010.

Big Bark. "3 New Mexicans Arrested at Copar Strip Mine." *Earth First! Journal* 9, no. 5 (May 1989): 20.

Bita, Natasha. "'Greenies' Lost Battle but Won the War." *Australian*, 10 June 2014, 2.

Blomley, Nicholas. "'Shut the Province Down': First Nations Blockades in British Columbia, 1984-1995." *BC Studies: The British Columbian Quarterly*, no. 111 (1996): 5–35.

Blumm, Michael C. "Ancient Forests and the Supreme Court: Issuing a Blank Check for Appropriation Riders." *Urban Law Annual; Journal of Urban and Contemporary Law* 43 (1993): 35–57.

Bocking, Mike. "Natives to Fight Logging but Won't Block Trucks." *Vancouver Sun*, 19 May 1988, B8.

Bohlen, Jim. *Making Waves: The Origins and Future of Greenpeace*. Montreal: Black Rose Books, 2001.

Bohn, Glenn. "Mine Road Blockade Organized." *Vancouver Sun*, 28 February 1987, A11.

— "Park Boundary Revision Outlines Mining Rights." *Vancouver Sun*, 30 January 1987, A12.

Bolan, Kim. "Strathcona Protestors Shrug Off Weekend Arrests." *Vancouver Sun*, 1 February 1988.

Bonnett, Mark, and Kurt Zimmerman. "Politics and Preservation: The Endangered Species Act and the Northern Spotted Owl." *Ecology Law Quarterly* 18 (1991): 105–71.

Bonyhady, Tim. *Places Worth Keeping: Conservationists, Politics and Law*. St Leonards: Allen & Unwin, 1993.

Bossin, Bob. The Clayoquot Women, Available [Online]: http://www3.telus.net/oldfolk/women.htm [Date Accessed 6 December 2016].

Braggs, Sierra. "Earth First!: The Rise of Eco-Action." Masters, Humboldt State University, 2012.

Branagan, Martin. "Art Alone Will Move Us: Nonviolence Developments in the Australian Eco-Pax Movement 1982-2003." University of New England, 2006.

— *Global Warming, Militarism and Nonviolence: The Art of Active Resistance*. Basingstoke: Palgrave Macmillan, 2013.

— "'We Shall Never Be Moved': Australian Developments in Nonviolence." *Journal of Australian Studies*, no. 80 (2004): 201–10.

Bralver, Peter, and Dan Strachan. "Los Angeles EF! Wins Fight for La's Last Wilderness." *Earth First! Journal* 9, no. 6 (June 1989): 1.

Brazil, Eric. "Tree Spikers Draw Sawmill Blood." *San Francisco Examiner*, 24 June 1987, D12.

Bridge, Carl and Kent Fedorowich, "Mapping the British World", *The Journal of Imperial and Commonwealth History*, 31 no. 2 (2003): 1–15.

Brinkman, Phil. "Spiking Incident Draws Investigation and Denials." *Southern Illinoisan*, 10 August 1990, 1–2.

— "Temporary Clamp Placed on Fairview Timber Sale." *Southern Illinoisan*, 4 July 1990, 1.

Brisman, Avi, and Nigel South. "A Green-Cultural Criminology: An Exploratory Outline." *Crime, Media, Culture: An International Journal* 9, no. 2 (2013): 115–35.

Brouillette, Kathy, Judy Baker, Adrian Donkers, and Michael Lockwood. "TWS Tactics: An NVA Scenario." *Groundswell: Newsletter for a New Society* 1, no. 3 (1983): 3–5.

Brouillette, Kathy, and Michael Lockwood. "Guide for Trainers." *Groundswell: Newsletter for a New Society* 1, no. 1 (October 1982): 4–5.

Brouwer, Stephen, ed. *The Message of Terania*. South Lismore: Terania Media, 1979.

Brown, Beverley. *In Timber Country*. Philadelphia: Temple University Press, 1995.

Brown, Bob, and Peter Singer. *The Greens*. Melbourne: Text Publishing, 1996.

Brown, Karen. "Bisti Circus." *Earth First! Journal* 3, no. 3 (March 1983): 5.

Brown, Nancy. "Park Activists Go Free." *Times-Colonist* 18 May 1989, 2.

— "Pillage of Oldest Park 'Will Continue'." *Times-Colonist*, 11 February 1987, B12.

Bruvere, Smokey. "Lubicon Lake Land Entitlement Breakdown in Negotiations," *Akwesasne Notes* Autumn (1986): 23.

Budworm, Bruce. "Battle for Fish Town Woods." *Earth First! Journal* 8, no. 4 (March 1988): 6.

Burg, Mary. How the Wet Tropics Was Won, Available [Online]: http://cafnec.org.au/about-cafnec/how-the-wet-tropics-was-won/ (Accessed 26 August 2014).

Burge, Mike. "13 Arrested for Blocking Entry to Timber Sale." *Register-Guard*, 27 March 1989, 1C.

— "Redwood Struggle Expected to Continue." *Ukiah Daily Journal*, 7 September 1990, 10.

Burgess, David. "Correspondence by email." 22 October 2020.

— "Police Interview Transcript: 13 November," (1989), 1–7.

Burgmann, Verity. "The Importance of Being Extreme," *Social Alternatives* 37 no. 2 (2011), 10–12.

Caffrey, Andy. 1980s Earth First!: The Antidote for Despair, Available [Online]: www.youtube.com/watch?v=DMxX7twsYNo (Accessed 6 April 2015).

'Canyon Wolf'. "Victory!!! Earth First! Saves Colorado Old Growth." *Earth First! Journal* 10, no. 2 (December 1989): 8.

Careless, Ric. *To Save the Wild Earth*. Vancouver: The Mountaineers, 1997.

Cassidy, Frank, and Norman Dale. *After Native Claims?: The Implications of Comprehensive Claims Settlements for Natural Resources in British Columbia*. Lantzville: Oolichan Books, 1988.

Chapman, Ron. " Fighting for the Forests: A History of the West Australian Forest Protest Movement, 1895-2001." Murdoch University, 2008.

Charles, Andrew. "Extinction Rebellion: A Short Critical Guide," *Overland*, October 2019, Available [Online]: https://overland.org.au/2019/10/extinction-rebellion-a-short-critical-guide/ (Accessed 2 November 2020).

Chase, Alston. *In a Dark Wood: The Fight over Forests and the Myths of Nature*. New Brunswick: Transaction, 2009.

Cherney, Daryl. Daryl Cherney Music, Available [Online]: http://asis.com/users/dced/hippieslyrics.htm (Accessed 14 June 2016).

— "Freedom Riders Needed to Save the Forest." *Earth First! Journal* 10, no. 5 (May 1990): 1, 6.

— "Triple Victory in Three Day Revolution." *Earth First! Journal* 9, no. 2 (December 1988): 1, 6.

Chornook, Kay. Telephone Interview, 14 February 2017, Monteverde, Costa Rica. 1 hour, 53 minutes.

Cianchi, John. *Radical Environmentalism: Nature, Identity and More-Than-Human Agency*. Basingstoke: Palgrave-Macmillan, 2015.

Circles, Lone Wolf, Karen Wood, and Moss. "Escalation! The Kalmiopsis 24." *Earth First! Journal* 9, no. 7 (August 1989): 6.

Clairmont, Don and Jim Potts, *For the Nonce: Policing and Aboriginal Occupations and Protests*, Ontario: Ipperwash Inquiry, 2006.

Claridge, Thomas. "Wilderness Society Wins First Round." *Globe and Mail* 24 March 1989, A5.

Clark, Hattie. "If a Tree Falls, John Seed Hears It." *Christian Science Monitor* (13 August 1987). http://search.proquest.com.ezp.lib.unimelb.edu.au/docview/1034959490?accountid=12372 (Accessed 22 May 2015).

Clearwater, Jericho. "Kalmiopsis Shutdown." *Earth First! Journal* 7, no. 7 (August 1987): 1, 6.

Coates, James. "Terrorists for Nature Proclaim Earth First!" *Chicago Tribune*, 2 August 1987, 21.

Coates, Peter A, and D.J.S. Morris. ""Support Your Right to Arm Bears (and Peccadillos)":The Higher Ground and Further Shores of American Environmentalism." *Journal of American Studies* 23, no. 3 (1989): 439–46.

Cockburn, Alexander. "Beat the Devil." *The Nation* (3 December 1990): 670–72.

Cohen, Ian. *Green Fire*. Sydney: Angus & Robertson, 1997.

Colombo, David. "Letter: Earth First! Had Part in Victory." *Southern Illinoisan*, 24 September 1990, 4.

Colong Committee. *How the Rainforest Was Saved*. Sydney: Colong Committee, 1983.

Connors, Libby, and Drew Hutton. *A History of the Australian Environmental Movement*. Cambridge: Cambridge University Press, 1999.

Coover, Virginia, Ellen Deacon, Charles Esser, and Christopher Moore. *Resource Manual for a Living Revolution*. Philadelphia: New Society Publishers, 1977.

Corsaletti, Louis. "Security Gantlet Thrown up to Discourage Logging Foes." *Seattle Times*, 6 August 1990.

Coulter, Karen, Telephone Interview, 9 May 2016, Fossil, USA. 2 hours, 17 minutes.

Cronon, Wilderness. "The Trouble with Wilderness: Or, Getting Back to the Wrong Nature." *Environmental History* 1, no. 1 (1996): 7–28.

Cruickshank, John. "3 Haidas Charged with Defying Order on Island Logging." *Globe and Mail*, 18 November 1985, A5.

—— "3 Haidas Charged with Defying Order on Island Logging." *Globe and Mail*, 18 November 1985, A5.

—— "Disobedience Very Civil at B.C. Rig." *Globe and Mail*, 15 February 1988, A4.

Cruickshank, John, and Jack Danylchuk. "Holds Investment in Forest Company, B.C. Minister Quits." *Globe and Mail*, 18 January 1986, A1.

Curley, John. "Protesters Vow to Bar Logging." *St. Louis Post-Dispatch*, 6 August 1990, 1,6.

Curran, Giorel. *21st Century Dissent: Anarchism, Anti-Globalization and Environmentalism*. Basingstoke: Palgrave Macmillan, 2006.

Danylchuk, Jack. "2 B.C. Ministers Meet Haida, Discuss Logging, Land Claims." *Globe and Mail*, 11 December 1985, A8.

Darnovsky, Marcy, Barbara Epstein, and Richard Flacks. "Introduction." In *Cultural Politics and Social Movements*, edited by Marcy Darnovsky, Barbara Epstein and Richard Flacks. vii–xxiii. Philadelphia: Temple Press, 1995.

Davenport, Paula. "Forest Service Cancels Sale of Shawnee Harvest." *St. Louis Post-Dispatch*, 29 September 1990, 6.

—— "Logging Opponents Dig in at Shawnee." *St. Louis Post-Dispatch*, 19 August 1990, 1,4.

—— "Protesters Hope to Save Shawnee Trees." *St. Louis Post-Dispatch*, 2 July 1990, 113.

—— "Rival Groups Join to Fight Shawnee Logging." *St. Louis Post-Dispatch*, 25 April 1990, 101.

—— "Timber Is Spared in Shawnee." *St Louis Dispatch*, 30 September 1990, D2.

Davis, John. "Ramblings." *Earth First! Journal* 10, no. 6 (June 1990): 2.

De Bare, Ilana. "A Tale of Two Owners: Old Redwoods, Traditions Felled in Race for Profits." *Los Angeles Times*, 20 April 1987, 1.

De Cleyre, Voltairine. *Direct Action*. New York: Mother Earth Publishing Association, 1912.

De Danan, Tuatha. "Is EF! Selling Out?" *Earth First! Journal* November (1983): 4.

De La Garza, Paul. "Hopes for Clear-Cutting Ban Rooted in Shawnee Dispute." *Chicago Tribune*, 22 October 1989, 16.

della Porta, Donatella. "Eventful Protest, Global Conflicts." *Distinktion* 17 (2008): 27–56.

della Porta, Donatella and Herbert Reiter. *Policing Protest: The Control of Mass Demonstrations in Western Democracies*. Minneapolis: University of Minnesota Press, 1998.

Department of Transport, *Roads for Prosperity*. London: Department of Transport, 1989.

Devall, Bill. "The Edge: The Ecology Movement in Australia." *Earth First! Journal* 4, no. 5 (May 1984): 12–13.

Dicanio, Margaret. *Encyclopedia of American Activism: 1960 to the Present*. Second ed. Lincoln: iUniverse Inc., 2005.

Dietrich, William. "New Law Forces Old-Growth Sale," *Seattle Times*, 21 November 1989, C3.

— *The Final Forest: The Battle for the Last Great Trees of the Pacific Northwest*. New York: Simon & Schuster, 1992.

Direct Action Manual Collective. *Earth First! Direct Action Manual*. Eugene: Cascadia Summer, 1997.

Doern, G Bruce, and Thomas Conway. *The Greening of Canada: Federal Institutions and Decisions*. Toronto: University of Toronto Press, 1994.

Doherty, Brian. "Manufactured Vulnerability: Protest Camp Tactics." In *Direct Action in British Environmentalism*, edited by Benjamin Seel, Matthew Paterson and Brian Doherty. 62–78. London: Routledge, 2000.

Dorst, Adrian. Email Interview, 1 September 2017, Tofino, Canada.

Douglas Shire Wilderness Action Group, *The Trials of Tribulation*, Port Douglas: Douglas Shire Wilderness Action Group, 1984.

Doyle, Timothy. "Direct Action in Environmental Conflict in Australia: A Re-Examination of Non-Violent Action." *Regional Journal of Social Issues*, no. 28 (1994): 1–13.

— *Environmental Movements in Minority and Majority Worlds*. New Brunswick: Rutgers University Press, 2005.

— *Green Power: The Environmental Movement in Australia*. Sydney: UNSW Press, 2000.

Draffan, George. "Cathedral Forest Action Group Fights for Oregon Old Growth." *Earth First! Journal* 4, no. 5 (June 1984): 4.

Draffan, George, and Mitch Freedman. "War in the Wenatchee." *Earth First! Journal* 6, no. 8 (1986): 8, 19.

Dubofsky, Melvyn. *We Shall Be All: A History of the Industrial Workers of the World*. Urbana, IL: University of Illinois Press, 2000.

Dumont, Clayton W. "The Demise of Community and Ecology in the Pacific Northwest: Historical Roots of the Ancient Forest Conflict." *Sociological Perspectives* 39, no. 2 (1996): 277–300.

Dunlap, Thomas. *Nature and the English Diaspora*. Cambridge: Cambridge University Press, 1999.

Dunn, Ross. "No Dams Campaign: It Depends on the Election." *Sydney Morning Herald*, 14 February 1983, 2.

— "High on Trees." *Nimbin News*, 20 August 1979, 2–4.

— "Lessons in Losing." *Maggies Farm*, October 1980, 11.

Durbin, Kathie. *Tree Huggers: Victory, Defeat & Renewal in the Northwest Ancient Forest Campaign*. Seattle, WA: Mountaineers, 1996.

Dutton, Don. "Ontario Indians Block Logging Road." *Toronto Star*, 3 June 1988, A22.

Dwyer, John. "Conflicts over Wilderness: Strathcona Provincial Park, British Columbia." Masters, Simon Fraser University, 1993.

Easthouse, Keith. EPIC Changes, Available [Online]: www.northcoastjournal.com/humboldt/epic-changes/Content?oid=2132384 (Accessed 5 November 2015).

Edwards, Bob. "With Liberty and Environmental Justice for All." In *Ecological Resistance Movements*, edited by Bron Taylor. 33–55. Albany, NY: SUNY, 1995.

Egan, Kelly. "Anti-Logging Road Activists Score Only Small Victories." *Ottawa Citizen*, 10 October 1989, A2.

— "Environmentalists Battle Loggers in Temagami." *Ottawa Citizen*, 10 October 1989, A1.

— "Wran Not Satisfied with Report on Logging." *Sydney Morning Herald*, 13 February 1982, 5.

Ellis, Elaine. "In Defense of Mother Earth." *Sun Sentinel*, 6 August 1989, 1E.

Epstein, Barbara. *Political Protest and Cultural Revolution: Nonviolent Direct Action in the 1970s and 1980s.* Berkeley: University of California Press, 1991.

Evans, Raymond. *A History of Queensland.* Cambridge: Cambridge University Press, 1997.

Extinction Rebellion UK. About Us, Available [Online]: https://extinctionrebellion.uk/the-truth/about-us/ (Accessed 1 November 2020).

Faithfull, Tony. "In Prison." *Groundswell: Newsletter for a New Society* 1, no. 2 (January 1983): 9.

Fattig, Paul. Future of Bald Mountain Remains Uncertain, Available [Online]: www.mailtribune.com/article/20030427/BIZ/304279999 (Accessed 5 November 2015).

Featherstone, Roger. "Grand Canyon Uranium Battle." *Earth First! Journal* 7, no. 4 (March 1987): 1.

Feigenbaum, Anna, Fabian Frenzel, and Patrick McCurdy. *Protest Camps.* London: Zed Books, 2013.

Fireman, Ken. "Summer of Heat over Redwoods' Fate." *Newsday*, 3 September 1990, 17.

Flynn, Tony. "Fishtown Protesters Concede." *Skagit Argus*, 9 February 1988, 3.

Foley, Griff. *Learning in Social Action: A Contribution to Understanding Informal Learning.* London: Zed, 1999.

Foreman, Dave. "Around the Campfire." *Earth First!* 3, no. 2 (December 1982): 2.

— "Around the Campfire." *Earth First!* 3, no. 6 (August 1983): 2.

— *Confessions of an Eco-Warrior.* New York: Crown, 1991.

— "Earth First!". *Earth First!* 2, no. 3 (February 1982): 4–5.

— "Editorial." *Earth First!* 1, no. 5 (June 1981): 1.

— "Editorial: The Lesson of Salt Creek." *Earth First! Journal* 3, no. 3 (March 1983): 2.

— "EF and Nonviolence: A Discussion." *Earth First! Journal* 3, no. 7 (September 1983): 11.

— "The Question of Growth in Earth First!". *Earth First! Journal* 8, no. 6 (June 1988): 32.

— "Welcome to EF." *Earth First! Journal* 5, no. 5 (May 1985): 16.

Foreman, Dave, and Bill Haywood, eds. *Ecodefense: A Field Guide to Monkeywrenching.* Tucson, AZ: Ned Ludd Books, 1987.

Foster, David. "Earth First! Takes Non-Violent Tack in California." *Phoenix Gazette*, 9 July 1990, B4.

Foster, Hamar. "Sannichton Bay Marina Case: Imperial Law, Colonial History and Competing Theories of Aboriginal Title, The." *University of British Columbia Law Review* 23 (1988): 629.

Foulkes, Brian. State V. Hund, Available [Online]: www.aclu-or.org/content/state-v-hund (Accessed 18 November 2015).

Fournier, Suzanne. "Court Imposes Deadline: Indians, Loggers Get 14 Days." *Province*, 3 November 1989, 43.

— "Gitksan Chiefs Jubilant over Court's Action." *Province*, 17 November 1989, 24.

— "Logging Drivers Reciprocate on Blockade." *Province*, 1 November 1989, 4.

— "Natives Blockade Loggers." *Province*, 31 October 1989, 1.

Freedman, Mitch. "Old Growth Strategy Revisited." *Earth First!* 9, no. 2 (December 1988): 7.

Freeman, Dave. Linn County Deputy Sheriff's Report, Available [Online]: www.penbay.org/ef/ronhuber_sheriffrp85.html [Date Accessed 17 November 2015].

Frideres, James "Circle of Influence Social Location of Aboriginals in Canadian Society." In *Aboriginal Peoples and Forest Lands in Canada*, edited by D.B. Tindall, Ronald Trosper and Pamela Perreault. 31–47. Vancouver: UBC Press, 2013.

Gamerman, Amy. "New Drug Bill Spells out Penalties for Sabotaging Logging Operations." *Colorado Springs Gazette*, 7 December 1988, B7.

Gamson, William. "Reflections on the Strategy of Social Protest." *Sociological Forum* 4, no. 3 (1989): 455–467.

Geniella, Mike. "Bari Juror Explains Verdicts, Marathon Deliberations." *Press Democrat*, 14 June 2002, A1.

— "Leadership Dispute Splits Earth First!" *Press Democrat*, 12 August 1990, A1.

George, Earl Maquinna. *Living on the Edge: Nuu-Chah-Nulth History from an Ahousaht Chief's Perspective.* Winlaw: Sono Nis Press, 2003.

George, Paul. *Big Trees Not Big Stumps.* Vancouver: Western Canada Wilderness Committee, 2006.

Gill, Ian. *All That We Say Is Ours.* Vancouver: Douglas & McIntyre, 2009.

Gitxsan History of Resistance. Available [Online]: www.gitxsan.com/culture/culture-history/gitxsan-history-of-resistance/ (Accessed 20 November 2016).

Glascott, Joseph. "43 Arrests at Sand Site," *Sydney Morning Herald*, 23 September 1980, 4.

— "Conservationist Quits as Adviser to Govt." *Sydney Morning Herald*, 10 September 1979, 2.

— "New Forest Studies Ordered." *Sydney Morning Herald*, 13 October 1979, 3.

— "Protesters Form Barrier in Trees." *Sydney Morning Herald*, 29 August 1979, 12.

— "B.C. Native Land Claim Freezes Vast Forest Area." *Vancouver Sun*, 17 December 1988, A1.

— "Blockade Marks Bid by Indians to Exercise 'Ownership' of Land." *Vancouver Sun*, 18 February 1988, A11.

— "Direct Action by Indians Urged." *Vancouver Sun*, 16 November 1988, A1.

— "Feast Ends Logging Truck Blockade." *Vancouver Sun*, 1 March 1988, B1.

— "Indians Block Logging Road." *Vancouver Sun*, 6 May 1988, B6.

— "Indians Consider 'Direct Action'." *Vancouver Sun*, 29 May 1990, B1.

— "Indians Fighting against Road Stand Firm Despite Injunctions." *Vancouver Sun*, 6 October 1988, A13.

— "Indians Hear Call to Battle Logging." *Vancouver Sun*, 12 April 1988, B5.

— "Indians Plan to Fight Sale of 20-Year Forest Licence." *Vancouver Sun*, 2 May 1988, A8.

— "Injunction Won't Curb Us, Native Vows." *Vancouver Sun*, 27 June 1987, A12.

— "Natives Say Wilderness Now a Rotting Heritage." *Vancouver Sun*, 14 June 1988, A1.

— "New Native Road Blockade Joins Four in Skeena Valley." *Vancouver Sun* 14 October 1989, C16.

— "Province Must Alter Policy, Westar Head Tells Meeting." *Vancouver Sun* 27 October 1989, A16.

— "Show of Force by RCMP Angers Kispiox Area Chief." *Vancouver Sun*, 2 March 1988, B1.

— Telephone Interview, 29 August 2017, Victoria, Canada. 1 hour, 31 minutes.

— "To the Barricades." *Vancouver Sun*, 21 July 1989, B1.

— "Tofino Protesters Halt Work on Logging Road." *Vancouver Sun*, 14 June 1988, F7.

— "Tsawout in Front Lines of B.C. Treaties Battle." *Vancouver Sun*, 4 November 1985, B1.

— "Westar Joins Northwest Timber Protest: Tired of Government Stance, Firm Reveals." *Vancouver Sun*, 23 February 1990, B3.

Glavin, Terry. "B.C. Blockade Resembles Lubicon Confrontation." *Vancouver Sun*, 20 October 1988, D15.

Goldberg, Kim. "Clayoquot Crackdown." In *Witness to Wilderness: The Clayoquot Sound Anthology*, edited by et al Howard Breen-Needham. 197–201. Vancouver: Arsenal Press, 1994.

Goldman, Patti and Kristen Boyles, "Forsaking the Rule of Law: The 1995 Logging without Laws Rider and Its Legacy," *Environmental Law* 27, no. 4 (1997): 1035–96.

Goldstick, Michael. *Wollaston: People Resisting Genocide.* Montreal: Black Rose, 1987.

Gorman, Tom. "Earth First! Tactics in Fight to Save Planet Anger Some, Tickle Others." *Los Angeles Times*, 14 August 1988, 1, 8–10.

Gram, Karen. "Environmentalist Says Life Threatened." *Vancouver Sun*, 5 August 1988, A1.

Grant, Donald. "Natives to Resume Logging Road Blockades." *Globe and Mail* 3 November 1989, A13.

Green, Roger. *Battle for the Franklin.* Sydney: Fontana/Australian Conservation Foundation, 1984.

Grenfell, Damian. "The State and Protest in Contemporary Australia: From Vietnam to S11." PhD, Monash, 2001.

Guujaw. "Correspondence by email." 6 June 2017.

Haines, Herbert. "Black Radicalization and the Funding of Civil Rights: 1957–1970." *Social Problems* 31, no. 2 (1984): 31–43.

Hamilton, Gordon. "Protesters' Oakalla Transfer Delayed." *Vancouver Sun*, 15 March 1988, B5.

Harden, Mark. "'Tree Sitters' Protest Logging Earth First! Environmentalists Camp out in Branches near Granby." *Denver Post*, 15 August 1989, 1.

Harper, Catherine. "Logging Halt Extended." *Sydney Morning Herald*, 5 September 1979, 1.

— "Public Inquiry into Terania Logging." *Sydney Morning Herald*, 26 September 1979, 2.

Harris, David. *The Last Stand.* New York: Times Books, 1996.

Harris, John. "City Halts Removal of Trees." *Austin American Statesman*, 9 February 1990, B2.

Hayes, Mark. "Greenies and Government at Loggerheads." *Nation Review*, 30 August 1979, 775.

Hayter, Roger, and John Holmes. "The Canadian Forest Industry: The Impacts of Globalization and Technological Change." In *Canadian Forest Policy: Adapting to Change*, edited by Michael Howlett. 127–56. Toronto: University of Toronto Press, 2001.

Health, Brian. "What Did You Expect to Accomplish Anyway?". *Earth First! Journal* 5, no. 1 (November 1984): 5.

Heatley, Dave. "The Early 80s: A Rising Momentum." In *For the Forests: A History of the Tasmanian Forest Campaigns*, edited by Helen Gee. 209–12. Hobart: The Wilderness Society, 2001.

Heikens, Norm. "Earth First! Vows Intensified Anti-Clearcutting Maneuvers." *Southern Illinoisian*, 28 August 1989.

Henry, Lisa. "Ecotrans – Collective & Individual Action." *Earth First! Journal* 11, no. 3 (February 1991): 6.

Henton, Darcy. "Band Evicts Environmentalists from Logging Road Ministry, Construction Company Also Told to Leave." *Toronto Star*, 1 November 1989, A12.

— "Band Members Making Last Stand Today in Bid to Halt Forest Logging Road." *Toronto Star*, 11 November 1989, D5.

— "Group Endures Trip through Rugged Bush in Bid to Save Forest." *Toronto Star*, 17 September 1989, A8.

— "Native Demonstrators Celebrate as Temagami Road Work Halted." *Toronto Star* 12 November 1989, A2.

— "Northerners Blast Toronto Activists for Meddling in Temagami Wilderness." *Toronto Star.*, 6 June 1989, A10.

— "OPP Bill Almost $1 Million in Temagami Logging Fight." *Toronto Star*, 4 December 1989, A10.

— "Police Arrest 5 as Group Blocks Logging Road." *Toronto Star*, 5 June 1989, A3.

Hinke, C.J. Telephone Interview, 26 October 2017, Bangkok, Thailand. 1 hour, 8 minutes.

— "The Prison Experience of a Tree Protector." *Journal of Prisoners on Prisons* 2, no. 2 (1990): 1–4.

Hinson, Dan. "Staying Put," *Orlando Sentinnel*, 13 July 1985, A16.

Hoberg, George, and Edward Morawski. "Policy Change through Sector Intersection: Forest and Aboriginal Policy in Clayoquot Sound." *Canadian Public Administration* 40, no. 3 (1997): 387–414.

Hodgins, Bruce. "Contexts of the Temagami Predicament." In *Temagami: A Debate on Wilderness*, edited by Mark Bray and Ashley Thomson. 123–39. Toronto: Dundurn Press, 1990.

— "Gary G. Potts." In *Blockades and Resistance: Studies in Actions of Peace and the Temagami Blockades of 1988-1989*, edited by Bruce W Hodgins, Ute Lischke and David T McNab. 19–23. Ontario: Wilfrid Laurier University Press, 2003.

— "The Temagami Blockades of 1989." In *Blockades and Resistance: Studies in Actions of Peace and the Temagami Blockades of 1988-1989*, edited by Bruce W Hodgins, Ute Lischke and David T McNab. 23–30. Ontario: Wilfrid Laurier Univ. Press, 2003.

Hoeben, Sophia, Email Interview, 12 April 2015, Nimbin, Australia.

Holloway, G. *The Wilderness Society: The Transformation of a Social Movement Organization.* Hobart: Department of Sociology, University of Tasmania, 1986.

Horsfield, Margaret, and Ian Kennedy. *Tofino and Clayoquot Sound: A History.* Madeira Park: Harbour Publishing, 2014.

Horter, Will. The Real Story Behind Gwaii Haanas, Available [Online]: https://dogwoodbc.ca/the-real-story-behind-gwaii-haanas/ (Accessed 1 June 2017).

Howe, Renate. "Nobody but a Bunch of Mothers': Grassroots Activism and Women's Leadership in 1970s Melbourne." In *Seizing the Initiative: Australian Women Leaders in Politics, Workplaces and Communities*, edited by Rosemary Francis, Patricia Grimshaw, and Ann Standish. 331–40. Melbourne: University of Melbourne, eScholarship Research Centre, 2012.

Howlett, Michael, and Jeremy Rayner. "The Business and Government Nexus: Principal Elements and Dynamics of the Canadian Forest Policy Regime." In *Canadian Forest Policy: Adapting to Change*, edited by Michael Howlett. 23–64. Toronto: University of Toronto Press, 2001.

Huber, Ron. "Giant Crane Attacks Tree Sitter." *Earth First! Journal* 5, no. 7 (August 1985): 1, 4–6.

— "Tree Climbing Hero." *Earth First! Journal* 5, no. 6 (June 1985): 1, 4.

— Yggdrasil Survives, Available [Online]: www.penbay.org/ef/yggdrasil_rebirth85.html (Accessed 17 November 2015).

Hume, Mark. "5 Arrests Expected in Logging Protest." *Vancouver Sun*, 12 August 1988, A1.

— "Blockade Stops Logging: Indians to Pick Another Target." *Vancouver Sun*, 16 February 1988, E8.

— "End to Mining, Logging in Park Wins Plaudits." *Vancouver Sun*, 2 September 1988, A1.

Humphries, David. "Government Halts Police Swoop on Logging Protest." *The Age*, 30 January 1984, 3.

Hungerford, Alice, ed. *Upriver: Untold Stories of the Franklin River Activists* Maleny: UpRiver Mob, 2013.

Hunter, Justine. "Indians Set up Road Blockade near Hazelton." *Vancouver Sun*, 29 September 1989, F10.

Hutchins, Brett, and Libby Lester. "Environmental Protest and Tap-Dancing with the Media in the Information Age." *Media, Culture & Society* 28, no. 3 (2006): 433–51.

India Block, Modular Boxes used by Extinction Rebellion are Protest Architecture, Available [Online]: www.dezeen.com/2019/10/17/extinction-rebellion-protest-architecture/ (Accessed 1 September 2020).

Ingalsbee, Timothy. "Earth First! Activism: Ecological Postmodern Praxis in Radical Environmentalist Identities." *Sociological Perspectives* 39, no. 2 (1996): 263–76.

Irvine, Graham. "Creating Communities at the End of the Rainbow." In *Belonging in the Rainbow Region: Cultural Perspectives on the NSW North Coast*, edited by Helen Wilson. 63–82. Lismore: Southern Cross University Press, 2003.

Israel, Bill. "'Tarzan and Jane' Try to Delay Loggers." *San Francisco Chronicle*, 3 September 1987, 25.

Israelson, David. "Just Plain Folks: The Environment's New Champions." *Toronto Star*, 15 November 1987, B1.

— "Province yet to Rule on Temagami Plan." *Toronto Star*, April 5 1990, A4.

— "Thousands March in Metro to Celebrate Earth Day." *Toronto Star*, 23 April 1990, A1.

Jager, Karin and Gero Rueter, "Hambach Forest: Germany's Sluggish Coal Phaseout sparks anger", *Deutsche Welle*, 19 January 2020, Available [Online]: www.dw.com/en/hambach-forest-germanys-sluggish-coal-phaseout-sparks-anger/a-52059845 (Accessed 31 July 2020).

Jakubal, Mikal. "Civilized Disobedience", http://civilizeddisobedience.com/2012/03/21/the-postcard-and-the-portaledge/. (Accessed 31 May 2015).

— Telephone Interview, 2 May 2016, Eureka, USA. 1 hour, 56 minutes.

Jardine, Jeff. "Water War of Yore Still Resonates with New Melones Protester." *Modesto Bee*, 8 July 2015, 15.

Järvikoski, Tino. "Alternative Movements in Finland: The Case of Koijarvi", *Acta Sociologica*, January (1981): 313–315.

Jasper, James. *The Art of Moral Protest: Culture, Biography, and Creativity in Social Movements.* Chicago: The University of Chicago Press, 1997.

— "Emotions and Social Movements: Twenty Years of Theory and Research." *Annual Review of Sociology* 37 (2011): 1–29.

Jasper, James, and Francesca Polletta. "Collective Identity and Social Movements." *Annual Review of Sociology* 27 (2001): 283–305.

Johnson, Hayley. "# NoDAPL: Social Media, Empowerment, and Civic Participation at Standing Rock." *Library Trends* 66, no. 2 (2017): 155–175.

Joiner, R. *Errinundra Plateau: Resolution of Conflict.* Melbourne: Government Printer, 1984.

Jones, Peter, Bryan Law, and Margaret Pestorius. "The Story of the Australian Nonviolence Network", Available [Online]: www.nonviolence.org.au/downloads/ann_story.pdf (Accessed 19 March 2010).

Judd, Mike. Telephone Interview, 24 September 2017, Pincher Creek, Canada. 31 minutes.

Kahn, P. (pseudonym). "Last Stand on Boggy Creek," *EFJ* 6, no. 3 (1986): 5.

Kaufman, Cynthia. *Ideas for Action: Relevant Theory for Change.* Cambridge: South End Press, 2003.

Kavanagh, Jean. "Jail Fails to Fall Logging Protest." *Vancouver Sun*, 10 August 1988, B5.

Kendell, Jeni, and Eddie Buivids. *Earth First.* Sydney: ABC Enterprises, 1987.

Kendell, Jeni, and Paul Tait. *Give Trees a Chance.* Gaia Films, 1980.

Kennedy, Peter. "State Moves to Curb Forest Protest." *Sydney Morning Herald*, 8 September 1979, 3.

Kerr, Andy. "Oregon Rare II Suit." *Earth First! Journal* September (1983): 1, 4.

Killan, Gerald. "The Development of a Wilderness Park System in Ontario, 1967–1990: Temagami in Context." In *Temagami: A Debate on Wilderness (Toronto, 1990)*, edited by Mark Bray and Ashley Tomson. 85–120. Tornoto: Dundurn Press, 1990.

King, Greg. "Freddies Set Their Sights High: Kalmiopsis Tree-Sitters Targeted." *Earth First! Journal* 8, no. 8 (September 1988): 1, 5.

— "New Battles in Maxxam Campaign." *Earth First! Journal* 8, no. 6 (June 1988): 3.

— "Old Growth Redwood." *Earth First! Journal* 7, no. 2 (December 1987): 9.

— "Redwood Tree Climbers." *Earth First! Journal* 7, no. 8 (September 1987): 1,6.

Kirkpatrick, Daniel. "Washington Old Growth Campaign." *Earth First!* 7, no. 8 (September 1987): 1, 4.

Kneen, Jamie. Web Post Regarding Wollaston Lake, Available [Online]: http://sisis.nativeweb.org/clark/nov1198can.html [Date Accessed 17 November 2016].

Knights, Sam. "The Story So Far." In *This Is Not A Drill: An Extinction Rebellion Handbook*, edited by Extinction Rebellion. 16–21. London: Penguin, 2019,

Koehler, Bart. "The Battle of Salt Creek." *Earth First!* 3, no. 2 (December 1982): 1.

— "Bisti Mass Trespass." *Earth First!* 3, no. 2 (December 1982): 11.

Kriesi, Hanspeter, Ruud Koopmans, Jan Willem Dyvendak, and Marco Giugni. *New Social Movements in Western Europe: A Comparative Analysis*. Minneapolis: University of Minnesota, 1995.

Krinsky, Ambrosia. "Mike Roselle, Earth First Founder 2013 Interview." www.youtube.com/watch?v=qfBQc6HO0Ys, 2013, 8:57.

Ladner, Kiera. "Aysaka'paykinit: Contesting the Rope around the Nations' Neck." In *Group Politics and Social Movements in Canada*, edited by Miriam Smith. 227–55. Toronto: University of Toronto Press, 2008.

Langelle, Orin. "EF!ers Face Jail for Defending Illinois Hardwoods." *Earth First! Journal* 10, no. 1 (November 1989): 7, 8.

— "Shawnee Saga Continues." *Earth First! Journal* 10, no. 8 (September 1990): 13.

— Telephone Interview, 13 May 2016, Buffalo, USA. 1 hour, 29 minutes.

Langelle, Orin, and John Wallace. "Showdown on the Shawnee." *Earth First! Journal* 10, no. 7 (August 1990): 1.

Langer, Valerie. "It Happened Suddenly (over a Long Period of Time)." In *Witness to Wilderness: The Clayoquot Sound Anthology*, edited by Howard Breen-Needham et al. 251–55. Vancouver: Arsenal Press, 1994.

Laronde, Mary. Interview. 20 September 2017, North Bay, Canada. 1 hour, 28 minutes.

Larson, Jeff. "Social Movements and Tactical Choice." *Sociology Compass* 7 (2013): 866–79.

Lawson, James. "Space, Strategy and Surprise." In *Blockades and Resistance: Studies in Actions of Peace and the Temagami Blockades of 1988-1989*, edited by Bruce W Hodgins, Ute Lischke and David T McNab. 157–92. Ontario: Wilfrid Laurier University Press, 2003.

Lawson, Susanne Hare. Email Interview, 15 October 2017, Tofino, Canada.

Lawrence, William. "Nuxalk People Obstruct Logging of ISTA Old-Growth Forest, 1995–1998", Available [Online]: https://nvdatabase.swarthmore.edu/content/nuxalk-people-obstruct-logging-itsa-old-growth-forest-1995-1998 (Accessed 1 November 2020).

Leaney, Rachel. "200 Environmentalists Given Training to Resist Temagami Logging." *Globe and Mail* 18 September 1989, A12.

Leggett, Dudley. "Green Alliance." *Sunshine News*, October 1980, 13–14.

— Telephone Interview, 5 February 2015, Byron Bay, Australia. 1 hour, 37 minutes.

Leonard, Jim. "Trees Cut on Mt. Graham." *Earth First! Journal* 11, no. 1 (November 1990): 13.

Liddiard, Brenda. Telephone Interview, 2 February 2015, Wellington, New Zealand. 58 minutes.

Lindeman, Tracey. "'Revolution Is Alive': Canada Protests Spawn Climate and Indigenous Rights Movement." *Guardian*, 28 February 2020, Available [Online]: www.theguardian.com/world/2020/feb/28/canada-pipeline-protests-climate-indigenous-rights (Accessed 1 November 2020).

Lindsay, Judy. "Separating Strathcona Antagonists Not Easy." *Vancouver Sun*, 18 February 1988, F1.

— "In Illinois Forest, Court Halts Cutting." *Chicago Tribune*, 6 September 1990, 17.

— "Two Arrested in Bid to Block Logging in Illinois." *Chicago Tribune*, 17 August 1990, 20.

— "Nonviolence and the Southwest." *Groundswell: Newsletter for a New Society* 1, no. 2 (January 1983): 6–8.

Lischke, Ute, and David T McNab. "Actions of Peace." In *Blockades and Resistance: Studies in Actions of Peace and the Temagami Blockades of 1988-1989*, edited by Bruce W Hodgins. 1–12. Ontario: Wilfrid Laurier University Press, 2003.

Loose Hip Circles (pseudonym). "Riotous Rendezvous Remembered." *Earth First! Journal* 9, no. 7 (August 1989): 19.

Love, Sam, and David Obst. *Ecotage*. New York: Pyramid, 1972.

Lukacs, Martin. From Queen Charlotte to Haida Gwaii, Available [Online]: www.dominionpaper.ca/articles/3248 (Accessed 27 October 2016).

Lunney, Daniel. "The Eden Woodchip Debate: Part II (1987-2004)" (paper presented at the 6th National Conference of the Australian Forest History Society, 2005).

McAdam, Doug. "The Framing Function of Movement Tactics: Strategic Dramaturgy in the American Civil Rights Movement." In *Comparative Perspectives on Social Movements: Political Opportunities, Mobilizing Structures and Cultural Meanings*, edited by Doug McAdam, John McCarthy and Mayer Zald. Cambridge: Cambridge University Press, 1996.

— "Tactical Innovation and the Pace of Insurgency." *American Sociological Review* 48 no. 6 (1983): 735–54.

McAdam, Doug, and Sidney Tarrow, "Ballots and Barricades: On the Reciprocal Relationship between Elections and Social Movements." *Perspectives on Politics* 8, no. 2 (2010): 529–542.

McAdam, Doug, Sidney Tarrow, and Charles Tilly. *Dynamics of Contention*. Cambridge: Cambridge University Press, 2001.

McCarthy, John D., and Mayer N. Zald. "Resource Mobilization and Social Movements: A Partial Theory." *American Journal of Sociology* 82, no. 6 (1977): 1212–41.

McClaren, Christie. "Indians Agree to End Blockade after Appeal Court Decision," *Globe and Mail*, 8 December 1988, A9.

— "New Conflict Looms on Logging Roads," *Globe and Mail* 22 November 1988, A12.

— "Ontario Government Starts Cutting Trees in Temagami Area," *Globe and Mail*, 21 March 1989, A5

— "Ontario Going to Court in Indian Dispute," *Globe and Mail*, 30 November 1988, A9

— "Plan Shares Logging Control with Indians," *Globe and Mail* 28 September 1988, A5.

— "South Moresby Logging Plan Sparks Ire." *Globe and Mail*, 21 October 1985, A5.

— "Temagami Indians Set to Block Logging Road," *Globe and Mail*, 24 March 1989, A5

McGregor, Adrian. "8 Arrests as Dozer Team Cuts Road." *Courier Mail*, 3 December 1983, 3.

— "Treetop Protest Stops Dozers." *Sunday Mail*, 4 December 1983, 5.

McGregor, Craig. "The Battle for Terania Creek." *National Times*, 28 August 1979, 7.

— "Picnic at Terania Creek." *National Times*, 9 June 1979, 14–15.

McInnes, Craig. "53 Parks to Be Created in Ontario." *Globe and Mail*, 18 May 1988, A1.

— "Hearing to Assess All Aspects of Forest Industry in Ontario." *Globe and Mail*, 10 May 1988, A15.

McIntyre, Iain. *Always Look on the Bright Side of Life: The AIDEX '91 Story*. Melbourne: Homebrew Press, 2008.

— Environmental Blockading Timeline, 1974–1997 Available [Online]: https://commonslibrary.org/environmental-blockading-in-australia-and-around-the-world-timeline-1974-1997/ [Date Accessed 8 November 2020]

McNab, David T. "Remembering an Intellectual Wilderness." In *Blockades and Resistance: Studies in Actions of Peace and the Temagami Blockades of 1988-1989*, edited by Bruce W Hodgins, Ute Lischke and David T McNab. 31–53. Ontario: Wilfrid Laurier University Press, 2003.

McQueen, James. *The Franklin: Not Just a River*. Ringwood: Penguin, 1983.

Maddison, Sarah, and Sean Scalmer. *Activist Wisdom: Practical Knowledge and Creative Tension in Social Movements* Sydney: UNSW Press, 2005.

Mack, Jacinda. *Remembering ISTA: Nuxalk Perspectives on Sovereignty and Social Change*. Toronto: York University, 2006.

Manes, Christopher. *Green Rage: Radical Environmentalism and the Unmaking of Civilization*. Boston, MA: Little Brown, 1990.

— "Overpopulation and Industrialism." *Earth First! Journal* 7, no. 4 (March 1987).

— "Population and AIDS." *Earth First! Journal* 7, no. 5 (May 1987).

Mann, Michaela. *Clearcut Conflict: Clayoquot Sound Campaign and the Moral Imagination*. Ontario: Saint Paul University, 2013.

Mansell, Michael. "Comrades or Trespassers on Aboriginal Land?". In *The Rest of the World Is Watching*, edited by Cassandra Pybus and Richard Flanagan. 101–6. Chippendale: Pan Macmillan, 1990.

Marr, Alec. "The Picton Blockades of 1986 and 1987." In *For the Forests: A History of the Tasmanian Forest Campaigns*, edited by Helen Gee. 174–76. Hobart: Wilderness Society, 2001.

Marsden, Steve. "Freddies Attack North Kalmiopsis Again." *Earth First! Journal* January (1985), 4.

— "The Weather from Bald Mounatin." *Earth First! Journal* August (1984): 9.

Marten, Robert. "A Hunting We Will Go." *Earth First! Journal* 12, no. 1 (November 1991): 26–27.

Martin, Douglas. Obituary: James Phillips, 70, Environmentalist Who Was Called the Fox, Available [Online]: www.nytimes.com/2001/10/22/us/james-phillips-70-environmentalist-who-was-called-the-fox.html?_r=0 (Accessed 16 December 2015).

Martin, Pauline. "Seven Protesters Arrested." *Campbell River Upper Islander*, 2 February 1988, 1.

Marty, Sid. *Leaning on the Wind: Under the Spell of the Great Chinook*. Victoria: Heritage House Publishing Co, 2011.

May, Elizabeth. *Paradise Won: The Struggle for South Moresby*. Toronto: McClelland and Stewart, 1990.

Meehan, Robyn. "On the Beach." *Maggies Farm*, October 1980, 12–13.

Melucci, Alberto. *Challenging Codes: Direct Action in the Information Age*. Cambridge: Cambridge University Press, 1996.

— *Nomads of the Present: Social Movements and Individual Needs in Contemporary Society*. London: Hutchison Radius, 1989.

Meredith, Peter. *Myles and Milo*. St Leonards: Allen & Unwin, 1999.

Meyer, David. "Protest and Political Opportunities." *Annual Review of Sociology* 30 (2004): 125–45.

— "Scholarship That Might Matter." In *Rhyming Hope with History: Activists, Academics and Social Movement Scholarship*, edited by David Croteau and William Hoynes. 191–205. Minneapolis: University of Minneapolis Press, 2005.

Michaelson, Alaina. "The Development of a Scientific Speciality of Diffusion through Social Relations: The Case of Role Analysis." *Social Networks* 15 (1993): 217–36.

Michelin, Lana. "Temagami Protesters Urged to Go." *Globe and Mail*, 6 November 1989, A12.

Middle Head Sand Mining Action Group. "Meeting Minutes." 22 September 1980, 1–3.

Mitchell, Robert, Angela Mertig, and Riley Dunlap. "Twenty Years of Environmental Mobilization: Trends among National Environmental Organizations." In *American Environmentalism,* edited by Riley Dunlap and Angela Mertig. New York: Taylor & Francis, 1992.

Moir, Rita. "CP Loses Permit to Spray B.C. Track with Herbicide." *Globe and Mail*, 23 July 1987, A4.

Molloy, Andy. "Activists Assaulted at Tract Pond." *Earth First! Journal* 11, no. 2 (December 1990): 1, 7.

Moore, Joseph G. "Two Struggles into One? Labour and Environmental Movement Relations and the Challenge to Capitalist Forestry in British Columbia, 1900-2000." PhD, McMaster University, 2002.

Morgan, Peter. "Contested Native Forests: A Theoretical and Empirical Study." PhD thesis, RMIT, 1997.

Morrison, Patt. "Terrorists or Saviors?" *LA Times*, 16 June 1991.

Morrow, Shayne. "Chief Councillor Remembers the First Days of a Long Battle on Meares." In *Ha-Shilth-Sa* (2014). Published electronically 25 April 2014. www.hashilthsa.com/news/2014-04-25/chief-councillor-remembers-first-days-long-battle-meares (Accessed 22 August 2016).

Morton, Nancy, and Dave Foreman. "Good Luck, Darlin'. It's Been Great.". *Earth First! Journal* 10, no. 8 (September 1990): 5.

Murphy, Simon. "Environmental Protesters Block Access to Parliament Square." *Guardian*, 24 November 2018, Available [Online]: www.theguardian.com/environment/2018/nov/24/environmental-protesters-block-access-to-parliament-square-extinction-rebellion (Accessed 1 September 2020).

Naess, Arne. "A Defence of the Deep Ecology Movement." *Environmental Ethics* 6, no. 3 (1984): 265–70.

National Policing Improvement Agency, *ACPO Manual of Guidance on Dealing with the Removal of Protestors: Faslane 365 2006 – 2007.* Bedfordshire: NPIA, 2007.

Native Forest Network. Temagami Chronology, Available [Online]: http://temagami.nativeweb.org/temagami-chronology.htm (Accessed 26 December 2016).

Negri, Sam. "Earth First! Founder Quits Group over Rhetoric." *Arizona Republic*, 15 August 1990, 20.

Nicholson, Nan. Telephone Interview, 3 August 2015, Terania Creek, Australia. 53 minutes.

— *Terania Creek*, Unpublished manuscript, 1982.

Nightcap Action Group. *Nightcap Handbook*. Nimbin: NAG, 1982.

— "Press Release: 4 August." 1982, 1.

Noreen, Barry. "Earth First! Members Divided by Differences in Philosophy." *Colorado Spring Gazette*, 14 October 1990, B1.

North Coast EF! "Press Release: Earth First! Responds to Timber Industry Propaganda Assault." July 1987, 1.

North East Forest Alliance. *Intercontinental Deluxe Guide to Blockading*. Lismore: North East Forest Alliance, 1993.

— "NEFA Meeting Minutes: 29 September," (NEFA, 1991).

— "NEFA Report: May", (NEFA, 1992)

Notzke, Claudia. *Aboriginal Peoples and Natural Resources in Canada*. North York: Captus Press, 1994.

NSW Department of Education. Australian Environmental Activism Timeline, Available [Online]: www.teachingheritage.nsw.edu.au/section03/timeenviron.php (Accessed 2 June 2014).

Olive, Andrea. *The Canadian Environment in Political Context*. Toronto: University of Toronto Press, 2016.

O'Neil, John D, Brenda D. Elias, and Annalee Yassi. "Situating Resistance in Fields of Resistance: Aboriginal Women and Environmentalism." In *Pragmatic Women and Body Politics*, edited by Margaret Lock and Patricia Kaufert. 260–86. Cambridge: Cambridge University Press, 1998.

Ongerth, Steve. Redwood Uprising Chapter 3: He Could Clearcut Forests Like No Other, Available [Online]: http://ecology.iww.org/texts/SteveOngerth/RedwoodUprising/3 (Accessed 30 March 2016).

— Redwood Uprising: Chapter 4: Maxxam's on the Horizon, Available [Online]: http://ecology.iww.org/texts/SteveOngerth/RedwoodUprising/4 (Accessed 20 March 2016).

— Redwood Uprising, Chapter 7: Way up High in the Redwood Giants, Available [Online]: http://ecology.iww.org/texts/SteveOngerth/RedwoodUprising/7 (Accessed 20 March 2016).

— Redwood Uprising, Chapter 11: I Knew Nothin' Till I Met Judi, Available [Online]: http://ecology.iww.org/texts/SteveOngerth/RedwoodUprising/11 (Accessed 20 May 2106).

— Redwood Uprising, Chapter 33: The Ghosts of Mississippi Will Be Watchin', Available [Online]: http://ecology.iww.org/texts/SteveOngerth/RedwoodUprising/33 (Accessed 20 May 2016).

Ord, Bill. "Cape Road Fight 'All Over'." *Courier Mail*, 14 December 1983, 3.

— "Forest Protester Tied to Cross." *Courier Mail*, 13 December 1983, 3.

— "Tenni: Forest Homework Fails." *Courier Mail*, 15 December 1983, 10.

O'Rizay, Mike. "Freddies Murder Millenium Grove." *Earth First! Journal* 6, no. 6 (June 1986): 13.

Osborne, Robin. "Confrontation at the Beach-Head," *National Times*, 13–19 July 1980, 5

Ostrow, Cecelia. "Letter from Oregon Jail." *Earth First! Journal* 5, no. 2 (December 1984): 7.

Paasonen, Karl-Erik. Telephone Interview, 7 October 2015, Canberra, Australia. 1 hour, 7 minutes.

Paehlke, Robert. "The Canadian Environmental Movement." In *Group Politics and Social Movements in Canada*, edited by Miriam Smith. 283–304. Toronto: University of Toronto Press, 2014.

Palmer, Tim. *Stanislaus: The Struggle for a River*. Berkeley: University of California Press, 1982.

Parfitt, Ben. "Blockade Observers at Loggerheads over Danger Posed by Man with Axe." *Vancouver Sun*, 6 October 1988, A13.

— "Tofino Police under Order to Nab Logging Protesters." *Vancouver Sun*, 28 June 1988, B7.

— "Indians' Trenches Shut Road to Park." *Vancouver Sun*, 13 July 1987, B12.

Parks, Andy. "Environmental Protest Songs of North East NSW 1979-1999." Honors, Southern Cross University, 1999.

Paterson, Matthew. "Swampy Fever: Media Constructions and Direct Action Politics," in *Direct Action in British Environmentalism*, 151–166.

Patterson, John, and Jean Ravine. "EF Shuts Down Uranium Mine." *Earth First! Journal* 7, no. 7 (August 1987): 1,4.

Peter, Ian. Telephone Interview, 8 June 2015, Sydney, Australia. 1 hour, 9 minutes.

Pickett, Karen. "Breaking up or Breaking Apart?". *Earth First! Journal* 11, no. 1 (November 1990).

— "Day of Outrage Shakes Forest Service Nationwide!". *Earth First! Journal* 8, no. 6 (June 1988): 1, 19.

— "Direct Action Fund 1990 Report." *Earth First! Journal* 12, no. 4 (March 1991): 34.

Pickett, Karen, and Woody Joe. "Redwood Summer Goes On!". *Earth First! Journal* 10, no. 6 (June 1990): 1.

Pickett, Karen, Mike Roselle, Doug Norlen, and Anonymous. "Blockade Personal Accounts." *Earth First! Journal* June (1983): 1, 6–7.

Pickett, Nelson. "Earth First! Padlock Fails to Halt Clear-Cut Start." *Oregonian*, 1 August 1990, B2.

Platiel, Rudy. "Ontario Native Band Blocks Planned Road into Disputed Region." *Globe and Mail*, 2 June 1988, N12.

Plows, Alexandra. "Praxis and Practice: The 'What, How and Why' of the U.K. Environmental Direct Action (E.D.A.) Movement in the 1990s." PhD thesis, University of Wales, 2002.

Polletta, Francesca. *It Was Like a Fever: Storytelling in Protest and Politics*. Chicago, IL: University of Chicago Press, 2006.

Pugh, Dailan. "Correspondence by email.", 16 November 2020

— Telephone Interview, 8 June 2015, Byron Bay, Australia. 56 minutes.

Rae, Michael. "Daintree Road." *Wilderness News* (April 1985): 3.

Rakestraw, Lawrence, and Mary Rakestraw. *History of the Willamette National Forest*. Eugene, OR: USDA Forest Service Pacific Northwest Region, 1991. www.foresthistory.org/ASPNET/Publications/region/6/willamette/chap6.htm.

Ramos, Howard. "What Causes Canadian Aboriginal Protest? Examining Resources, Opportunities and Identity, 1951-2000." *The Canadian Journal of Sociology* 31, no. 2 (2006): 211–34.

Rawls, John. *A Theory of Justice*. Cambridge MA: Belknap Press, 1999.

RCMP. Organizational Structure, Available [Online]: www.rcmp-grc.gc.ca/about-ausujet/organi-eng.htm (Accessed 7 February 2017).

Reclaim The Streets, Propaganda, Available [Online]: http://rts.gn.apc.org/prop15.htm (Accessed 10 October 2010).

Reinsborough, Patrick. "Post Issue Activism." In *Globalize Liberation: How to Uproot the System and Build a Better World*, edited by David Solnit. 161–212. San Francisco, CA: City Lights Books, 2004.

Restless, Randall. "First American Tripod is Constructed in Cove/Mallard," *EFJ* 21, no. 1 (2000): 56–57.

Revkin, Andrew. *The Burning Season: The Murder of Chico Mendes and the Fight for the Amazon Rainforest*. Washington, DC: Island Press, 2004.

Ricketts, Aiden. "Om Gaia Dudes: The North East Forest Alliance's Old Growth Forest Campaign." Chap. Om Gaia Dudes: The North East Forest Alliance's Old Growth Forest Campaign In *Belonging in the Rainbow Region: Cultural Perspectives on the N.S.W. North Coast*, edited by Helen Wilson. 121–48. Lismore, NSW: Southern Cross University Press, 2003.

Road Alert. *Road Raging: Top Tips for Wrecking Roadbuilding*. London: Road Alert!, 1997.

Robinson, Michael. "The Battle for Murrelet Grove." *Earth First! Journal* 10, no. 8 (September 1990): 15.

Rodgers, Kathleen. *Welcome to Resisterville: American Dissidents in British Columbia*. Vancouver: UBC Press, 2014.

Rodrigues, Gomercindo. *Walking The Forest with Chico Mendes: Struggle for Justice in the Amazon*. Austin: University of Texas, 2007.

Rogers, Nicole. "Law, Order and Green Extremists." In *Green Paradigms and the Law*, edited by Nicole Rogers. 142–62. Lismore, NSW: Southern Cross University, 1998.

Roggeband, Conny. "Translators and Transformers: International Inspiration and Exchange in Social Movements." *Social Movement Studies* 6, no. 3 (2007): 245–59.

Roland, Paul. "Breitenbush Blockade Draws National Attention to Ancient Forests." *Earth First! Journal* 9, no. 5 (May 1989): 6.

Rootes, Christopher. "Acting Locally: The Character, Contexts and Significance of Local Environmental Mobilisations." *Environmental Politics* 16, no. 3 (2007): 722–741.

— "Exemplars and Influences: Transnational Flows in the Environmental Movement." *Australian Journal of Politics and History* 61 no. 3 (2015): 414–431.

— "From Local Conflict to National Issue: When and How Environmental Campaigns Succeed in Transcending the Local." *Environmental Politics* 22, no. 1 (2013): 95–114.

Rosen, Robert. Telephone Interview, 29 January 2015, Sydney, Australia. 1 hour, 35 minutes.

Roselle, Mike. "Earth First! Takes Regional Forester's Office." *Earth First! Journal* November (1984): 1.

— "Guest Editorial: Nomadic Action Group." *Earth First! Journal* 7, no. 8 (September 1987): 3.

— "Meares Island: Canada's Old Growth Struggle." *Earth First! Journal* February (1985): 1, 5.

— "Middle Santium Struggle Continues." *Earth First! Journal* August (1984): 4.

— "Oregon Trials." *Earth First! Journal* 5, no. 2 (December 1984): 3–4.

— "Roadkill." *Earth First! Journal* 10, no. 3 (February 1990): 27–28.

— "Sinkyone Struggle Continues." *Earth First! Journal* 4, no. 1 (December 1983): 10.

— "Tree Huggers Save Redwoods." *Earth First! Journal* 4, no. 1 (November 1983): 1, 4.

— *Tree Spiker: From Earth First! To Lowbagging, My Struggles in Radical Environmental Action.* New York: St. Martin Press, 2009.

Roselle, Mike, and Karen Pickett. "Direct Action Fund: The Year in Review." *Earth First! Journal* 9, no. 4 (March 1989): 26.

Runciman, Claire, Harry Barber, Linda Parlane, Gill Shaw, and John Stone. *Effective Action for Social Change.* East Ringwood, Victoria, Australia: ACF Books, 1986.

Ryberg, Erik. "Lessons from Mt. Graham." *Earth First! Journal* 10, no. 3 (February 1990): 8.

Scalmer, Sean. *Gandhi in the West: The Mahatma and the Rise of Radical Protest.* Cambridge: Cambridge University Press, 2011.

— "Translating Contention: Culture, History, and the Circulation of Collective Action." *Alternatives: Global, Local, Political* 25, no. 4 (2000): 491–514.

Scarce, Rik. *Eco-Warriors: Understanding the Radical Environmental Movement.* Chicago, IL: Noble Press, 1990.

Scholtz, Chrissa. *Negotiating Claims: The Emergence of Indigenous Land Claim Negotiation in Australia, Canada, New Zealand and the United States.* New York: Routledge, 2006.

Schweingruber, David. "Mob Sociology and Escalated Force: Sociology's Contribution to Repressive Police Tactics." *The Sociological Quarterly* 41, no. 3 (2000): 371–89.

Seed, John. "Australia Reports In!". *Earth First!* 2, no. 8 (1982): 9.

— "Introduction." *World Rainforest Report*, no. 2 (August 1984): 1.

Seel, Benjamin, Matthew Patterson, and Brian Doherty. "Direct Action in British Environmentalism." In *Direct Action in British Environmentalism*, edited by Benjamin Seel, Matthew Patterson, and Brian Doherty. 1–24. London: Routledge, 2000.

Seel, Benjamin, and Alexandra Plows. "Coming Live and Direct: Strategies of Earth First!". In *Direct Action in British Environmentalism*, edited by Benjamin Seel, Matthew Paterson, and Brian Doherty. 112–32. London: Routledge, 2000.

Sewell Jr, William H. *Logics of History: Social Theory and Social Transformation.* Chicago, IL: University of Chicago Press, 2009.

Sewell, W.R.D., P. Dearden, and J. Dumbrell. "Wilderness Decisionmaking and the Role of Environmental Interest Groups: A Comparison of the Franklin Dam, Tasmania and South Moresby, British Columbia Cases." *Natural Resources Journal (USA)* 29, no. 1 (1989).

Shabecoff. "The Acre by Acre Effort to Save the Environment." *New York Times*, 30 December 1984, A2.

Sharp, Gene. *The Politics of Nonviolent Action.* Boston, MA: Extending Horizons Books, 1973.

Sickler, Linda. "Activists Fight the Sale of Timber." *Southern Illinoisan*, 16 July 1990, 3.

Silvaggio, Anthony. "The Forest Defense Movement, 1980-2005: Resistance at the Point of Extraction, Consumption and Production." PhD thesis, University of Oregon, 2005.

Simper, Errol. "17 Arrested as Police Break Forest 'Sit-In'." *The Australian*, 18–19 August 1979, 3.

— "Sit-in Blockade by Forest Lovers to Stop Loggers." *The Australian*, 17 August 1979, 2.

Sister Extraterrestial (pseudonym). "Earth First! Activists Conference." *Earth First! Journal* 8, no. 4 (March 1988): 9.

Smith, Dan. "Temporary Truce Calms Haida-B.C. Dispute." *Toronto Star*, 19 December 1985, A18.

Smith, Marlene. Telephone Interview, 24 November 2017, Comox Valley, Canada. 1 hour, 10 minutes.

Smith, Wes. "Tangled Woods " *Chicago Tribune*, 17 May 1990, 1.

Socratrees (pseudonym). "Tactical Thoughts on the Maxxam Protests." *Earth First! Journal* 7, no. 6 (June 1987): 5.

Sollund, Ragnhild Aslaug "Introduction: Critical Green Criminology – an Agenda for Change." In *Green Harms and Crimes: Critical Criminology in a Changing World*, edited by Ragnhild Aslaug Sollund. 1–26. Basingstoke: Palgrave Macmillan, 2015.

Somerville, James. *Saving the Rainforest: The NSW Campaign 1973-1984*. Narrabeen: James Somerville, 2005.

Soule, Sarah. "Diffusion Processes within and across Movements." In *Blackwell Companion to Social Movements*, edited by David Snow, Sarah Soule, and Hanspeter Kriesi. 294–310. Oxford: Blackwell, 2004.

South East Conservation Working Group. "Media Release: 18 September," (SECWG, 1989).

—"Press Release 20 March 1989," (SECWG, 1989)

South East Forest Alliance. "SEFA Activity Sheet: 9–14 August 1989," (SEFA, 1989).

— South East Forest Campaign Strategy Paper, (SEFA, 1989).

— *South-East Forests Campaign Handbook*, (SEFA, 1988).

Spain, David. "David Spain's Terania." *Nimbin News*, 3 September 1979, 10–11.

— "Report from Terania Basin." *Nimbin News*, 27 August 1979, 18–19.

Speece, Darren. "Defending Giants: The Battle over Headwaters Forest and the Transformation of American Environmental Politics, 1850 to 1999." Master's thesis, University of Maryland, 2010.

Speirs, Rosemary. "Parks Issue Catches Peterson in a Crossfire." *Toronto Star*, 16 January 1988, D1.

Sprig. "Our Kind Live in Forests to Save Forests from Our Kind." *Earth First! Journal* November–December (2000): 28–29.

Stammer, Larry. "Environment Radicals Target of Probe into Lumber Mill Accident." *Los Angeles Times*, 15 May 1987, 3.

Starr, Amory. "'… (Excepting Barricades Erected to Prevent Us from Peacefully Assembling)': So-Called 'Violence' in the Global North Alterglobalization Movement." *Social Movement Studies* 5, no. 1 (2006): 61–81.

— *Global Revolt: A Guide to the Movements against Globalization*. London, New York: Zed Books, 2005.

Steel, Suzanne. "Logging Ended in Prime Temagami Area." *Gazette*, 24 April 1990, B1.

Stentz, Zack. "Osprey Grove Falls." *Earth First! Journal* 11, no. 1 (November 1990): 9–10.

Stephens, Peter. "Group Puts up Fight to Save Rainforest." *Australian*, 8 May 1979.

Stevens, Christi. "Daybreak Dozer Occupation." *Earth First! Journal* 8, no. 8 (September 1988): 7.

Stevens, Christi, and Barbara Dugelby. "Cavebugs Saved from Oblivion!". *Earth First! Journal* 9, no. 1 (September 1988): 1,4.

Stickells, Lee "Negotiating Off-Grid," *Fabrications: The Journal of the Society of Architectural Historians, Australia and New Zealand* 25, no. 1 (2015): 104–29.

Stiehm, Judith. "Nonviolence Is Two." *Sociological Inquiry* 38, no. 1 (1968).

Streetly, Joanna. No Pasaran: The Fight for Sulphur Passage, Available [Online]: www.tofino-bc.com/geography/clayoquot-protests-sulphur-passage.php (Accessed 22 August 2016).

Swanson, Peter. "EF!: Violence or Non-Violence?". *Earth First!* 3, no. 7 (September 1983): 12.

Swedlow, Brendon. "Scientists, Judges, and Spotted Owls: Policymakers in the Pacific Northwest." *Duke Environmental Law and Policy Forum* 13, no. 1 (2002): 187–278.

Szveteca, Annie. "Activists· Defend Hawaii's Last Rainforest." *Earth First! Journal* 10, no. 7 (August 1990): 10.

T.A. "Blockade #6." *Earth First! Journal* (August 1983): 9.

Tahoma (pseudonym), "Cascadia Rising!," *EFJ* November-December: 14–15, 73.

Tarrow, Sidney. *Power in Movement: Social Movements and Contentious Politics*. Third ed. Cambridge: Cambridge University Press 2011.

Taylor, Bron. "Earth First! And Global Narratives of Popular Ecological Resistance." In *Ecological Resistance Movements*, edited by Bron Taylor. 11–34. Albany, NY: SUNY Press, 1995.

Taylor, Diane. "Extinction Rebellion Activists Claim Victory in HS2 Tree Protest", *Guardian*, 27 April 2019, Available [Online]: www.theguardian.com/environment/2019/apr/27/extinction-rebellion-activists-scale-trees-in-anti-hs2-protest (Accessed 1 September 2020).

Taylor, Matthew. "The Evolution of Extinction Rebellion," *Guardian*, 4 August 2020; Available [Online]: www.theguardian.com/environment/2020/aug/04/evolution-of-extinction-rebellion-climate-emergency-protest-coronavirus-pandemic (Accessed 1 November 2020).

Taylor, Matthew, and Damien Gayle. "Dozens Arrested After Climate Protest Blocks Five London Bridges." *Guardian*, 17 November 2018, Available [Online]: www.theguardian.com/environment/2018/nov/17/thousands-gather-to-block-london-bridges-in-climate-rebellion." (Accessed 1 September 2020).

— "Tourism, Lumber Industries Clash over Plan to Build Logging Road in Northern Ontario Wilderness." *Toronto Star*, 5 May 1988, A24.

Taylor, Verta, and Nella Van Dyke. "'Get up, Stand Up': Tactical Repertoires of Social Movements." In *The Blackwell Companion to Social Movements*, edited by David Snow, Sarah Soule, and Hanspeter Kriesi. 262–93. Malden: Blackwell Publishing Ltd, 2004.

Terpstra, Jan. "Policing Protest and the Avoidance of Violence: Dilemmas and Problems of Legitimacy." *Journal of Criminal Justice and Security* 8, no. 3–4 (2006): 203–12.

Thomas, Chant. "Return to Bald Mountain." *Earth First! Journal* 7, no. 4 (March 1987): 1, 4.

Thompson, Peter. *Bob Brown of the Franklin River* Sydney: George Allen & Unwin, 1984.

Tilly, Charles. *Contentious Performances*. Cambridge: Cambridge University Press, 2008.

— *From Mobilization to Revolution*. Reading, UK: Addison-Wesley, 1978.

— *Regimes and Repertoires*. Chicago, IL: University of Chicago, 2006.

Tindall, D.B., Ronald Trosper, and Pamela Perreault. "The Social Context of Aboriginal Peoples and Forest Land Issues." In *Aboriginal Peoples and Forest Lands in Canada*, edited by DB Tindall, Ronald Trosper, and Pamela Perreault. 3–14. Vancouver: UBC Press, 2013.

TK (pseudonym). Telephone Interview, 17 June 2017. 1 hour, 29 minutes.

Trussell, Kent. "Save Middle Head Beach." *Maggies Farm*, April 1980, 8.

— "Montana Earth First!Ers Get Federal Subpoenas." *Earth First! Journal* 10, no. 1 (November 1989): 1.

— "So You Got a Death Threat...". *Earth First! Journal* 10, no. 6 (June 1990): 6.

Turner, Ralph, and Turner Killian. *Collective Behaviour*. Sydney: Prentice-Hall, 1987.

Turvey, Nigel. *Terania Creek: Rainforest Wars*. Cairndale: Glasshouse Books, 2006.

Underdog', 'Savannah. "Texas Earth First!, Locks up Outer Loop Construction." *Earth First! Journal* 10, no. 4 (March 1990): 12–13.

United Press International (UPI). "Logging Protester Arrested," *Washington Post*, 22 May 1985, A13

— "Oregon Logging Sends Protester up a Tree." *Chicago Tribune*, 13 July 1985, 4.

— Protesters Remain in Jail, Available [Online]: www.upi.com/Archives/1983/07/01/Protesters-remain-in-jail/6156425880000/ (Accessed 4 November 2015).

Van Huizen, Philip. "'Panic Park'": Environmental Protest and the Politics of Parks in British Columbia's Skagit Valley." *BC Studies*, no. 170 (2011): 67.

Vanderziel, Kathleen M. "Hatfield Riders & Environmental Preservation: What Process Is Due." *Boston College Environmental Affairs Law Review* 19, no. 2 (1991): 431–79.

Vaughn, Peter. "Police Interview Transcript: 5 January," (1990).

Veenker, Matt. "Blockaders Roughed up in Middle Santium." *Earth First! Journal* August (1984): 1.

Vercellotti, Tim. "'Lock and Block' Tactic Is Spreading Amongst Groups Protesting Abortions " *Pittsburg Press*, 25 June 1989.

Vucetic, Srdjan. "The Anglosphere: A Genealogy of an Identity in International Relations." PhD Thesis. The Ohio State University, 2008.

Wade, Valeri. "Kalimopsis Kangaroo Court." *Earth First! Journal* 7, no. 8 (September 1987): 9.

—— Telephone Interview, 18 May 2016, Bellingham, Washington. 1 hour, 28 minutes.

Wagner, Travis. "Reframing Ecotage as Ecoterrorism: News and the Discourse of Fear." *Environmental Communication* 2, no. 1 (2008): 25–39.

Walker, Cam. Telephone Interview, 28 September 2015, Castlemaine, Australia. 59 minutes.

Walker, Martin, and David Rowan. "Shock Troops in the Eco-War." *Guardian*, 17 August 1990, 21.

Walkom, Thomas. "Action Speaks Louder Than Words " *Globe and Mail*, 26 October 1988, A7.

— "At Loggerheads over Logging." *Globe and Mail* 28 September 1987, A7.

Wall, Derek. *Earth First! And the Anti-Roads Movement: Radical Environmentalism and Comparative Social Movements*. London: Routledge, 1999.

Wallner, Astrid. "The Aboriginal Peoples' Position in Land-Use Conflicts in British Columbia, Canada." *Bulletin* 62 (1998): 61–66.

Wang, Dan, and Alessandro Piazza. "The Use of Disruptive Tactics in Protest as a Trade-Off: The Role of Social Movement Claims." *Social Forces* 94, no. 4 (2016): 1675–710.

Wang, Dan, and Sarah Soule. "Tactical Innovation in Social Movements: The Effects of Peripheral and Multi-Issue Protest." *American Sociological Review* 81, no. 3 (2016): 517–48.

Ward, Doug. "Arrests Clear Way for Park Drilling." *Vancouver Sun*, 3 February 1988, B1.

— "Showdown a War of Nerves." *Vancouver Sun*, 2 February 1988, A1.

Ward, Susan, and Kitty van Vuuren. "Belonging to the Rainbow Region: Place, Local Media, and the Construction of Civil and Moral Identities Strategic to Climate Change Adaptability." *Environmental Communication* 7, no. 1 (2013): 63–79.

Watson, John. "Terania Mania." *Nimbin News*, 24 September 1979, 15–16.

Watson, Ian. *Fighting over the Forests*. North Sydney: Angus and Robertson, 1990.

Watson, Imogen. "After a Soul Searching Lockdown Extinction Rebellion Plots Its Next Move," *The Drum*, 20 July 2020, Available [Online]: www.thedrum.com/news/2020/07/20/after-soul-searching-lockdown-extinction-rebellion-plots-its-next-move (Accessed 1 November 2020).

Watson, Paul. "In Defense of Tree Spiking." *Earth First! Journal* 10, no. 8 (September 1990): 8–9.

— Letter to Ha-Shilth-Sa Reporter, Available [Online]: http://forestcouncil.org/paul-watson-on-clayoquot-tree-spiking/ (Accessed 11 November 2016).

Waud, Pam, and Robin Tindale, eds. *The Franklin Blockade*. Hobart: The Wilderness Society, 1983.

Weber, Thomas. "Nonviolence Is Who? Gene Sharp and Gandhi," *Peace & Change* 28, no. 2 (2003): 250–65.

Weber, Thomas and Robert Burrowes, "Nonviolence: An Introduction," *Peace Dossier* no. 27 (1991): 1–10.

White, Rob, and Diane Heckenberg. *Green Criminology: An Introduction to the Study of Environmental Harm*. London, New York: Routledge, 2014.

Widick, Richard. *Trouble in the Forest: California's Redwood Timber Wars* Minneapolis: University of Minnesota Press, 2009.

Wilkes, Rima. "A Systematic Approach to Studying Indigenous Politics: Band-Level Mobilization in Canada, 1981–2000." *The Social Science Journal* 41, no. 3 (2004): 447–58.

— "The Protest Actions of Indigenous Peoples: A Canadian-U.S. Comparison of Social Movement Emergence." *American Behavioral Scientist* 40, no.4 (2006): 510–525.

Wilkes, Rima, and Tamara Ibrahim. "Direct Action over Forests and Beyond." In *Aboriginal Peoples and Forest Lands in Canada*, edited by D.B. Tindall, Ronald Trosper and Pamela Perreault. 74–88. Vancouver: BCM Press, 2013.

Wilkie, Bill. Telephone Interview, 3 September 2014, Mossman, Australia. 2 hours, 22 minutes.

— *The Daintree Blockade*. Mossman: Four Mile Books, 2017.

Willems-Braun, Bruce. "Colonial Vestiges: Representing Forest Landscapes on Canada's West Coast." In *Troubles in the Rainforest* edited by Trevor Barnes and Roger Hayter. 99–130. Victoria: Western Geographical Press, 1997.

Willow, Marcy. "Foresticide in Middle Santium." *Earth First! Journal* 5, no. 5 (May 1985): 10.

— "Last Stand for the Last Stand." *Earth First! Journal* 4, no. 5 (May 1984): 7.

Willsin, Kelpie, and George Shook. "Forest Action Workshop." *Earth First! Journal* 11, no. 6 (June 1991): 25.

Wilson, Hap. Telephone Interview, 17 January 2017, Rousseau, Canada. 1 hour, 8 minutes.

Wilson, Jeremy. *Talk and Log: Wilderness Politics in BC, 1965-96*. Vancouver: University of British Columbia Press, 1998.

Wilson, Roger. *From Manapouri to Aramoana*. Auckland: Earthworks Press, 1982.

Winfield, Mark. *Blue-Green Province: The Environment and the Political Economy of Ontario*. Vancouver: UBC Press, 2012.

Woestendiek, John. "Anti-Logging Protest Still Spreading Its Roots." *Philadelphia Inquirer*, 12 August 1990, A2.

Wolfe, Patrick. *Settler Colonialism and the Transformation of Anthropology*. London: Cassell, 1999.

Wolke, Howie. "The Grizzly Den." *Earth First! Journal* 3, no. 7 (September 1983): 12.

Wilson, Deborah. "Police Break up Blockade of Temagami Logging Road," *Globe and Mail* 6 June 1989, A18.

Wilson, Pip. "Impressions of Middle Head." *Maggies Farm*, October 1980, 9.

Winstanley, Paul. Telephone Interview, 14 November 2017, Vancouver, USA. 2 hours, 3 minutes.

Wood, Karen, Telephone Interview, 27 June 2016, Pittsburgh, USA. 1 hour, 12 minutes.

Wood, Lesley. *Direct Action, Deliberation and Diffusion: Collective Action After the WTO Protests in Seattle*. Cambridge: Cambridge University Press, 2012.

Workman, Bill. "San Bruno Mountain Condos: Chained Protesters Cut Free." *San Francisco Chronicle*, 25 August 1987, 5.

Wren, Christopher. "Canadian Battle Rages over Lovely Timbered Isle." *New York Times*, 17 May 1985, A2.

Wyant, Dan. "15 Protesters Arrested on Logging Road." *Register-Guard*, 5 June 1984, 1C.

— "Protesters Vow to Continue Fight." *Register-Guard*, 8 June 1984, 1B.

Wyland, Scott. "Demonstrators Delay Clear-Cutting near Forest Park." *Oregonian*, 4 May 1990, C4.

Yeates, Lisa, Telephone Interview, 29 January 2015, Launceston, Australia. 3 hours, 12 minutes.

York, Geoffrey, and Loreen Pindera. *People of the Pines: The Warriors and the Legacy of Oka.* Toronto: Little, Brown, 1991.

Zable, Benny. Email Interview, 25 March 2015, Nimbin, Australia.

Zakin, Susan. *Coyotes and Town Dogs: Earth First! And the Radical Environmental Movement.* New York: Viking, 1993.

Zierenberg, Nancy. "Time to Move On." *Earth First!* 10, no. 8 (September 1990): 2–3.

Zelko, Frank. "Making Greenpeace: The Development of Direct Action Environmentalism in British Columbia." *BC Studies*, no. 142/143 (2004): 197–240.

"Zig Zag", *BC Native Blockades and Direct Action.* Unknown: Warrior Publications, 2006.

INDEX

For Product Safety Concerns and Information please contact our EU
representative GPSR@taylorandfrancis.com
Taylor & Francis Verlag GmbH, Kaufingerstraße 24, 80331 München, Germany